Mathematical Theory of Optimization

Nonconvex Optimization and Its Applications

Volume 56

Mathematical Theory of Optimization

Edited by

Ding-Zhu Du

Department of Computer Science and Engineering,
University of Minnesota, Minneapolis, U.S.A.

Panos M. Pardalos

Center for Applied Optimization,
ISE Department, University of Florida, Gainesville, U.S.A.

and

Weili Wu

Department of Computer Science and Engineering,
University of Minnesota, Minneapolis, U.S.A.

KLUWER ACADEMIC PUBLISHERS
DORDRECHT / BOSTON / LONDON

A C.I.P. Catalogue record for this book is available from the Library of Congress.

ISBN 1-4020-0015-4

Published by Kluwer Academic Publishers,
P.O. Box 17, 3300 AA Dordrecht, The Netherlands.

Sold and distributed in North, Central and South America
by Kluwer Academic Publishers,
101 Philip Drive, Norwell, MA 02061, U.S.A.

In all other countries, sold and distributed
by Kluwer Academic Publishers,
P.O. Box 322, 3300 AH Dordrecht, The Netherlands.

Printed on acid-free paper

Printed in the Netherlands.

“And it was then that all these kinds of things thus established received their shapes from the Ordering One, through the action of Ideas and Numbers.”

- Plato, Timaeus

“Since the fabric of the world is the most perfect and was established by the wisest Creator, nothing happens in this world in which some reason of maximum or minimum would not come to light.”

- Euler

Contents

Preface

Optimization is of central importance in all sciences. Nature inherently seeks optimal solutions. For example, light travels through the "shortest" path and the folded state of a protein corresponds to the structure with the "minimum" potential energy. In combinatorial optimization, there are numerous computationally hard problems arising in real world applications, such as floorplanning in VLSI designs and Steiner trees in communication networks. For these problems, the exact optimal solution is not currently real-time computable. One usually computes an approximate solution with various kinds of heuristics. Recently, many approaches have been developed that link the discrete space of combinatorial optimization to the continuous space of nonlinear optimization through geometric, analytic, and algebraic techniques. Many researchers have found that such approaches lead to very fast and efficient heuristics for solving large problems. Although almost all such heuristics work well in practice there is no solid theoretical analysis, except Karmakar's algorithm for linear programming. With this situation in mind, we decided to teach a seminar on nonlinear optimization with emphasis on its mathematical foundations. This book is the result of that seminar.

During the last decades many textbooks and monographs in nonlinear optimization have been published. Why should we write this new one? What is the difference of this book from the others? The motivation for writing this book originated from our efforts to select a textbook for a graduate seminar with focus on the mathematical foundations of optimization. We tried to find a textbook containing more theoretical components and found that no one met our needs. In fact, there are many very fundamental theoretical results that have been published during the last fifteen years, not yet included in textbooks. It is our objective to make these recent theoretical developments the main topic of the present book. We focus on the mathematical theory of optimization, especially, various convergence problems in nonlinear optimization and the theoretical bridges between continuous and discrete optimization problems.

This book contains three major parts. The first part consists of chapters 2-6. In this part, we introduce the convergence theory which was established successfully for solving a long-standing open problem on the global convergence of Rosen's gradient projection method. We will try to explain this theory in an intuitive way. In particular, we would like to point out to the reader that the theory is very powerful to deal with situations where the sequence of points generated by an algorithm has a cluster point but the sequence does not converge to the cluster point.

The second part consists of chapters 9-10. This part essentially is dedicated to Powell's conjecture on the global convergence of quasi-Newton methods. There still exist many important open problems about convergence of quasi-Newton methods, for example, the global convergence of various quasi-Newton methods with/without exact line search, including Powell's conjecture. The main obstacle for solving these open problems appears to be the case where the sequence of points generated by an algorithm has a cluster point but the sequence does not converge to the cluster point. Could we borrow some ideas from theory in the first part to solve these open problems on quasi-Newton methods? We hope so!

The material after chapter 11 constitutes the third part of the book which is about combinatorial and global optimization. As we mentioned above, it is one of our goals to train students with sufficient knowledge working in the interface of continuous and discrete optimization problems. In fact, almost all hard combinatorial optimization problems are looking for a global optimal solution instead of local optimal solution. To build a bridge between continuous and discrete optimization problems, it is important to understand the theory of global optimization. A recent success applying a continuous nonlinear optimization approach to a discrete optimization problem is the proof of the Gilbert-Pollak conjecture on the Steiner ratio. We introduce this work in chapter 11.

During the past decade several interesting results have appeared on the subject of global optimization. In chapter 15, we provide an introduction to global optimization of general quadratic problems. The concept of the convex envelope of a function is introduced and the Pardalos-Rosen algorithm for concave minimization problems is presented. In addition, we discuss the formulation of discrete optimization problems as continuous quadratic programming problems. As an example, the maximum clique problem in general graphs is formulated as the maximization of a nonconcave quadratic function over the unit simplex (Motzkin-Strauss formulation).

We wish to express our thanks to the graduate students and colleagues who have direct or indirect contributions in the process of writing this book. Finally, we wish to thank Professor Eugene Shragowitz for allowing us to use some material from his unpublished work in chapter 1.

DingZhu Du, Panos M. Pardalos and Weili Wu
University of Minnesota and University of Florida
November 1998

Chapter 1

Optimization Problems

Most practical questions can be reduced to problems of largest and smallest magnitudes ... and it is only by solving these problems that we can satisfy the requirements of practice which always seeks the best, the most convenient.

P. L. ČEBYŠEV

Accurate modeling of scientific problems often leads to the formulation of optimization problems involving continuous and/or discrete variables. In recent years optimization has seen a dramatic increase in activities. This is a natural consequence of new algorithmic developments and the increased power of computers. Many of these problems can be very large, although what is large in optimization reflects not only the size but also the inherent complexity of a problem.

The main subject studied in this book is the mathematical theory of nonlinear optimization. We first study some examples. These examples will tell us that nonlinear optimization techniques are very useful in solving optimization problems in the real world, in particular in connection with recent technological developments. Nonlinear optimization is also very important in the development of new techniques for solving combinatorial optimization problems, including linear programming.

1.1 Introduction

Did you ever watch how a spider catches a fly or a mosquito? Usually, a spider hides at the edge of its net. When a fly or a mosquito hits the net, the spider will pick up each line in the net to choose the tense one and then goes rapidly along the line to its prey. Why does the spider chooses the tense line? Some biologists explain that the line gives the shortest path from the spider to its prey.

Did you hear the following story about a wise general? He had a duty to capture a town behind a mountain. When he and his soldiers reached the top of the mountain, he found that his enemy had already approached the town very closely from another way. His dilemma was how to get in the town before the enemy arrive. It was a challenging problem for the general. The general solved the problem by asking each soldier to roll down the mountain in a blanket. Why is this faster? Physicists tell us that a free ball rolling down a mountain always chooses the most rapid way.

Do you know the tale of a horse match of Tian Gi? It is a story set in B.C. time. Tian Gi was a general in one of several small counties of China, called Qi. The King of Qi knew that Tian Gi had several good horses and ordered Tian Gi to have a horse match with him. The match consisted of three rounds. In each round, each side chose a horse to compete with the other side. Tian Gi knew that his best horse could not compete with the best one of the King, his second best horse could not compete with the second best one of King, and his third best horse could not compete with the third best one of the King. Therefore, he did not use his best horse against the best horse of the King. Instead, he put his third best horse in the first round against the best one of the King, his best horse in the second round against the second best one of the King, and his second best horse in the third round against the third best one of the King. The final result was that although he lost the first round of the match, he won the last two rounds. Tian Gi's strategy was the best to win this match. Today, economists tell us that many economic systems and social systems can be modeled into games. Each contestant in the game tries to maximize certain benefits.

Optimality is a fundamental principle, establishing natural laws, ruling biologic behaviors, and conducting social activities. Therefore, optimization started from the earliest stages of human civilization. Of course, before mathematics was well established, optimization could be done only by simulation. One may find many wise men's stories in the human history about it. For example, to find the best way to get out of a mountain, someone followed a stream, and to find the best way to get out from a desert, someone set an old horse free and followed the horse's trace.

In the last century or even today, simulation is still used for optimizing something. For example, to find a shortest path on a network, one may make a net with rope in a proportional size and pull the net tightly between two destinations. The tense rope shows the shortest path. To find an optimal location of a school for three villages, one may drill three holes on a table and put a piece of rope in each hole. Then tie three rope-ends above the table together and hang a one-kg-weight on each rope-end under the table. When this mechanical system is balanced, the knot of the three rope-pieces points out the location of the school.

The history of optimization in mathematics can be divided into three periods.

In the first period, one did not know any general method to find a maximum/minimum point of a function. Only special techniques were found to maximize/minimize some special functions. A typical function is the quadratic function of one variable

$$y = ax^2 + bx + c.$$

The study of quadratic functions was closely related to the study of constantly-accelerating movement. What is the highest point that a stone is thrown out with certain initial speed and certain angle? What is the farthest point where a stone thrown with certain initial speed can reach when throwing angle varies? These were questions considered by some physicists and generals. In fact, the stone-throwing machine was an important weapon in military.

Today, computing maximum/minimum points of a quadratic function is still, an important technique of optimization, existing in elementary mathematics books. The technique had been also extended to other functions such as

$$y = \frac{x^2 + x + 1}{x^2 + 2x + 3}.$$

Actually, multiplying both sides by $x^2 + 2x + 3$ and simplifying, we obtain

$$(y - 1)x^2 + (2y - 1)x + (3y - 1) = 0.$$

Since x is a real number, we must have

$$(2y - 1)^2 - 4(y - 1)(3y - 1) \geq 0.$$

Therefore,

$$-8y^2 + 12y - 3 \geq 0,$$

that is,

$$\frac{3 - \sqrt{3}}{4} \leq y \leq \frac{3 + \sqrt{3}}{4}.$$

It is interesting to note that with this technique we obtained the global maximum and minimum of y.

A new period started in 1646 by Pierre de Fermat. He proposed, in his paper *Treatises on Maxima and Minima*, a general approach to compute local maxima/minima points of a differentiable function, that is, setting the derivative of the function to be zero. Today, this approach is still included in almost all textbooks of Calculus as an application of differentiation. In this period, optimization existed scattered and disorderly in mathematics. Because optimization had not become an important branch of applied mathematics, some mathematicians did not pay so much attention to results on optimization and some contributions were even not put in any publication. This left many mysteries in the history of optimization.

For example, who is the first person who proposed the Steiner tree? It was such a mystery. To obtain a clear view, let us explain it in a little detail.

In the same paper mentioned above, Fermat also studied a problem of finding a point to minimize the total distance from it to three given points in the Euclidean plane. Suppose three given points are (x_1, y_1), (x_2, y_2), and (x_3, y_3). Then the total distance from a point (x, y) to these three points is

$$f(x,y) = \sum_{i=1}^{3} \sqrt{(x - x_i)^2 + (y - y_i)^2}.$$

By Fermat's general method, the minimum point of $f(x, y)$ must satisfy the following equations

$$\frac{\partial f}{\partial x} = \sum_{i=1}^{3} \frac{x - x_i}{\sqrt{(x - x_i)^2 + (y - y_i)^2}} = 0,$$
$$\frac{\partial f}{\partial y} = \sum_{i=1}^{3} \frac{y - y_i}{\sqrt{(x - x_i)^2 + (y - y_i)^2}} = 0.$$

However, obtaining x and y from this system of equations seems hopeless. Therefore, Fermat mentioned this problem again in a letter to A. Mersenne that it would be nice if a clear solution could be obtained for this problem.

Torricelli, a student of Galilei, obtained a clever solution with a geometric method. He showed that if three given points form a triangle without an angle of at least 120°, then the solution is a point at which three segments from it to three given points produce three angles of 120°. Otherwise, the solution is the given point at which the triangle formed by the three given points has an angle of at least 120°. This result can also be proved by the mechanic system described at the beginning of this article. In the first case, the knot of the three rope-pieces stays not at any given point and hence the balance condition of the three forces of equal magnitude yields the condition on the angles. In the second case, the knot falls in one of the three holes, and the condition on the angle guarantees that the knot would not move away from the hole.

Fermat's problem was extensively studied later and was generalized to four points by J. Fr. Fagnano in 1775 and to n points by Tedenat and L'Huiller in 1810. Fagnano pointed out that it is very easy to find the solution of Fermat's problem for four points. When four given points form a convex quadrilateral, the solution of Fermat's problem is the intersection of two diagonals, i.e., the intersection of two diagonals minimizes the total distance from one point to four given points. Otherwise, there must be one of the given points lying inside the triangle formed by the other three given points; this given point is the solution.

On March 19, 1836, Schumacher wrote a letter to Gauss. In his letter, he mentioned a paradox about Fermat's problem: Consider a convex quadrilateral $ABCD$. It has been known that the solution of Fermat's problem for four points A, B, C, and D is the intersection E of diagonals AC and BD. Suppose extending DA and CB can obtain an intersection F. Now, moving A and B to F. Then E will also be moved to F. However, when the angle at F is less than 120°, the point F cannot be the solution of Fermat's problem for three given points F, D, and C. What happens? (Fig. 1.1.)

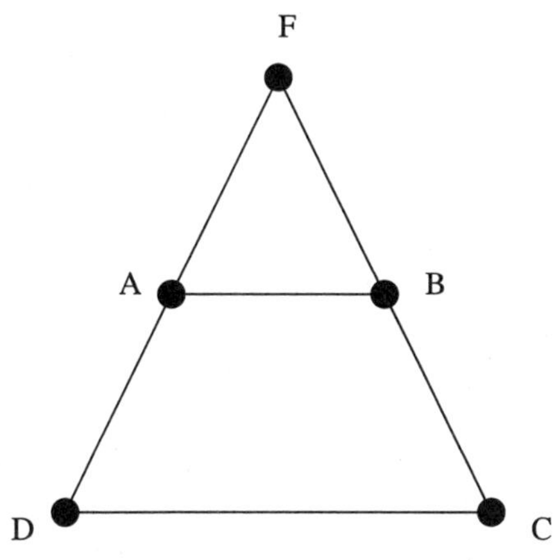

Figure 1.1: Schumacher's paradox.

On March 21, 1836, Gauss wrote a letter to Schumacher in which he explained that the mistake of Schumacher's paradox occurs at the place where Fermat's problem for four points A, B, C, and D is changed to Fermat's problem for three points F, C, and D. When A and B are identical to F, the total distance from E to four points A, B, C, and D equals $2EF+EC+ED$, not $EF+EC+ED$. Thus, the point E may not be the solution of Fermat's problem for F, C, and D. More importantly, Gauss proposed a new problem. He said that it is more interesting to find a shortest network rather than a point. Gauss also presented several possible connections of the shortest network for four given points.

Unfortunately, Gauss' letter was discovered only in 1986. From 1941 to 1986, many publications have followed R. Courant and H. Robbins who in their popular book *What is mathematics?* (published in 1941) called Gauss' problem as the Steiner tree problem. The Steiner tree has become a popular and important name. If you search "Steiner-tree" with *yahoo.com* on the internet, then you will receive a list of more than 1000 web-pages on Steiner trees. We have no way to change back the name from Steiner trees to Gauss trees. It may be worth mentioning that Jakob Steiner, a geometrician in nineteen century whose name is used for the shortest networks, has not been found so far to have any significant contribution to Steiner trees.

George B. Dantzig who first proposed the simplex method to solve linear programming in 1947 stated in his article *Linear Programming: The Story about How It Began*: "What seems to characterize the pre-1947 era was lack of any interest in trying to optimize." Due to the lack of interests in optimization, many important works appeared before 1947 were ignored. This happened not only for Steiner trees, but also to other areas of optimization, including some important contributions in linear and nonlinear programming.

The discovery of linear programming started a new age of optimization. However, in the same article mentioned above, Dantzig made the following comments: Linear programming was unknown prior to 1947. This is not quite correct; there were some late exceptions. Fourier (of Fourier series fame) in 1823 and the well-known Belgian mathematician de la Vallée Poussin in 1911 each wrote a paper about it. Their work had as much influence on post-1947 developments as would finding in an Egyptian tomb an electronic computer built in 3000 BC. Leonid Kantorovich's remarkable 1939 monograph on the subject was also neglected for ideological reasons in the USSR. It was resurrected two decades later after the major developments had already taken place in the West. An excellent paper by Hitchcock in 1941 on the transportation problem was also overlooked until after others in the late 1940's and early 1950's have independently rediscovered its properties.

He also recalled how he made his discovery. "My own contribution grew out of my World War II experience in the Pentagon. During the war period (1941-45), I had become an expert on programming-planning methods using desk calculators. In 1946 I was Mathematical Advisor to the US Air Force Comptroller in the Pentagon. I had just received my PhD (for research I had done mostly before the war) and was looking for an academic position that would pay better than a low offer I had received from Berkeley. In order to entice me to not take another job, my Pentagon colleagues, D. Hitchcock and M. Wood, challenged me to see what I could do to mechanize the planning process. I was asked to find a way to more rapidly compute a time-staged development, training and logistical supply program. In those days 'mechanizing' planning meant using analog devices or punch-card equipment. There were no electronic computers."

This challenge problem made Dantzig discover his great work in linear programming without electronic computer. But, we have to point out that it is due to the rapid development of computer technology that applications of linear programming can be made so wide and so great, and areas of optimization can have so fast growing.

In 1951, Albert Tucker and his student Harold Kuhn published the Kuhn-Tucker conditions. This is considered as an initial point of nonlinear programming. However, A. Takayama in his book *Mathematical Economics* has an interesting comment on these conditions: "Linear programming aroused interest in constraints in the form of inequalities and in the theory of linear inequalities and convex sets. The Kuhn-Tucker study appeared in the middle of this interest with a full recognition of such developments. However, the theory of nonlinear programming when constraints are all in the form of equalities has been known for a long time – in fact, since Euler and Lagrange. The inequality constraints were treated in a fairly satisfactory manner already in 1939 by Karush. Karush's work is apparently under the influence of a similar work in the calculus of variations by Valentine. Unfortunately, Karush's work has been largely ignored." Yet, this is another work first appeared before 1947 and it was ignored. In the 1960's, G. Zoutendijk, J.B. Rosen, P. Wolfe, M.J.D. Powell, and others published a number of algorithms for solving nonlinear optimization problems. These algorithms form the main body of nonlinear programming.

Today, optimization has become a very large interdisciplinary area between mathematics, computer science, industrial engineering, and management science. The techniques in nonlinear optimization are used not only for solving nonlinear programming problems, but also for solving other problems, such as combinatorial optimization problems. Actually, from yesterday-and-today's fast growth of computer technology, optimization can be expected to continue its great speed of developments tomorrow. These developments may contain deep understanding of the successful heuristics for combinatorial optimization problems with nonlinear programming approaches. As many mysteries and open problems still exist, it will be an area receiving a great attention.

Two examples presenting in the next two sections are discrete problems in nature. We show how to solve them with nonlinear optimization approaches.

1.2 Floorplanning

Electronic Chip Design is one of the current top technologies in the world. In fact, the chip is the heart of many electronic devices and has become a symbol of the level of advancement of technology in a country. The chip design can be divided into several levels and each level has several steps. One of the most consequential steps in each level is floorplanning, because decisions made on this step will be retained through all levels of a design hierarchy.

In floorplanning, the input contains a set of rectangles, called *modules*. The internal structure of each module is designed in higher levels, and possibly is also a result of another floorplanning problem. Some modules are

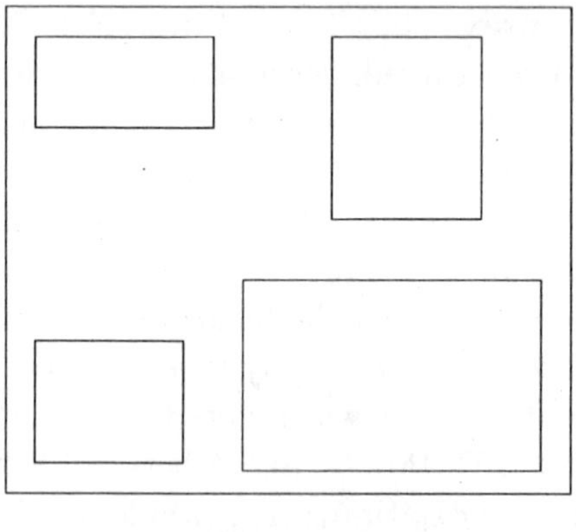

Figure 1.2: Floorplanning.

flexible and others are rigid. Let M_1 be the set of flexible modules and M_2

the set of rigid modules. Rigid modules have their dimensions fixed, i.e. the width w_m and the height h_m ($m \in M_2$) are defined. It is also assumed that a rotation of rigid modules is allowed. Usually, for rigid modules one allows coordinates of pins for external interconnects to be fixed. These coordinates are given as $x_{m,i}$ and $y_{m,i}$, where m is an index for a module and i is the pin number for the module. The coordinates of the pins are defined with respect to the lower left corner of the fixed module. For the flexible cells the positions of the pins for external interconnects initially are assigned to the center of the flexible cell and then re-assigned to the position specified by the structure of the flexible block. For each flexible block m ($m \in M_1$) area S_m is given as a number, i.e. $w_m h_m = S_m$, where w_m and h_m are the unknown width and height of the module m. For flexible modules and for a whole chip constraints can be imposed on the aspect ratio, i.e. $a_m \leq \frac{w_m}{h_m} \leq b_m$. One also allows the number of input/output pads on the periphery of the chip to be fixed to the sides and positions or to be decided in the process of floorplanning.

On each level of hierarchical floorplanning external interconnections are defined by a netlist where for each module a set of nets connected to it is given, i.e. each "external" net is defined by modules and pins that this net is connected to. On each net one considers a timing delay which is defined by a function of coordinates of terminals, parameters of interconnects and drivers. Depending on the technology, a computational model for approximate evaluation of the timing delay may change, but initially given in the form of the quadratic function of a length of the distance between the driver and a receiver in the form $t(mi, nj) = c_0 + c_1 d_{mi,nj} + c_2 d^2_{mi,nj}$, where $d_{mi,nj}$ is the Manhatten distance between the pin i from the module m and the pin j from the module n and the coefficients are selected to fit the experimental data.

Several objective functions for minimization could be applied in the design process. It is not required that the same objective function be used on consecutive hierarchical levels.

1. The first objective function is the area of the minimal covering rectangle R for a set of M nonoverlapping rectangles presenting modules and spaces between them if necessary for routing of buses.

2. The second objective function is the total length of edges of interconnects.

3. The third one is the maximal timing delay on interblock connections.

All of the objectives stated above could be applied to the partial solution when a part of a floorplan is frozen and only a part of it is a subject of optimization. Such a problem definition allows us to fix a part of the floorplan if required and evaluate a difference between the floorplan redesigned from the changed requirements and the partially redesigned floorplan.

The first objective function can be written as

$$\max_m x_m \cdot \max_m y_m.$$

By introducing two new variables x and y, instead of minimizing the function $\max_m x_m \cdot \max_m y_m$, one can minimize

$$xy$$

subject to

$$\begin{aligned} x_m &\leq x \quad \text{for all } m \\ y_m &\leq y \quad \text{for all } m. \end{aligned}$$

The second objective function is in the form of

$$\sum_i |f_i|$$

where f_i's are linear functions. For each i, one introduces a new variable z_i. Instead of minimizing the function with absolute values, one can equivalently minimize

$$\sum_i z_i$$

subject to

$$\begin{aligned} f_i &\leq z_i \quad \text{for all } i \\ -f_i &\leq z_i \quad \text{for all } i. \end{aligned}$$

The third objective function can be treated similarly as the first objective function. The resulting new objective function is a quadratic function and some new quadratic constraints are introduced.

In summary, the objective function in the mathematical programming model can be considered as a linear or quadratic function in any case.

The constraints on a floorplan first of all should prevent overlapping of the modules. It means that for any pair of the modules at least one of the following linear inequalities holds (here all coordinates are related to the lower left corner):

$$x_m + w_m \leq x_n \tag{1.1}$$
$$x_n + w_n \leq x_m \tag{1.2}$$
$$y_m + h_m \leq y_n \tag{1.3}$$
$$y_n + h_n \leq y_m \tag{1.4}$$

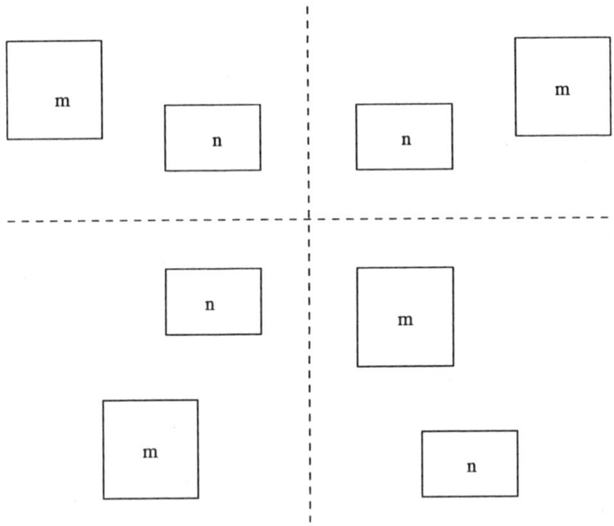

Figure 1.3: Four possible positions between two modules.

Usually, in an optimization problem, constraints are given by a system of equalities and inequalities; they are required to hold simultaneously. However, the nonoverlapping condition is described by four inequalities which are not required to hold simultaneously. One of the approaches to handle this unusual situation is to introduce two new integer (binary) variables. For example, one introduces integer variables u and v with $u^2 = 1$ (i.e. $u = 1$ or -1) and $v^2 = 1$ and then reduces the above module-nonoverlapping condition to the following constraints:

$$\begin{aligned} x_m + w_m - x_n &\leq (2 - u - v)L \\ -x_m + w_n + x_n &\leq (2 - u + v)L \end{aligned}$$

$$
\begin{aligned}
y_m + h_m - y_n &\leq (2 + u - v)L \\
-y_m + h_n + y_n &\leq (2 + u + v)L
\end{aligned}
$$

where L is a sufficiently large constant (for instance, choose $L = \sum_m (w_m + h_m)$). The choice of L guarantees that L is larger than every left side in the four inequalities, so that the inequality holds nontrivially only if the coefficient of L becomes zero.

Since two integer variables will be introduced for each pair of modules, the number of integer variables will be quite large. This is one of the major difficulties for solving the resulting integer programming. We now introduce a new method based on nonlinear optimization.

Instead of two integer variables, we introduce one nonnegative variable $u_{mn} \geq 0$. Note that the function

$$
\begin{aligned}
&(x_m + w_m + u_{mn} - x_n)^2 (x_n + w_n + u_{mn} - x_m)^2 \\
&\cdot (y_m + h_m + u_{mn} - y_n)^2 (y_n + h_n + u_{mn} - y_m)^2
\end{aligned}
$$

achieves the minimum value 0 with $u_{mn} \geq 0$ if and only if one of the four inequalities (1.2)-(1.4) holds, i.e. the two modules do not overlap. Thus, we can include this function into the objective function and only $u_{mn} \geq 0$ into constraints.

Each flexible module m has two constraints:

$$
\begin{aligned}
w_m h_m &= S_m \\
a_m &\leq \frac{w_m}{h_m} \leq b_m
\end{aligned}
$$

Define

$$
r_m = \sqrt{\frac{w_m}{h_m}}.
$$

Then

$$
\begin{aligned}
w_m &= r_m \sqrt{S_m} \\
h_m &= \frac{\sqrt{S_m}}{r_m} \\
\sqrt{a_m} &\leq r_m \leq \sqrt{b_m}.
\end{aligned}
$$

Now, we want to replace variable w_m by r_m. Meanwhile, relax $h_m = \frac{\sqrt{S_m}}{r_m}$ to

$$
h_m \geq \frac{\sqrt{S_m}}{r_m}. \tag{1.5}
$$

This relaxation does not change the optimal solution for the first objective function. However, for the second and third objective functions, this relaxation may change their optimal solutions. But, the optimal solution of the relaxation still suggests a good feasible solution of the original problem and the effect of relaxation can be reduced through adjustment in the next level of the hierarchical designs.

Now, one can linearize the nonlinear constraint (1.5) in several ways.

A naive way is to use a tangent line to replace a curve. But, for our problem this is not a good way since it introduces nonfeasible solutions and the nonlinear curve cannot be approximated very well.

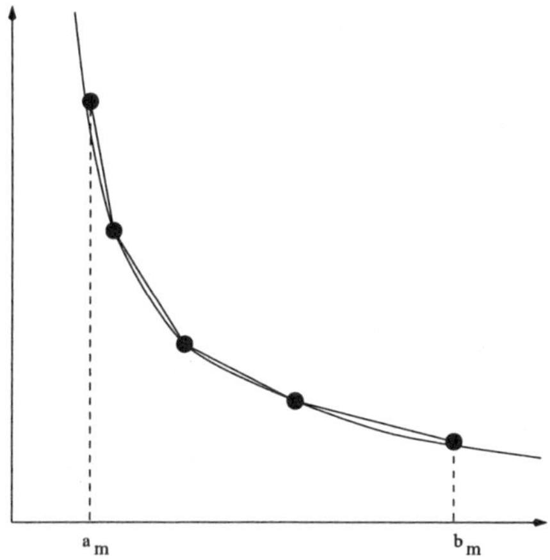

Figure 1.4: Replace the curve by chords.

Actually, one uses several chords on the curve to replace the curve. For example, choose n points $(h_m^1, r_m^1), (h_m^2, r_m^2), \cdots, (h_m^n, r_m^n)$, where $r_m^1 = a_m < r_m^2 < \cdots < r_m^n = b_m$, on the curve $h_m = \frac{\sqrt{S_m}}{r_m}$. The line through chords, between points (h_m^i, r_m^i) and (h_m^{i+1}, r_m^{i+1}) is the following

$$\frac{h_m - h_m^i}{r_m - r_m^i} = \frac{h_m^{i+1} - h_m^i}{r_m^{i+1} - r_m^i}.$$

When the curve $h_m = \frac{\sqrt{S_m}}{r_m}$ is replaced by the $n-1$ chords through the n points, the constraint (1.5) is replaced by $n-1$ linear constraints

$$h_m - h_m^i \geq \frac{h_m^{i+1} - h_m^i}{r_m^{i+1} - r_m^i}(r_m - r_m^i)$$

for $i = 1, 2, \cdots, n-1$. Since $h_m = \frac{\sqrt{S_m}}{r_m}$ is a convex function, this replacement would not introduce any nonfeasible solution. In fact, the curve $h_m = \frac{\sqrt{S_m}}{r_m}$ can be approximated very closely if we choose n sufficiently large.

Of course, one does not like to have a large number of linear constraints. To keep n small and make the approximation to the curve more accurate, one may place more points around point $(\sqrt[4]{S_m}, \sqrt[4]{S_m})$. In fact, placing two or three points around $(\sqrt[4]{S_m}, \sqrt[4]{S_m})$ can already make the approximation work very well.

Based on different requirements, one can formulate different optimization models for floorplanning. The following are two examples.

Area Optimization under Timing and Other Constraints In this model, the goal is to minimize the area of the covering rectangle which introduces a quadratic term in the objective function and several linear constraints. Other terms in the objective function and constraints come from considerations on module-nonoverlapping, flexible modules, and timing limits. When time delay is computed approximately by a linear function, all constraints can be reduced to linear constraints. Thus, this model is a nonlinear minimization problem with linear constraints.

Timing Optimization under Area and Other Constraint When timing delay is approximated by a linear function, this goal contributes a linear term in the objective function. All constraints except the one introduced from the area limits are linear. The constraint introduced from the area is in the form

$$xy \leq S.$$

This is a nonlinear constraint. One might use this technique for dealing with flexible modules. If such a technique is used, one should replace this nonlinear constraint by a condition that one of several linear constraints holds. This is a similar situation in module nonoverlapping and hence, a nonlinear term in the objective function and a linear constraint would be introduced.

Another way to deal with this concave constraint is to use a linear constraint to approximate the concave constraint locally. In each iteration, one uses the tangent line of the nonlinear curve $xy = S$ at the current point to create the linear constraint. The only disadvantage is that this linear constraint has to be updated at the beginning of each iteration.

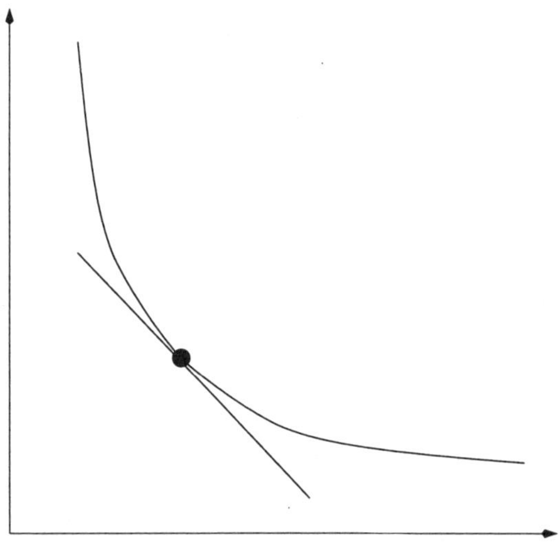

Figure 1.5: Replace a concave constraint by a linear constraint

1.3 Satisfiability

A company wants to employ a student from a university. The job is to program in Java. There are three candidates John, Steve, and Bill. The company received the following comments from three Professors at the university:

"John and Steve know Java very well."

"Steve and Bill are excellent programmers in Java."

"I know Bill is good in Java, but I don't think that Steve knows Java well."

If each comment is at least half correct, then whom should the company choose?

Let x_1, x_2, x_3 be three Boolean variables such that

$$x_1 = \begin{cases} 1 & \text{John knows Java well} \\ 0 & \text{otherwise} \end{cases}$$

$$x_2 = \begin{cases} 1 & \text{Steve knows Java well} \\ 0 & \text{otherwise} \end{cases}$$

$$x_3 = \begin{cases} 1 & \text{Bill knows Java well} \\ 0 & \text{otherwise.} \end{cases}$$

Since each of three comments is at least half correct, we have

$$x_1 \vee x_2 = x_2 \vee x_3 = x_3 \vee \bar{x}_2 = 1$$

i.e.

$$(x_1 \vee x_2) \wedge (x_2 \vee x_3) \wedge (x_3 \vee \bar{x}_2) = 1. \tag{1.6}$$

Expanding it, we obtain that

$$(x_1 \wedge x_2 \wedge x_3) \vee (x_1 \wedge x_3) \vee (x_2 \wedge x_3) = 1.$$

This means that there are three possible solutions of the logic equation 1.6. However, every solution of the logic equation 1.6 contains $x_3 = 1$. Therefore, the company should choose Bill.

A *conjunctive normal form* (CNF) is a conjunction of m clauses $C_1 \wedge C_2 \wedge \cdots \wedge C_m$ where each clause is a disjunction of several literals (i.e. Boolean variables and negation of Boolean variables). Given a CNF formula F, the satisfiability problem (SAT) is to determine whether there exists an assignment to Boolean variables to make F have value 1 (true).

SAT is a very important problem in computational complexity theory. It was the first problem discoved to be NP-complete. Thus, an efficient algorithm for SAT has a small likelihood to exist. One of the techniques for solving SAT is to transform it to a nonlinear (global) optimization problem.

Let $x_1, x_2, \cdots, x_n$ be Boolean variables and $\boldsymbol{x}$ the vector $(x_1, x_2, \cdots, x_n)$ in $\{0,1\}^n$. Let $y_1, y_2, \cdots, y_n$ be real variables and $\boldsymbol{y}$ the vector $(y_1, y_2, \cdots, y_n)$ in R^n. Define $f_1(\boldsymbol{y}) : R^n \to R$ by

$$f_1(\boldsymbol{y}) = \sum_{i=1}^{m} c_i(\boldsymbol{y})$$

where the clause function $c_i(\boldsymbol{y})$ is the product of n literal functions $q_{ij}(y_j)$ as follows:

$$c_i(\boldsymbol{y}) = \prod_{j=1}^{n} q_{ij}(y_j),$$

where

$$q_{ij}(y_j) = \begin{cases} (y_j - 1)^2 & \text{if } x_j \text{ is in clause } C_i \\ (y_j + 1)^2 & \text{if } \bar{x}_j \text{ is in clause } C_i \\ 1 & \text{neither } x_j \text{ nor } \bar{x}_j \text{ is in } C_i. \end{cases}$$

The correspondence between $\boldsymbol{x}$ and $\boldsymbol{y}$ is included in the following:

$$x_j = \begin{cases} 1 & \text{if } y_j = 1 \\ 0 & \text{if } y_j = -1 \\ \text{undefined} & \text{otherwise.} \end{cases}$$

A CNF formula $C_1 \wedge C_2 \wedge \cdots \wedge C_m$ is satisfiable if and only if the minimization problem

$$\min f_1(\boldsymbol{y})$$

has the minimum value 0. Define

$$f(\boldsymbol{y}) = f_1(\boldsymbol{y}) + f_2(\boldsymbol{y})$$

where

$$f_2(\boldsymbol{y}) = \sum_{j=1}^{n} (y_j^2 - 1)^2.$$

Then it is still true that A CNF formula $C_1 \wedge C_2 \wedge \cdots \wedge C_m$ is satisfiable if and only if the minimization problem

$$\min f(\boldsymbol{y})$$

has the minimum value 0. The purpose of including the function $f_2(\boldsymbol{y})$ is to improve the convexity at the minimum point.

The current objective function $f(\boldsymbol{y})$ has the following two interesting properties:

1. There exists $\boldsymbol{y}$ such that $f(\boldsymbol{y}) = 0$ if and only if there exists $\boldsymbol{y}$ such that $f(\boldsymbol{y}) < 1$.

2. At every minimum point $\boldsymbol{y}^*$, $f(\boldsymbol{y})$ is strictly convex.

To prove the first property, suppose that $f(\boldsymbol{y}) < 1$. It follows that for every i, $c_i(\boldsymbol{y}) < 1$. Thus, each $c_i(\boldsymbol{y})$ contains a factor $(y_j - 1)^2$ or $(y_j + 1)^2$ whose value is less than 1. Therefore, either $c_i(\boldsymbol{y})$ has a factor $(y_j + 1)^2$ with $y_j < 0$ or $c_i(\boldsymbol{y})$ has a factor $(y_j - 1)^2$ with $y_j > 0$. Define

$$y_j^* = \begin{cases} 1 & \text{if } y_j \geq 0 \\ -1 & \text{if } y_j < 0. \end{cases}$$

Then we have that $f(\boldsymbol{y}^*) = 0$.

To prove the second property, suppose $\boldsymbol{y}^*$ is a minimum point of $f(\boldsymbol{y})$. We first note that $f_2(\boldsymbol{y})$ is strictly convex at $\boldsymbol{y}^*$ since

$$f_2(\boldsymbol{y}^* + \alpha \boldsymbol{d}) = 4(d_1^2 + d_2^2 + \cdots + d_n^2)\alpha^2 + o(\alpha^2).$$

Since $\boldsymbol{y}^*$ is also a global minimum point of $f_1(\boldsymbol{y})$, $f_1(\boldsymbol{y})$ must be convex at $\boldsymbol{y}^*$. Therefore, the sum of f_1 and f_2 must be strictly convex at $\boldsymbol{y}^*$.

The first property means that once a point with f's value < 1 is found the computation may stop. The second property encourage us to minimize the function f with nonlinear optimization algorithms.

Exercises

1. Formulate the following logic puzzles into satisfiability problems and solve them:

   (a) Three men named Lewis, Miller, and Nelson fill the positions of accountant, cashier, and clerk in a department store. If Nelson is the cashier, Miller is the clerk. If Nelson is the clerk, Miller is the accountant. If Miller is not the cashier, Lewis is the clerk. If Lewis is the accountant, Nelson is the clerk. What is each man's job?

   (b) Messrs. Spinnaker, Buoy, Luff, Gybe, and Windward are yacht owners. Each has a daughter, and each has named his yacht after the daughter of one of the others. Mr. Spinnaker's yacht, the *Iris*, is named after Mr. Buoy's daughter. Mr. Buoy's own yacht is the *Daffodil*; Mr. Windward's yacht is the *Jonquil*; Mr. Gybe's, the *Anthea*. Daffodil is the daughter of the owner of the yacht which is named after Mr. Luff's daughter. Mr. Windward's daughter is named Lalage. Who is Jonquil's father?

   (c) Six men, Andrews, Bllaine, Colter, Doister, Ebert, and Fisher, are the only members eligible for the offices of president, vice-president, and secretary in a certain organization. If Andrews won't be an officer unless Ebert is president, Blaine won't serve if he outranks Colter, Blaine won't serve with Fisher under any conditions, Colter won't serve with both Ebert and Fisher, Colter won't serve if Fisher is president or Blaine is secretary, Doister won't serve with Colter or Ebert unless he outranks them, Ebert won't be vice-president, Ebert won't be secretary if Doister is an officer, Ebert won't serve with Andrews unless Fisher serves too, Fisher won't serve unless either he or Colter is president, how can the three offices be filled?

(d) Four men, one of whom was known to have committed a certain crime, made the following statements when questioned by the police. Archie:"Dave did it." Dave:"Tony did it." Gus:"I didn't do." Tony:"Dave lied when he said I did it." If only one of these four statements is true, who was the guilty man? On the other hand, if only one of these four statements is false, who was the guilty man?

2. Consider an undirected graph $G = (V, E)$ where $V = \{1, 2, \cdots, n\}$. Each edge (i, j) in E has a nonnegative cost c_{ij}. Given a subset R of vertices, find a minimum cost tree interconnecting all vertices in R. This is called the *network Steiner tree* problem. Show the following.

(a) Let $\mathcal{P}$ be the family of partitions (I, J) of vertex set V such that $I \cap R \neq \emptyset$ and $J \cap R \neq \emptyset$. Then the network Steiner tree problem is equivalent to the following:

$$\begin{array}{rll} \text{minimize} & \sum_{(i,j)\in E} c_{ij}x_{ij} & \\ \text{subject to} & \sum_{i\in I, j\in J} x_{ij} \geq 1 & (\text{for } (I, J) \in \mathcal{P} \\ & x_{ij} \in \{0, 1\} & \text{for } (i, j) \in E. \end{array}$$

(b) Add a vertex o to G and assign to each edge (o, i) for $i \in V - R$ with cost $c_{oi} = 0$. Choose a vertex $r \in R$ and assign to edge (o, r) also with cost $c_{or} = 0$. Denote

$$\begin{aligned} E^* &= E \cup \{(o, i) \mid i \in (V - (R - \{r\}))\}, \\ V^* &= V \cup \{o\}. \end{aligned}$$

Then the network Steiner tree problem is equivalent to the following:

$$\begin{array}{rll} \text{minimize} & \sum_{(i,j)\in E^*} c_{ij}x_{ij} & \\ \text{subject to} & \sum_{i\in I, j\in J} x_{ij} \geq 1 & \text{for all partition } (I, J) \text{ of } V^* \\ & x_{oi} + x_{pq} \leq 1 & \text{for } i \in V^* - R \\ & & \text{and } (i = p \text{ or } q), \\ & x_{ij} \in \{0, 1\} & \text{for } (i, j) \in E^*. \end{array}$$

3. Given n points in the Euclidean plane, find a shortest tree interconnecting the n points. This is called the *Euclidean Steiner tree* problem. Let $\boldsymbol{x}_1, \boldsymbol{x}_2, \cdots, \boldsymbol{x}_n$ be the n given points. Let $\boldsymbol{x}_{n+1}, \boldsymbol{x}_{n+2}, \cdots, \boldsymbol{x}_{2n-2}$ be $n-2$ points varying in the Euclidean plane. Denote $P = 1, 2, \cdots, n$ and $S = \{n+1, n+2, \cdots, 2n-2\}$. Show that the Euclidean Steiner tree problem is equivalent to the following:

$$\begin{array}{lll} \text{minimize} & \sum_{1 \leq i < j \leq 2n-2} \|\boldsymbol{x}_i - \boldsymbol{x}_j\|_2 \cdot y_{ij} & \\ \text{subject to} & \boldsymbol{x}_i \in R^2 & \text{for } i \in S \\ & \sum_{j \in S} y_{ij} = 1 & \text{for } i \in P \\ & \sum_{j \in P \cup (S - \{i\})} y_{ij} = 3 & \text{for } i \in S \\ & y_{ij} & 1 \leq i < j \leq 2n-2, \end{array}$$

where $y_{ij} = y_{ji}$.

4. Consider a directed graph $G = (V, E)$ where $V = \{1, 2, \cdots, n\}$. Each edge (i, j) in E has a nonnegative cost c_{ij}. Given a vertex r and and a subset R of vertices, find a minimum cost tree containing paths from root r to all vertices in R. This is called the *directed network Steiner tree* problem. Show the following.

   (a) Suppose there are $|R|$ units of flow from r to R and each vertex in R demands one unit. Let y_{ij} be the amount of flow on edge (i, j), and x_{ij} a binary variable corresponding to edge (i, j). Then the directed network Steiner tree problem can be formulated as follows:

$$\begin{array}{lll} \text{minimize} & \sum_{(i,j) \in E} c_{ij} x_{ij} & \\ \text{subject to} & \sum_{(i,j) \in O(i)} y_{ij} - \sum_{(j,i) \in I(i)} y_{ji} = -1 & \text{for } i \in R \\ & \sum_{(r,j) \in O(r)} y_{rj} - \sum_{(j,r) \in I(r)} y_{jr} = |R| & \\ & \sum_{(i,j) \in O(i)} y_{ij} - \sum_{(j,i) \in I(i)} y_{ji} = 0 & \text{for } i \in V - R \\ & & -\{r\} \\ & 0 \leq y_{ij} \leq |R| x_{ij} & \text{for } (i,j) \in E \\ & x_{ij} \in \{0, 1\} & \text{for } (i,j) \in E, \end{array}$$

   where $O(i)$ $(I(i))$ is the set of out-edges (in-edges) at vertex i.

   (b) Suppose there are $|R|$ units of flow from r to R and each vertex in R demands one unit. Let y_{ij} be the amount of flow on edge

(i,j). Let $\varepsilon > 0$. Then the solution of the following concave minimization problem goes to the solution of the directed network Steiner tree problem as ε goes to zero.

$$
\begin{array}{lll}
\text{minimize} & \sum_{(i,j)\in E} c_{ij} \frac{y_{ij}}{y_{ij}+\varepsilon} & \\
\text{subject to} & \sum_{(i,j)\in O(i)} y_{ij} - \sum_{(j,i)\in I(i)} y_{ji} = -1 & \text{for } i \in R \\
& \sum_{(r,j)\in O(r)} y_{rj} - \sum_{(j,r)\in I(r)} y_{jr} = |R| & \\
& \sum_{(i,j)\in O(i)} y_{ij} - \sum_{(j,i)\in I(i)} y_{ji} = 0 & \text{for } i \in V - R \\
& & -\{r\} \\
& 0 \le y_{ij} & \text{for } (i,j) \in E.
\end{array}
$$

Chapter 2

Linear Programming

If there are obstacles,
the shortest line between two points
may be the crooked line.
BERTOLT BRECHT

As we stated in introduction (Section 1.1), linear programming initiated the modern theory of optimization. Indeed, many algorithms for nonlinear optimization use linear programming as a subroutine. Therefore, we first briefly introduce linear programming as a preparation for later developments, especially we introduce Karmarkar algorithm as a bridge between linear and nonlinear optimization.

2.1 Simplex Method

An optimization problem is called a *linear programming* if its objective function is linear and its feasible region is constrained by linear inequalities and/or linear equalities. The following is an example.

$$\begin{array}{ll} \min & x_1 + x_2 \\ \text{subject to} & x_1 - x_2 \leq 1 \\ & x_1 \geq 0, x_2 \geq 0. \end{array}$$

In this example, the feasible region is constrained by three linear inequalities, which is a two-dimensional polyhedron as shown in Figure 2.1. Its objective function $x_1 + x_2$ reaches the minimum at point $(0, 0)$ which is a vertex of the polyhedron.

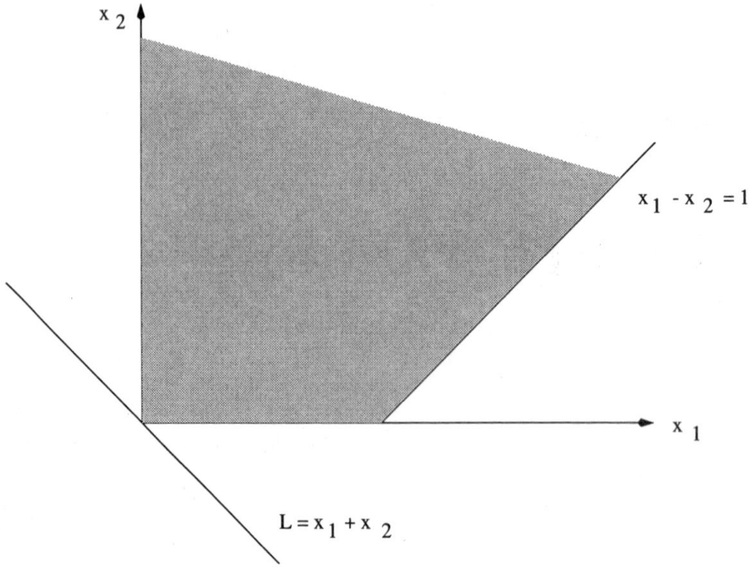

Figure 2.1: An example of linear programming

Linear programming has a standard form as follows:

$$
\begin{array}{lrl}
(LP) & \min & \boldsymbol{c}^T \boldsymbol{x} \\
 & \text{subject to} & A\boldsymbol{x} = \boldsymbol{b} \\
 & & \boldsymbol{x} \geq \boldsymbol{0}
\end{array}
$$

where A is an $m \times n$ matrix, $\boldsymbol{c}, \boldsymbol{x} \in R^n$, and $\boldsymbol{b} \in R^m$. Every linear programming can be transformed into an equivalent linear programming in the standard form. In fact, if a variable x_i is not non-negative, then we can use two non-negative variables to replace it, i.e., set $x_i = u_i - v_i$, $u_i \geq 0$, $v_i \geq 0$. If there is an inequality containing at least two variables, then we can introduce a new non-negative variable to make the equality sign holds. For example, the above linear programming can be transformed to the following standard form.

$$
\begin{array}{rl}
\min & x_1 + x_2 \\
\text{subject to} & x_1 - x_2 + w = 1 \\
 & x_1 \geq 0, x_2 \geq 0, w \geq 0.
\end{array}
$$

In the standard form of linear programming, we usually assume $rank(A) = m$. In fact, if the feasible domain is not empty, $rank(A) < m$ means the existence of some useless constraints which can be deleted to make that the rank of coefficient matrix equals the number of rows.

There are three important families of algorithms for solving linear programming, simplex methods, ellipsoid methods, and interior-point methods. In this section, we introduce the simplex method.

Simplex method is motivated from one important observation on the linear programming in the standard form, that is, if optimal solution exists, it can be found among vertices of feasible region.

What is a vertex of feasible region? Note that the feasible region of every linear programming is a polyhedron. A point $\boldsymbol{x}$ in a polyhedron Ω is called a *vertex* or an *extreme point* if it has property that

$$\left.\begin{array}{l} \boldsymbol{x} = \frac{\boldsymbol{y}+\boldsymbol{z}}{2} \\ \boldsymbol{y}, \boldsymbol{z} \in \Omega \end{array}\right\} \Rightarrow \boldsymbol{x} = \boldsymbol{y} = \boldsymbol{z}.$$

With this definition, let us first give a formal proof to our observation.

Lemma 2.1.1 *Let* $\Omega = \{\boldsymbol{x} \mid A\boldsymbol{x} = \boldsymbol{b}, \boldsymbol{x} \geq 0\}$. *If* $\min_{\boldsymbol{x}\in\Omega} \boldsymbol{c}^T\boldsymbol{x}$ *has an optimal solution, then it can be found in its vertices.*

Proof. Consider an optimal solution $\boldsymbol{x}^*$ with maximum number of zero components among optimal solutions. We will show that $\boldsymbol{x}^*$ is a vertex of Ω. By contradiction, suppose $\boldsymbol{x}^*$ is not, that is, there exist $\boldsymbol{y}, \boldsymbol{z} \in \Omega$ such that $\boldsymbol{x}^* = \frac{\boldsymbol{y}+\boldsymbol{z}}{2}$ but $\boldsymbol{x}^*$, $\boldsymbol{y}$, and $\boldsymbol{z}$ are distinct (note: if two of them are equal, then they are all equal). Since $\boldsymbol{c}^T\boldsymbol{x}^* \leq \boldsymbol{c}^T\boldsymbol{y}$, $\boldsymbol{c}^T\boldsymbol{x}^* \leq \boldsymbol{c}^T\boldsymbol{z}$, and $\boldsymbol{c}^T\boldsymbol{x}^* = (\boldsymbol{c}^T\boldsymbol{y} + \boldsymbol{c}^T\boldsymbol{z})/2$, we must have $\boldsymbol{c}^T\boldsymbol{x}^* = \boldsymbol{c}^T\boldsymbol{y} = \boldsymbol{c}^T\boldsymbol{z}$. This means that $\boldsymbol{y}$ and $\boldsymbol{z}$ are also optimal solutions. It follows that all feasible points on line $\boldsymbol{x}^* + \alpha(\boldsymbol{y} - \boldsymbol{x}^*)$ are optimal solutions. However, Ω does not contain any line. Thus, the line $\boldsymbol{x}^* + \alpha(\boldsymbol{y} - \boldsymbol{x}^*)$ must have a point $\boldsymbol{x}'$ not in Ω, that is, $\boldsymbol{x}'$ violates at least one constraint.

Note that for any α, $A(\boldsymbol{x}^* + \alpha(\boldsymbol{y} - \boldsymbol{x}^*)) = \boldsymbol{b}$. Thus, $\boldsymbol{x}'$ cannot violate constraint $A\boldsymbol{x} = \boldsymbol{b}$. Moreover, for $x_i^* = 0$, since $x_i^* = (y_i + z_i)/2$ and $y_i, z_i \geq 0$, we must have $z_i = y_i = x_i = 0$. Therefore, the ith component of $\boldsymbol{x}^* + \alpha(\boldsymbol{y} - \boldsymbol{x}^*)$ is equal to 0 for any α. This means that x' cannot violate constraint $x_i \geq 0$. Hence, x' must violate a constraint $x_j \geq 0$ for some j with $x_j^* > 0$. Now, we can easily find an optimal solution between $\boldsymbol{x}'$ and $\boldsymbol{x}^*$, which has one more zero-component than $\boldsymbol{x}^*$, a contradiction. □

Simplex method starts from a vertex and moves from one to another vertex at each iteration while the objective function value is going down. To describe it in detail, let us first give a sufficient and necessary condition for a feasible point to be a vertex.

Lemma 2.1.2 *A feasible point $\boldsymbol{x} \in \Omega$ is a vertex if and only if there exist m column-indices $j_1, j_2, \cdots, j_m$ such that columns $\boldsymbol{a}_{j_1}, \boldsymbol{a}_{j_2}, \cdots, \boldsymbol{a}_{j_m}$ of A are linearly independent and $x_j = 0$ for $j \notin I = \{j_1, j_2, \cdots, j_m\}$.*

Proof. For "if" part, suppose $\boldsymbol{x} = (\boldsymbol{y} + \boldsymbol{z})/2$ and $\boldsymbol{y}, \boldsymbol{z} \in \Omega$. Note that $x_j = 0$ implies $y_j = z_j = 0$. Thus, $(x_{j_1}, x_{j_2}, \cdots, x_{j_m})$, $(y_{j_1}, y_{j_2}, \cdots, y_{j_m})$, and $(z_{j_1}, z_{j_2}, \cdots, z_{j_m})$ are all solutions of the following system of linear equations

$$\boldsymbol{a}_{j_1} x_{j_1} + \boldsymbol{a}_{j_2} x_{j_2} + \cdots + \boldsymbol{a}_{j_m} x_{j_m} = b,$$

which has unique solution since $\boldsymbol{a}_{j_1}, \boldsymbol{a}_{j_2}, \cdots, \boldsymbol{a}_{j_m}$ are linearly independent. Thus, $(x_{j_1}, x_{j_2}, \cdots, x_{j_m}) = (y_{j_1}, y_{j_2}, \cdots, y_{j_m}) = (z_{j_1}, z_{j_2}, \cdots, z_{j_m})$. Hence, $\boldsymbol{x} = \boldsymbol{y} = \boldsymbol{z}$. Therefore, $\boldsymbol{x}$ is a vertex.

For "only if" part, suppose $\boldsymbol{x}$ is a vertex with m' nonzero components $x_{j_1}, x_{j_2}, \cdots, x_{j_{m'}}$. We claim that the following system of linear equations has unique solution

$$\boldsymbol{a}_{j_1} x_{j_1} + \boldsymbol{a}_{j_2} x_{j_2} + \cdots + \boldsymbol{a}_{j_{m'}} x_{j_{m'}} = \boldsymbol{b}. \tag{2.1}$$

In fact, for otherwise, suppose the system of linear equations has another solution $(x'_{j_1}, x'_{j_2}, \cdots, x'_{j_{m'}}) \neq (x_{j_1}, x_{j_2}, \cdots, x_{j_{m'}})$. Set

$$x'_j = 0 \text{ for } j \notin \{j_1, j_2, \cdots, j_{m'}\}.$$

Then $A\boldsymbol{y} = \boldsymbol{b}$ and $\boldsymbol{x} + \alpha(\boldsymbol{x}' - \boldsymbol{x}) \geq 0$ for $|\alpha|$ sufficiently small. Choose $\alpha > 0$ such that $\boldsymbol{y} = \boldsymbol{x} + \alpha(\boldsymbol{x}' - \boldsymbol{x}) \geq 0$ and $\boldsymbol{z} = \boldsymbol{x} - \alpha(\boldsymbol{x}' - \boldsymbol{x}) \geq 0$. Then $\boldsymbol{y}, \boldsymbol{z} \in \Omega$, $\boldsymbol{x} = (\boldsymbol{y} + \boldsymbol{z})/2$, and $\boldsymbol{x} \neq \boldsymbol{y}$, contradicting the fact that $\boldsymbol{x}$ is a vertex. Thus, our claim is true. It follows that $\boldsymbol{a}_{j_1}, \boldsymbol{a}_{j_2}, \cdots, \boldsymbol{a}_{j_{m'}}$ are linearly independent. Hence, $m' \leq m$. In the case of $m' < m$, we can put more columns to form a maximum independent subset of columns including $\boldsymbol{a}_{j_1}, \boldsymbol{a}_{j_2}, \cdots, \boldsymbol{a}_{j_{m'}}$, which meets our requirement. □

A vertex of the feasible region is also called a *feasible basic solution.* The index subset $I = \{j_1, j_2, \cdots, j_m\}$ associated with a feasible basic solution in Lemma 2.1.2 is called a *feasible basis.* Denote $A_I = (\boldsymbol{a}_j, j \in I)$. Then an index subset I is a feasible basis if and only if $rank(A_I) = m = |I|$ and $A_I^{-1}\boldsymbol{b} \geq 0$. In fact, the vertex associated with I is the x with $x_I = A_I^{-1}\boldsymbol{b}$ and $x_{\bar{I}} = 0$ where $\bar{I} = \{1, 2, \cdots, n\} - I$.

Now, suppose $\boldsymbol{x}$ is a feasible basic solution associated with feasible basis I. How do we find another feasible basic solution $\boldsymbol{x}^+$ with feasible basis I^+ such that $\boldsymbol{c}^T\boldsymbol{x} > \boldsymbol{c}^T\boldsymbol{x}^+$? To answer this question, let us study the objective function $\boldsymbol{c}^T\boldsymbol{x}$. From $A\boldsymbol{x} = \boldsymbol{b}$, we know that $\boldsymbol{x}_I = A_I^{-1}(\boldsymbol{b} - A_{\bar{I}}\boldsymbol{x})$. Thus,

$$\begin{aligned} \boldsymbol{c}^T\boldsymbol{x} &= \boldsymbol{c}_I^T\boldsymbol{x}_I + \boldsymbol{c}_{\bar{I}}^T\boldsymbol{x}_{\bar{I}} \\ &= \boldsymbol{c}_I^T A_I^{-1}\boldsymbol{b} + (\boldsymbol{c}_{\bar{I}}^T - \boldsymbol{c}_I^T A_I^{-1} A_{\bar{I}})\boldsymbol{x}_{\bar{I}} \end{aligned}$$

If $\boldsymbol{c}_{\bar{I}}^T - \boldsymbol{c}_I^T A_I^{-1} A_{\bar{I}} \geq 0$, then $\boldsymbol{c}^T\boldsymbol{x} \geq \boldsymbol{c}_I A_I^{-1}\boldsymbol{b}$ and, moreover, the equality sign holds if $\boldsymbol{x}_{\bar{I}} = 0$. This means that the current feasible basic solution x associated with feasible basis I is optimal. If $\boldsymbol{c}'_{\bar{I}} = \boldsymbol{c}_{\bar{I}}^T - \boldsymbol{c}_I^T A_I^{-1} A_{\bar{I}}$ has a negative component, say $c'_{j'} < 0$ for some $j' \in \bar{I}$, then increasing the value of $x_{j'}$ may decrease the value of $\boldsymbol{c}^T\boldsymbol{x}$. Thus, using j' to replace an index in I may result in a better feasible basis. When can we do so? We next study this problem.

Denote $(a'_{ij}) = A' = A_I^{-1}A$ and $\boldsymbol{b}' = A_I^{-1}\boldsymbol{b}$. First, please note that if j' can be used to replace the i'th index $j_{i'}$ in I, then we must have $a'_{i'j'} > 0$. In fact, denote $I^+ = (I - \{j_{i'}\}) \cup \{j'\}$. Then $(A_{I^+}^{-1}A, A_{I^+}^{-1}\boldsymbol{b})$ can be obtained from $(A', \boldsymbol{b}')$ by dividing the i'th row with $a'_{i'j'}$ and then subtracting $a'_{ij'}$ times the i'th row from each row $i \neq i'$. This means that if I^+ is a feasible basis, then $x_{j'}$ in the feasible basic solution associated with I^+ equals $b'_{i'}/a'_{i'j'}$. Thus, we must have $a'_{i'j'} > 0$ in order to have $b'_{i'}/a'_{i'j'} \geq 0$. Moreover, for $i \neq i'$, the ith component of $A_{I^+}^{-1}b$ is equal to $b'_i - a'_{ij'}b'_{i'}/a'_{i'j'}$. For $a'_{ij'} \leq 0$, it is clear that $b'_i - a'_{ij'}b'_{i'}/a'_{i'j'} \geq 0$. However, for $a'_{ij'} > 0$, $b'_i - a'_{ij'}b'_{i'}/a'_{i'j'} \geq 0$ requires another condition on the choice of i' such that $b'_i/a'_{ij'} > b'_{i'}/a_{i'j'}$, that is,

$$\frac{b'_{i'}}{a'_{i'j'}} = \min\{\frac{b'_i}{a'_{ij'}} \mid a'_{ij'} > 0\}. \tag{2.2}$$

If there is a positive $a'_{ij'}$, then such a choice always exists. If $a'_{ij'} \leq 0$ for all i, then the linear programming has no optimal solution. In fact, set $x_j = 0$ for all $j \in \bar{I} - \{j'\}$ and $\boldsymbol{x}_I = \boldsymbol{b}' - (a'_{1j'}x_{j'}, \cdots, a'_{mj'}x_{j'})^T$. This is a feasible solution. The objective function on this feasible solution is equal to $\boldsymbol{c}_I^T A_I^{-1}\boldsymbol{b} + c'_{j'}x_{j'}$. Since $\boldsymbol{c}'_{j'} < 0$, $\boldsymbol{c}_I^T A_I^{-1}\boldsymbol{b} + c'_{j'}x_{j'}$ goes to the negative infinity as $x_{j'}$ goes to the positive infinity.

The above discussion indicates that if $c'_{\bar{I}}$ has a component $c'_{j'} < 0$, then either the linear programming has no optimal solution or we can choose i' by

(2.2) and move our attention to a new feasible basis $I^+ = (I - \{j'\}) \cup \{j_{i'}\}$. This movement is called a *pivot*. Essentially, the simplex method consists of a sequence of pivots as follows:

Simplex Method

- *Initially, a feasible basis I is found in some way.*
- *In each iteration, compute $\mathbf{c}'^T = \mathbf{c}^T - \mathbf{c}_I^T A_I^{-1} A$, $A' = A_I^{-1} A$, and $\mathbf{b}' = A_I^{-1}\mathbf{b}$. Those values are usually put in a* simplex *table as follows:*

$$\begin{array}{c|c} -L & \mathbf{c}' \\ \hline \mathbf{b}' & A' \end{array}$$

where $L = \mathbf{c}_I^T A_I^{-1}\mathbf{b}$. If $\mathbf{c}' \geq \mathbf{0}$, then stop; the current feasible basic solution associated with I is optimal. If $\mathbf{c}'$ has a component $c'_{j'} < 0$, then find an i' such that (2.2) holds. If such an i' does not exist, then stop; the linear programming has no optimal solution. If such an i' exists, then perform a pivot at $a'_{i'j'}$.

The following is an example.

Example 2.1.3

$$\begin{array}{ll} \min & -2x_2 + x_3 \\ s.t. & x_1 + x_2 + x_3 \leq 9 \\ & -x_1 + 2x_2 - x_3 \leq 4 \\ & 2x_1 - x_2 - x_3 \leq 5 \\ & x_1, x_2, x_3 \geq 0 \end{array}$$

Solution. First, we introduce three variables x_4, x_5, x_6 to transform it into the standard form:

$$\begin{array}{ll} \min & -2x_2 + x_3 \\ s.t. & x_1 + x_2 + x_3 + x_4 = 9 \\ & -x_1 + 2x_2 - x_3 + x5 = 4 \\ & 2x_1 - x_2 - x_3 + x_6 = 5 \\ & x_1, x_2, x_3, x_4, x_5, x_6 \geq 0 \end{array}$$

Clearly, x_4, x_5, x_6 form an initial feasible basis with the following simplex table:

0	0	−2	1	0	0	0
9	1	1	1	1	0	0
4	−1	2	−1	0	1	0
5	2	−1	−1	0	0	1

Choose x_2 to enter in basis.

0	0	−2	1	0	0	0
9	1	1	1	1	0	0
2	−1/2	1	−1/2	0	1/2	0
5	2	−1	−1	0	0	1

Perform a pivot at a'_{22}.

4	−2	0	0	0	1	0
7	3/2	0	3/2	1	−1/2	0
2	−1/2	1	−1/2	0	1/2	0
7	3/2	0	−3/2	0	1/2	1

Choose x_1 to enter in basis.

4	−2	0	0	0	1	0
14/3	1	0	1	2/3	−1/3	0
2	−1/2	1	−1/2	0	1/2	0
14/3	1	0	−1	0	1/3	2/3

Perform a pivot at a'_{13}.

40/3	0	0	−2	0	5/3	4/3
0	0	0	2	2/3	−2/3	−2/3
13/3	0	1	−1	0	2/3	1/3
14/3	1	0	−1	0	1/3	2/3

Put x_3 into basis.

40/3	0	0	0	2/3	1	2/3
0	0	0	1	1/3	−1/3	−1/3
13/3	0	1	0	1/3	1/3	0
14/3	1	0	0	1/3	0	1/3

Finally, we obtained a minimum solution $x_1 = 14/3, x_2 = 13/3, x_3 = 0$. □

A remaining problem is whether the objective function value is going down after a pivot. Note that the objective function value on the feasible basic solution associated with I^+ is $\boldsymbol{c}_I^T A_I^{-1}\boldsymbol{b} + \boldsymbol{c}'_{j'}\boldsymbol{b}'_{i'}/a'_{i'j'}$. This value compared with $\boldsymbol{c}_I^T A_I^{-1}\boldsymbol{b}$, the objective function value on the feasible basic solution associated with I, is smaller when $b'_{i'} > 0$. Therefore, if we make the following assumption, then each pivot will make the objective function value go down and hence a new feasible basic solution will be reached. After finitely many pivots, the algorithm will terminate since the number of feasible basic solutions is finite.

Nondegeneracy Assumption: For any feasible basis I, $A_I^{-1}\boldsymbol{b} > 0$.

What would happen if Nondegeneracy Assumption does not hold? The simplex algorithm described as above may fall in a cycle. In order to prevent the algorithm from going in a cycle, we need to employ more complicated rules for making the choice of j' and i'. One of such rules is based on lexicographical ordering in vectors. Consider two vectors $\boldsymbol{x}$ and $\boldsymbol{y}$. We say that $\boldsymbol{x}$ is smaller than $\boldsymbol{y}$ in lexicographical ordering, denoted by $\boldsymbol{x} <_L \boldsymbol{y}$, if $x_1 = y_1, \cdots, x_{k-1} = y_{k-1}$, and $x_k < y_k$ for some k.

Lexicographical Ordering Method

- *Initially, a feasible basis I is found in some way and rearrange variable labels such that $I = \{1, 2, \cdots, m\}$.*

- *In each iteration, compute $\mathbf{c}'^T = \mathbf{c}^T - \mathbf{c}_I^T A_I^{-1} A$, $A' = A_I^{-1}A$, and $\mathbf{b}' = A_I^{-1}\mathbf{b}$. If $\mathbf{c}' \geq \mathbf{0}$, then stop; the current feasible basic solution associated with I is optimal. If $\mathbf{c}'$ has a component $c'_{j'} < 0$, then find an i' such that*

$$\left(\frac{b'_{i'}}{a'_{i'j'}}, \frac{a'_{i'1}}{a'_{i'j'}}, \cdots, \frac{a'_{i'n}}{a'_{i'j'}}\right)$$

is the smallest among

$$\left\{\left(\frac{b'_i}{a'_{ij'}}, \frac{a'_{i1}}{a'_{ij'}}, \cdots, \frac{a'_{in}}{a'_{ij'}}\right) \mid a_{ij'} > 0\right\},$$

in lexicographic ordering. If such an i' does not exist, then stop; the linear programming has no optimal solution. If such an i' exists, then perform a pivot at $a'_{i'j'}$.

Those rules guarantee the following:

- Initially, for every $i = 1, \cdots, m$, $(b'_i, a'_{i1}, \cdots, a'_{in}) >_L \mathbf{0}$, due to the variable relabeling.
- In each iteration, it is preserved that for every $i = 1, \cdots, m$,

$$(b'_i, a'_{i1}, \cdots, a'_{in}) >_L \mathbf{0},$$

  due to the choice of i'.
- In each iteration, $(-L, \boldsymbol{c}')$ increases in lexicographical ordering. Therefore, the algorithm visit each feasible basis at most once and the objective function value L is nonincreasing.

Hence, the algorithm either finds nonexistence of optimal solution or find an optimal feasible basic solution with feasible basis I such that $\mathbf{c}^T - \mathbf{c}_I^T A_I^{-1} A \geq 0$. This concludes the following.

Theorem 2.1.4 *If the linear programming (LP) has an optimal solution, then it has an optimal feasible basic solution associated with feasible basis I such that $\mathbf{c}^T - \mathbf{c}_I^T A_I^{-1} A \geq 0$. Moreove, if a feasible basis I satisfies $\mathbf{c}^T - \mathbf{c}_I^T A_I^{-1} A \geq 0$, then the feasible basic solution associated with I is optimal.*

2.2 Duality

An important part in theory of linear programming is about duality. Consider the following linear programming

$$\begin{array}{lrl} (DLP) \qquad & \max & \boldsymbol{y}^T \boldsymbol{b} \\ & \text{subject to} & \boldsymbol{y}^T A \leq \mathbf{c}^T. \end{array}$$

(DLP) is called the *dual* problem of (LP). In fact, they have a very interesting relation.

Theorem 2.2.1 *Suppose $\boldsymbol{x}$ and $\boldsymbol{y}$ are feasible solutions of (LP) and (DLP), respectively. Then $\mathbf{c}^T\boldsymbol{x} \geq \boldsymbol{y}^T\boldsymbol{b}$.*

Proof. $\mathbf{c}^T\boldsymbol{x} \geq (\boldsymbol{y}^T A)\boldsymbol{x} = \boldsymbol{y}^T\boldsymbol{b}$. □

From the proof of Theorem 2.2.1, it is easy to see that $\boldsymbol{c}^T\boldsymbol{x} = \boldsymbol{y}^T\boldsymbol{b}$ if and only if

$$(\boldsymbol{c}^T - \boldsymbol{y}^T A)\boldsymbol{x} = 0.$$

This condition is called the *complementary-slackness* condition.

Corollary 2.2.2 *(a) (LP) has feasible solutions but no optimal solution if and only if (DLP) has no feasible solution. (b) (DLP) has feasible solutions but no optimal solution if and only if (LP) has no feasible solution. (c) (LP) has an optimal solution if and only if (DLP) has an optimal solution.*

Corollary 2.2.3 *Suppose x and y are feasible solutions of (LP) and (DLP), respectively. If $(\mathbf{c}^T - \boldsymbol{y}^T A)\boldsymbol{x} = 0$, then x and y are optimal solutions of (LP) and (DLP), respectively.*

Theorem 2.2.4 *Suppose $\boldsymbol{x}^*$ and $\boldsymbol{y}^*$ are optimal feasible solutions of (LP) and (DLP), respectively. Then $\mathbf{c}^T\boldsymbol{x}^* = (\boldsymbol{y}^*)^T\boldsymbol{b}$.*

Proof. It suffices to show that if (LP) and (DLP) have optimal solutions, then there exist feasible solutions $\boldsymbol{x}$ and $\boldsymbol{y}$ for (LP) and (DLP), respectively, such that $(\mathbf{c}^T - \boldsymbol{y}^T A)\boldsymbol{x} = 0$. By Theorem 2.1.4, (LP) has a feasible basis I such that $\mathbf{c}^T - \mathbf{c}_I^T A_I^{-1} A \geq 0$. Suppose $\boldsymbol{x}$ is the feasible basic solution of (LP) associated with I and $\boldsymbol{y} = \mathbf{c}_I A_I^{-1}$. Then $\boldsymbol{y}$ is a feasible solution of (DLP). In addition, we have

$$(\boldsymbol{c}^T - \boldsymbol{y}^T A)\boldsymbol{x} = 0$$

since

$$\boldsymbol{x}_{\bar{I}} = 0 \quad \text{and} \quad \boldsymbol{c}_I^T - \boldsymbol{y}^T A_I = 0.$$

□

Corollary 2.2.5 (Farkas Lemma) *Let A be a given $m \times n$ matrix and $\mathbf{b}$ a given m-dimensional vector. Then there exists $\boldsymbol{x} \geq \mathbf{0}$ such that $A\boldsymbol{x} = \mathbf{b}$ if and only if $A^T\boldsymbol{y} \geq \mathbf{0}$ implies $\mathbf{b}^T\boldsymbol{y} \geq 0$.*

Proof. Consider (LP) with $\boldsymbol{c} = \boldsymbol{0}$. If there exists $\boldsymbol{x} \geq \boldsymbol{0}$ such that $A\boldsymbol{x} = \boldsymbol{b}$, i.e., (LP) has feasible solution, then (LP) has a minimum solution. In fact, every feasible solution of (LP) is its minimum solution with minimum value 0. If $A^T\boldsymbol{y} \geq \boldsymbol{0}$, then $-\boldsymbol{y}$ is a feasible solution of (DLP). By Theorem 2.2.4, $\boldsymbol{b}^T\boldsymbol{y} \geq 0$.

Conversely, suppose $A^T\boldsymbol{y} \geq \boldsymbol{0}$ implies $\boldsymbol{b}^T\boldsymbol{y} \geq 0$. If (DLP) has no feasible solution, then by Corollary 2.2.2 (LP) has a feasible solution. If (DLP) has a feasible solution, then it must have a maximum solution with maximum value ≤ 0. Hence, (LP) has a feasible solution. □

2.3 From Linear to Nonlinear

The most important development of linear programming in the 1980's was the discovery of interior-point methods. The key idea of these methods is to introduce a nonlinear potential function so that nonlinear optimization techniques are employed in solving linear programming. To explain this idea explicitly, let us present the first interior-point algorithm, Karmarkar's algorithm, in this section.

Karmarkar's algorithm deals with the problem in the following form.

$$\begin{array}{ll} \text{minimize} & \boldsymbol{c}^T\boldsymbol{x} \\ \text{subject to} & A\boldsymbol{x} = \boldsymbol{0}, \\ & \boldsymbol{e}^T\boldsymbol{x} = 1, \\ & \boldsymbol{x} \geq \boldsymbol{0}, \end{array} \tag{2.3}$$

where $\boldsymbol{e} = (1, \cdots, 1)^T$. Suppose that the minimum value of problem (2.3) is non-negative. Karmarkar's algorithm determines whether the minimum value is zero and finds an optimal point if the minimum value is zero.

Before introducing Karmarkar's algorithm, we explain how to transfer the usual form of linear programming to the one in form (2.3). Consider the following usual form:

$$\begin{array}{ll} \text{maximize} & \boldsymbol{c}^T\boldsymbol{x} \\ \text{subject to} & A^T\boldsymbol{x} \leq \boldsymbol{b}. \end{array} \tag{2.4}$$

First, note that the dual of (2.4) is as follows.

$$\begin{array}{ll} \text{minimize} & \boldsymbol{b}^T\boldsymbol{y} \\ \text{subject to} & A\boldsymbol{y} = \boldsymbol{c}, \\ & \boldsymbol{y} \geq \boldsymbol{0}. \end{array} \tag{2.5}$$

By the duality theorem of linear programming, (2.4) has an optimal solution if and only if the optimal value of the following problem is zero.

$$\begin{array}{ll} \text{minimize} & \boldsymbol{b}^T\boldsymbol{y} - \boldsymbol{c}^T\boldsymbol{x} \\ \text{subject to} & A\boldsymbol{y} = \boldsymbol{c}, \\ & A^T\boldsymbol{x} \leq \boldsymbol{c}, \\ & \boldsymbol{y} \geq \boldsymbol{0}. \end{array} \tag{2.6}$$

Introducing slack variables and writing $\boldsymbol{x}$ as the difference of two nonnegative vectors, (2.6) can be reduced to (2.3) except that $\boldsymbol{e}^T\boldsymbol{x} = 1$ may not hold. Furthermore, we can obtain (2.3) by changing all variables $x_1, \cdots, x_n$ in the following way.

$$\begin{aligned} z_i &= x_i/(1 + x_1 + \cdots + x_n),\ i = 1, 2, \cdots, n, \\ z_{n+1} &= 1/(1 + x_1 + \cdots + x_n). \end{aligned}$$

Instead of linear objective function, Karmarkar considered the following potential function:

$$g(\boldsymbol{x}) = \frac{(\boldsymbol{c}^T\boldsymbol{x})^n}{x_1 \cdots x_n}.$$

He then applied a gradient method to this nonlinear objective function with linear constraints. In this gradient method, a search direction is determined based on the gradient at the current point and then a line search is performed along the direction in order to find a new point with better value of the potential function. A very technical method is also employed to do the line search.

In order to determine the search step length, he first introduced a projection transformation which maps the current point $\boldsymbol{x}_k$ to the center of the simplex $S = \{\boldsymbol{x} \,|\, \boldsymbol{e}^T\boldsymbol{x} = 1, \boldsymbol{x} \geq \boldsymbol{0}\}$:

$$y_i = \frac{(x_i/x_{ki})}{\sum_{j=1}^n (x_j/x_{kj})}.$$

Under this transformation, the potential function becomes

$$\hat{g}(\boldsymbol{y}) = \frac{(\hat{\boldsymbol{c}}^T\boldsymbol{y})^n}{y_1 \cdots y_n} = g(\boldsymbol{x}),$$

and the simplex S maps to itself. Note that the gradient of the denominator of $\hat{g}(\boldsymbol{y})$ at the center $\boldsymbol{y}_k = (\frac{1}{n}, \cdots, \frac{1}{n})$ is $(1, \cdots, 1)$ which is perpendicular to the simplex. The steepest descent feasible direction is the gradient projection of numerator into the tangent plane $\{\boldsymbol{d} \mid \boldsymbol{e}^T\boldsymbol{d} = 0, \hat{A}\boldsymbol{d} = \boldsymbol{0}\}$ where

$$\hat{A} = A \begin{pmatrix} x_{n1} & & 0 \\ & \ddots & \\ 0 & & x_{kn} \end{pmatrix}.$$

Let $\boldsymbol{d}$ denote this projection. Let $\boldsymbol{y}_{k+1} = \boldsymbol{y}_k + \lambda\boldsymbol{d}/\|\boldsymbol{d}\|$ where $\lambda = \frac{1}{2\sqrt{n(n-1)}}$. Note that the minimum distance from the center $\boldsymbol{y}_k$ to the boundary of the simplex is 2λ. Thus, $\boldsymbol{y}_{k+1}$ is a feasible point. The new point $\boldsymbol{x}_{k+1}$ is selected to be the ordinary image of the point $\boldsymbol{y}_{k+1}$ under the projective transformation, namely,

$$\boldsymbol{x}_{k+1,i} = \frac{x_{ki}y_{k+1,i}}{\sum_{j=1}^n x_{kj}y_{k+1,j}}.$$

Next, we estimate the ratio $g(\boldsymbol{x}_{k+1})/g(\boldsymbol{x}_k) = \hat{g}(\boldsymbol{y}_{k+1})/\hat{g}(\boldsymbol{y}_k)$. Since $\boldsymbol{c}^T\boldsymbol{y}$ is linear, it decreases in a constant rate along a direction. Let $\boldsymbol{x}^*$ be an optimal point of (2.3) and $\boldsymbol{y}^*$ the image of $\boldsymbol{x}^*$ under the projective transformation. It is clear that $\hat{\boldsymbol{c}}^T\boldsymbol{y}^* = 0$. Note that the distance between $\boldsymbol{y}_k$ and $\boldsymbol{y}^*$ is not greater than that between $\boldsymbol{y}_k$ and point $(1, 0, \cdots, 0)$. The latter distance is $\sqrt{(n-1)/n}$. Thus, the descent rate of $\hat{\boldsymbol{c}}^T\boldsymbol{y}$ along the direction $\boldsymbol{y}^* - \boldsymbol{y}_k$ is not smaller than $\hat{\boldsymbol{c}}^T\boldsymbol{y}_k/\sqrt{(n-1)/n}$. Since $\boldsymbol{d}$ is the steepest descent feasible direction of $\hat{\boldsymbol{c}}^T\boldsymbol{y}$ at $\boldsymbol{y}_k$, the descent rate of $\hat{\boldsymbol{c}}^T\boldsymbol{y}$ along $\boldsymbol{d}$ is at least $\hat{\boldsymbol{c}}^T\boldsymbol{y}_k/\sqrt{(n-1)/n}$. Therefore,

$$\hat{\boldsymbol{c}}^T\boldsymbol{y}_{k+1} \le \hat{\boldsymbol{c}}^T\boldsymbol{y}_k - \lambda\hat{\boldsymbol{c}}^T\boldsymbol{y}_k\sqrt{\frac{n}{n-1}} = \hat{\boldsymbol{c}}^T\boldsymbol{y}_k\left(1 - \frac{1}{2(n-1)}\right).$$

So,

$$\frac{\text{the numerator of } \hat{g}(\boldsymbol{y}_{k+1})}{\text{the numerator of } \hat{g}(\boldsymbol{y}_k)} \le \left(1 - \frac{1}{2(n-1)}\right)^n.$$

Since the denominator of $\hat{g}(\boldsymbol{y}_k) = (\frac{1}{n})^n$, the remainder is to find a lower bound for the denominator of $\hat{g}(\boldsymbol{y}_{k+1})$. Write $\boldsymbol{y}_{k+1} = (\frac{1}{n} + \varepsilon_1, \cdots, \frac{1}{n} + \varepsilon_n)^T$ where $\varepsilon_1, \cdots, \varepsilon_n$ satisfy

$$\sum_{i=1}^n \varepsilon_i = \sum_{i=1}^n (y_{k+1,i} - y_{ki}) = 0,$$

$$\sum_{i=1}^n \varepsilon^2 = \|\boldsymbol{y}_{k+1} - \boldsymbol{y}_k\|^2 = \frac{1}{4n(n-1)}.$$

We study the following optimization problem:

$$\text{minimize} \quad \phi(\varepsilon_1, \cdots, \varepsilon_n) = \prod_{i=1}^{n}(1 + n\varepsilon_i) \tag{2.7}$$

$$\text{subject to} \quad \sum_{i=1}^{n} \varepsilon_i = 0,$$

$$\sum_{i=1}^{n} \varepsilon_i^2 = \frac{1}{4n(n-1)}.$$

By Lagrange's method, an optimal solution $(\varepsilon_1^*, \cdots, \varepsilon_n^*)$ of (2.7) must satisfy equations

$$\frac{1}{1 + n\varepsilon_i^*} = \xi \cdot \varepsilon_i^* + \eta \text{ for } i = 1, \cdots, n,$$

where ξ and η are Lagrange multiplers. That is, ε_i^*'s are all solutions of quadratic equation

$$n\xi\varepsilon^2 + (\xi + n\eta)\varepsilon + \eta - 1 = 0.$$

Thus, ε_i^*'s have at most two possible values. Since $\sum_{i=1}^{n} \varepsilon_i^* = 0$, all ε_i^*'s being equal implies $\varepsilon_i^* = 0$ for all i, contradicting the second constraint. Hence, ε_i^* have exactly two values; one is positive and another is negative. We claim that only one of ε_i^*'s can be negative. In fact, suppose to the contrary that $\varepsilon_1^* = \varepsilon_2^* < 0$. Without loss of generality, assume $\varepsilon_3^* > 0$. Denote $a = \varepsilon_1^* + \varepsilon_2^* + \varepsilon_3^*$ and $b = (\varepsilon_1^*)^2 + (\varepsilon_2^*)^2 + (\varepsilon_3^*)^2$. Clearly,

$$\begin{aligned}(1 + n\varepsilon_1^*)(1 + n\varepsilon_2^*)(1 + n\varepsilon_3^*) &= 1 + na + \frac{n^2}{2}(a^2 - b) + n^3\varepsilon_1^*\varepsilon_2^*\varepsilon_3^* \\ &> 1 + na + \frac{n^2}{2}(a^2 - b).\end{aligned}$$

Select $\bar{\varepsilon}_1 < \bar{\varepsilon}_2 = \bar{\varepsilon}_3$ such that $\bar{\varepsilon}_1 + \bar{\varepsilon}_2 + \bar{\varepsilon}_3 = a$ and $\bar{\varepsilon}_1^2 + \bar{\varepsilon}_2^2 + \bar{\varepsilon}_3^2 = b$. Similarly, we can obtain $(1 + n\bar{\varepsilon}_1)(1 + n\bar{\varepsilon}_2)(1 + n\bar{\varepsilon}_3) < 1 + na + \frac{n^2}{2}(a^2 - b)$. Therefore, $\phi(\bar{\varepsilon}_1, \bar{\varepsilon}_2, \bar{\varepsilon}_3, \varepsilon_4^*, \cdots, \varepsilon_n^*) < \phi(\varepsilon_1^*, \cdots, \varepsilon_n^*)$, contradicting the optimality of $(\varepsilon_1^*, \cdots, \varepsilon_n^*)$. Now, assuming $\varepsilon_1^* < \varepsilon_2^* = \cdots = \varepsilon_n^*$, it is easy to see that $\varepsilon_1^* = -1/(2n)$ and $\varepsilon_2^* = \cdots = \varepsilon_n^* = 1/(2n(n-1))$. Thus,

$$\prod_{i=1}^{n}(1 + n\varepsilon_i) \geq \frac{1}{2}\left(1 + \frac{1}{2(n-1)}\right)^{n-1}.$$

Hence

$$\begin{aligned}
\frac{\hat{g}(\boldsymbol{y}_{k+1})}{\hat{g}(\boldsymbol{y}_k)} &= \frac{(\hat{\boldsymbol{c}}^T \boldsymbol{y}_{k+1}/\hat{\boldsymbol{c}}^T \boldsymbol{y}_k)^n}{\prod_{i=1}^n (1+n\varepsilon_i)} \\
&\leq \frac{\left(1-\frac{1}{2(n-1)}\right)^n}{\frac{1}{2}\left(1+\frac{1}{2(n-1)}\right)^{n-1}} \\
&= 2\left(1-\frac{1}{2(n-1)}\right)\cdot\frac{1}{\left(1+\frac{1}{n-1.5}\right)^{n-1}} \\
&\leq \frac{2}{e}.
\end{aligned}$$

Here, we note that $(1+\frac{1}{x})^{x+0.5}$ converges decreasingly to e. Therefore,

$$\boldsymbol{c}^T\boldsymbol{x}_k \leq \frac{\boldsymbol{c}^T\boldsymbol{x}_k}{(x_{k1}\cdots x_{kn})^{1/n}} \leq \left(\frac{2}{e}\right)^{k/n}\frac{\boldsymbol{c}^T\boldsymbol{x}_1}{(x_{11}\cdots x_{1n})^{1/n}} = \left(\frac{2}{e}\right)^{k/n}\boldsymbol{c}^T\boldsymbol{x}_1.$$

Clearly, if (2.3) has a positive optimal value, then $\boldsymbol{c}^T\boldsymbol{x}_k > \left(\frac{2}{e}\right)^{k/n}\boldsymbol{c}^T\boldsymbol{x}_1$. Suppose that all coefficients are integers of absolute values less than T. We know that each extreme point of polytope $\{\boldsymbol{x} \mid A\boldsymbol{x} = \boldsymbol{0}, \boldsymbol{e}^T\boldsymbol{x} = 1, \boldsymbol{x} \geq \boldsymbol{0}\}$ is a rational vector whose components are all of denominator less than n^nT^n. Therefore, if $k = \lceil (n^2\log_2 n + n^2\log_2 T - \log_2(\boldsymbol{c}^T\boldsymbol{x}_1))/\log_2(\frac{e}{2})\rceil = O(n^2\log_2 n)$, then $\boldsymbol{c}^T\boldsymbol{x}_k > (\frac{2}{e})^{k/n}\boldsymbol{c}^T\boldsymbol{x}_1$. This means that in a polynomial number of iterations the algorithm can find whether the optimal value of (2.3) is zero. If the optimal value is zero, then $\boldsymbol{c}^T\boldsymbol{x}_k \leq (\frac{2}{e})^{k/n}\boldsymbol{c}^T\boldsymbol{x}_1 < n^{-n}T^{-n}$. Find an extreme point $\boldsymbol{x}^*$ of the polytope such that $\boldsymbol{c}^T\boldsymbol{x}^* \leq \boldsymbol{c}^T\boldsymbol{x}_k$. Then $\boldsymbol{x}^*$ is an optimal point. Note that a linear programming with rational coefficients can be transformed to the one with integer coefficients by multiplying a number. Therefore, Karmarkar's algorithm is a polynomial-time algorithm for linear programming with rational coefficients.

Karmarkar's algorithm is the first interior-point algorithm for linear programming. After the publication of Karmarkar's algorithm, many interior-point methods have been designed for linear and convex optimization problems. All these methods view combinatorial objects as limiting cases of nonlinear continuous objects and exploit the topological and geometric properties of the continuous space.

Exercises

1. Use simplex method to solve the following:

$$\begin{aligned} \min \quad & -2x_1 + x_2 \\ \text{subject to} \quad & x_1 + x_2 \le 3 \\ & 2x_1 + x_2 \ge 1 \\ & x_1 - x_2 \le 2 \\ & x_1 \ge 0 \\ & x_2 \ge 0. \end{aligned}$$

2. Solve the following problem with the lexicographic ordering method:

$$\begin{aligned} \min \quad & -\tfrac{3}{4}x_4 + 20x_5 - \tfrac{1}{2}x_6 + 6x_7 \\ \text{s.t.} \quad & x_1 + \tfrac{1}{4}x_4 - 8x_5 - x_6 + 9x_7 = 0 \\ & x_2 + \tfrac{1}{2}x_4 - 12x_5 - \tfrac{1}{2}x_6 + 3x_7 = 0 \\ & x_3 + x_6 = 1 \\ & x_1, \cdots, x_7 \ge 0 \end{aligned}$$

3. Show the following:

   (a) In every iteration of the simplex method,

$$\begin{pmatrix} 1 & \mathbf{c} \\ \mathbf{0} & A \end{pmatrix} \boldsymbol{y} = \mathbf{0} \Leftrightarrow \begin{pmatrix} 1 & \mathbf{c}' \\ \mathbf{0} & A' \end{pmatrix} \boldsymbol{y} = \mathbf{0}.$$

   (b) Suppose $I = \{j_1, \cdots, j_m\}$ is a basis and $s \notin I$. Define $(n+1)$-dimensional vector $\boldsymbol{y}$ by

$$y_j = \begin{cases} c'_s & \text{if } j = 0 \\ -1 & \text{if } j = s \\ a'_{is} & \text{if } j = j_i \text{ where } 1 \le i \le m \\ 0, & \text{otherwise.} \end{cases}$$

   where $(\mathbf{c}')^T = \mathbf{c}^T - \mathbf{c}_I^T A_I^{-1} A$ and $A' = A_I^{-1} A$. Then

$$\begin{pmatrix} 1 & \mathbf{c} \\ \mathbf{0} & A \end{pmatrix} \boldsymbol{y} = \mathbf{0}.$$

Furthermore, suppose I' is another basis and $(\boldsymbol{c}'')^T = \boldsymbol{c}^T - \boldsymbol{c}_{I'}^T A_{I'}^{-1} A$. Then

$$c''_s - c'_s = \sum_{j \in I \subseteq I'} c''_j y_j.$$

(c) Suppose the simplex method fails into a cycle. Let t be the maximum index of a variable which gets in and out basis during the cycle. Suppose I is a feasible basis with which x_t gets out from basis and x_s gets in basis, and I' is a feasible basis with which x_t gets in basis. Then one of the following must occur:

- $c'_s < 0$,
- $c''_j < 0$ for some $j < t$,
- $a'_{is} > 0$ for some $j_i \in I \subseteq I'$.

(d) In the simplex method, we may choose i' and j' in the following way:

$$\begin{aligned} j' &= \min\{j \mid c'_j < 0\} \\ j_{i'} &= \min\{j_r \mid \frac{b'_r}{a'_{rj'}} = \min\{\frac{b'_i}{a'_{ij'}} \mid a_{ij'} > 0\}\} \end{aligned}$$

where j_r is the index of variable in basis such that $a'_{rj_r} = 1$. This rule is called *Bland rule*. Show that the simplex method with Bland rule halts in finitely many iterations without nondegeneracy assumption.

4. There is a direct proof for Farkas lemma as follows: "If there exists $\boldsymbol{x} \geq \boldsymbol{0}$ such that $A\boldsymbol{x} = \boldsymbol{b}$, then it is obvious that $A^T\boldsymbol{y} \geq \boldsymbol{0}$ implies $\boldsymbol{b}^T\boldsymbol{y} = \boldsymbol{x}^T A^T \boldsymbol{y} \geq 0$. Now, suppose that there is no $\boldsymbol{x} \geq \boldsymbol{0}$ such that $A\boldsymbol{x} = \boldsymbol{b}$. We need to find a vector $\boldsymbol{y}$ such that $A^T\boldsymbol{y} \geq \boldsymbol{0}$ and $\boldsymbol{b}^T\boldsymbol{y} < 0$. Let C be the convex cone generated by the column vectors of A. Then $\boldsymbol{b} \notin C$. Suppose that $\boldsymbol{b}^*$ is the element in C such that $\|\boldsymbol{b} - \boldsymbol{b}^*\|$ reaches the minimal value. (Such a vector $\boldsymbol{b}^*$ exists since $\min\{\|\boldsymbol{b} - \boldsymbol{c}\| \mid \boldsymbol{c} \in C\} = \min\{\|\boldsymbol{b} - \boldsymbol{c}\| \mid \boldsymbol{c} \in C, \|\boldsymbol{c}\| \leq 2\|\boldsymbol{b}\|\}$ and the latter minimization is over a compact region.) Define $\boldsymbol{y} = \boldsymbol{b}^* - \boldsymbol{b}$. Clearly, $A^T\boldsymbol{y} \geq \boldsymbol{0}$ and $\boldsymbol{b}^T\boldsymbol{y} < 0$."

(a) Use Farkas lemma to show Theorem 2.2.4.

(b) Give a geometric explanation to Farkas lemma.

5. A company has m department stores and n storages. To prepare Christmas sales, the m department stores want to input $a_1, a_2, \cdots, a_m$ units of certain goods, respectively. The n storages have $b_1, b_2, \cdots, b_n$ units of the goods and the total amount just meet the demand, i.e., $\sum_{i=1}^m a_i = \sum_{j=1}^n b_j$. Suppose the cost for the ith storage to send a unit of goods to the jth store is $c_{ij} > 0$. Find a flow to minimize total cost. This problem can be formulated as the following linear programming:

$$\begin{array}{lll} \min & \sum_{i=1}^m \sum_{j=1}^n c_{ij} x_{ij} & \\ \text{s.t.} & \sum_{i=1}^m x_{ij} = b_j & \text{for } 1 \le j \le n \\ & \sum_{j=1}^n x_{ij} = a_i & \text{for } 1 \le i \le m \\ & x_{ij} \ge 0 & \text{for } 1 \le i \le m, 1 \le j \le n. \end{array}$$

   (a) Show that a subset X of variables is a basis if and only if $|X| = m + n - 1$ and the graph $G(X)$ with vertex set X and edge set $\{(x_{ij}, x_{hk}) \in X \times X \mid (i = h \text{ or } j = k) \text{ and } x_{ij} \ne x_{hk}\}$ is a tree.

   (b) Suppose X is a feasible basis. Show that for each variable $x_{ij} \notin X$, the graph $G(X \cup \{x_{ij}\})$ contains exactly one cycle $(x_{ij}, x_{ij_1}, x_{i_1 j_1}, \cdots, x_{i_k j_k}, x_{i_k j})$. Define $c'_{ij} = c_{ij} - c_{ij_1} + c_{i_1 j_1} - \cdots + c_{i_k j_k} - c_{i_k j}$. Show that if $c'_{ij} \ge 0$ for all $x_{ij} \notin X$, then the feasible basic solution associated with basis X is minimum.

   (c) Compute p_i for $1 \le i \le m$ and q_i for $1 \le j \le n$ in the following way: Initially, pick up an i such that the ith row contains only one variable in basis X, and assign $p_i = 0$. Then, for each $x_{ij} \in X$ do the following until all p_i's and q_j's have been assigned: if p_i is assigned and q_j is not assigned, then set $q_j = c_{ij} - p_i$; if p_i is not assigned and q_j is assigned, then set $p_i = c_{ij} - q_j$. Show that for every $x_{ij} \notin X$, $c'_{ij} = c_{ij} - p_i - q_j$.

6. Show that any linear programming can be transformed to a one with known feasible solution and the same optimal value of objective function.

Chapter 3

Blind Man's Method

There is a budding morrow in midnight;
There is a triple sight in blindness keen.
KEATS

A blind man who sees is better than
a seeing man who is blind.
PERSIAN PROVERB

Finding a maximum point of a function is like climbing a mountain. When a blind man climbs a mountain, how does he do it? He is unable to see the global direction and may collect only local information with his stick. A naive way for him is to choose the steepest ascent direction at his current point (Figure 3.1) and along this direction climbs to a new point. In this chapter, we discuss this method of a blind man.

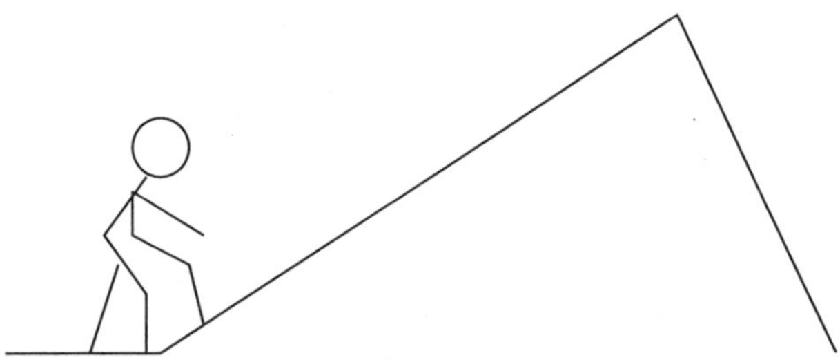

Figure 3.1: A blind man climbs a mountain.

3.1 Unconstrained Optimization

Consider an unconstrained optimization problem

$$\max_{\boldsymbol{x}} f(\boldsymbol{x})$$

where $f(\boldsymbol{x})$ is a continuously differentiable function in n-dimensional Euclidean space R^n. Each point in R^n is considered as a column vector.

A scalar x and a vector $\boldsymbol{x}$ can be easily distinguished from their printing. For the vector $\boldsymbol{x}$, x_i denotes the ith component of $\boldsymbol{x}$ and $\|\boldsymbol{x}\|$ is the Euclidean norm of $\boldsymbol{x}$, i.e. $\|\boldsymbol{x}\| = \sqrt{\sum x_i^2}$. As usual, $\boldsymbol{x}^T\boldsymbol{y}$ is the inner product of vectors $\boldsymbol{x}$ and $\boldsymbol{y}$, i.e. $\boldsymbol{x}^T\boldsymbol{y} = \sum x_i y_i$.

To design an algorithm for solving the above maximization problem, we choose an initial point $\boldsymbol{x}_1$ and study the kth iteration. Suppose at the beginning of the kth iteration, a blind man stands at the point $\boldsymbol{x}_k$. Which way should he go? Since he can only collect local information with his stick, a naive choice is the steepest ascent (Figure 3.1). What is the steepest ascent at $\boldsymbol{x}_k$? Let us look at an example.

Consider a function of two variables, $f(x, y) = 1 - x^2 - 2y^2$. It can be described by a family of curves $c = 1 - x^2 - 2y^2$. Suppose $(x, y)^T$ is on the curve $c = 1 - x^2 - 2y^2$. Clearly, the steepest ascent at (x, y) is in the direction given by the normal vector of the tangent line at $(x, y)^T$, which is $(f_x, f_y)^T$. This vector is called the *gradient* of f, denoted by $\nabla f(x, y) = (f_x, f_y)^T$.

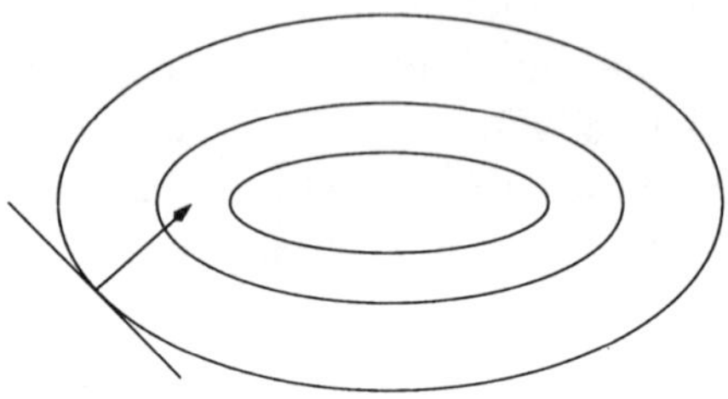

Figure 3.2: Gradient.

In general, suppose $\boldsymbol{x} = (x_1, x_2, \cdots, x_n)^T$. Then the gradient of $f(\boldsymbol{x})$ at the point x is $\nabla f(\boldsymbol{x}) = (f_{x_1}, f_{x_2}, \cdots, f_{x_n})^T$. The steepest ascent at $\boldsymbol{x}$ is in the direction given by the gradient at $\boldsymbol{x}$ (Figure 3.2).

Now, the blind man follows the gradient direction to search a new point. Where should he stop? The easiest way for a blind man is to stop at the first turning point, that is, the first point starting lower down. However, if the first turning point is too far or does not exist, then the blind man will get in trouble. Thus, a maximum step length may be welcome. This gives the following algorithm.

Blind Man's Method: *Choose an initial point* $\boldsymbol{x}_1$ *and a maximum step length* $\bar{\lambda}$. *At the kth iteration, carry out the following steps.*

1. *Compute gradient* $\boldsymbol{d}_k = \nabla f(\boldsymbol{x}_k)$.

2. *If* $\boldsymbol{d}_k = \boldsymbol{0}$, *then stop (*$\boldsymbol{x}_k$ *is a local maximal point). Else, if for* $\lambda \in [0, \bar{\lambda}]$

$$\nabla f(\boldsymbol{x}_k + \lambda \boldsymbol{d}_k)^T \boldsymbol{d}_k > 0$$

then set $\lambda_k = \bar{\lambda}$; *else. find the smallest positive number* λ_k *in* $(0, \lambda]$ *such that*

$$\nabla f(\boldsymbol{x}_k + \lambda_k \boldsymbol{d}_k)^T \boldsymbol{d}_k = 0.$$

Set

$$\boldsymbol{x}_{k+1} = \boldsymbol{x}_k + \lambda_k \boldsymbol{d}_k.$$

Note that $\nabla f(\boldsymbol{x}_k)^T \boldsymbol{d}_k = \|\nabla f(\boldsymbol{x}_k)\|^2 > 0$ when $\boldsymbol{d}_k \neq \boldsymbol{0}$. Thus, for $0 \leq \lambda < \lambda_k$,

$$\frac{df(\boldsymbol{x}_k + \lambda \boldsymbol{d}_k)}{d\lambda} = \nabla f(\boldsymbol{x}_k + \lambda \boldsymbol{d}_k)^T \boldsymbol{d}_k > 0.$$

It follows that $f(\boldsymbol{x}_k + \lambda \boldsymbol{d}_k)$ is strictly increasing with respect to λ in $[0, \lambda_k]$. Therefore,

$$f(\boldsymbol{x}_k) < f(\boldsymbol{x}_{k+1}).$$

That is, the Blind Man's method produces an ascendant sequence of points.

3.2 Global Convergence

In Blind Man's method, when $\boldsymbol{d}_k = \boldsymbol{0}$, we have

$$\nabla f(\boldsymbol{x}_k) = \boldsymbol{0}.$$

This condition cannot guarantee that $\boldsymbol{x}_k$ is a maximum point of $f(\boldsymbol{x})$. However, it is a necessary condition. In fact,

$$f(\boldsymbol{x}) = f(\boldsymbol{x}_k) + \nabla f(\boldsymbol{x}_k)^T (\boldsymbol{x} - \boldsymbol{x}_k) + o(\|\boldsymbol{x} - \boldsymbol{x}_k\|).$$

If $\nabla f(\boldsymbol{x}_k) \neq \mathbf{0}$, then for sufficiently small λ, we will have

$$f(\boldsymbol{x}_k + \lambda \boldsymbol{d}_k) = f(\boldsymbol{x}_k) + \lambda \|\nabla f(\boldsymbol{x}_k)\|^2 + o(\lambda) > f(\boldsymbol{x}_k).$$

Thus, $\boldsymbol{x}_k$ cannot be a maximum point.

A point $\boldsymbol{x}$ is called a *stationary* point of a function f if $\nabla f(\boldsymbol{x}) = \mathbf{0}$. An algorithm is said to be *globally convergent* if it either stops at a stationary point or generates an infinite sequence of points in which every cluster point is a stationary point.

Theorem 3.2.1 *Blind Man's method is globally convergent.*

Proof. Clearly, if the algorithm stops within finitely many iterations, then it stops at a stationary point. Suppose the algorithm generates an infinite sequence of points $\boldsymbol{x}_1, \boldsymbol{x}_2, \cdots$. Consider a cluster point $\boldsymbol{x}^*$ of this sequence. There exists a subsequence, $\boldsymbol{x}_{k_1}, \boldsymbol{x}_{k_2}, \cdots$, convergent to $\boldsymbol{x}^*$.

First, we assume that the sequence of numbers $\lambda_{k_1}, \lambda_{k_2}, \cdots$ converges to 0. Thus, the sequence of points $\boldsymbol{x}_{k_1+1}, \boldsymbol{x}_{k_2+1}, \cdots$ is also convergent to $\boldsymbol{x}^*$ since $\boldsymbol{x}_{k+1} = \boldsymbol{x}_k + \lambda_k \nabla f(\boldsymbol{x}_k)$ and f is continuously differentiable. Without loss of generality, we may also assume that $\lambda_{k_i} < \bar{\lambda}$ for all i. It follows that

$$\nabla f(\boldsymbol{x}_{k_i+1})^T \boldsymbol{d}_{k_i} = 0$$

for all $i = 1, 2, \cdots$. Letting i go to infinity, we obtain

$$\|\nabla f(\boldsymbol{x}^*)\| = 0.$$

Hence, $\nabla f(\boldsymbol{x}^*) = \mathbf{0}$.

Next, we assume that the sequence of numbers $\lambda_{k_1}, \lambda_{k_2}, \cdots$ is not convergent to 0. Hence, there exists a number $\zeta > 0$ such that $\lambda_{k_i} > \zeta$ for infinitely many i's. Without loss of generality, we may assume $\lambda_{k_i} > \zeta$ for all i. Then for every $\lambda \in (0, \zeta)$,

$$f(\boldsymbol{x}_{k_i+1}) = f(\boldsymbol{x}_k + \lambda_k \boldsymbol{d}_k) \geq f(\boldsymbol{x}_k + \lambda \boldsymbol{d}_k).$$

Therefore, for $k = k_1, k_2, \cdots$, we have

$$\begin{aligned} f(\boldsymbol{x}_{k+1}) - f(\boldsymbol{x}_k) &\geq f(\boldsymbol{x}_k + \lambda \boldsymbol{d}_k) - f(\boldsymbol{x}_k) \\ &= \lambda \nabla f(\boldsymbol{x}_k + \lambda'_k \boldsymbol{d}_k)^T \boldsymbol{d}_k \end{aligned} \tag{3.1}$$

where $\lambda'_k \in (0, \lambda)$. Clearly, the sequence of numbers $\lambda'_{k_1}, \lambda'_{k_2}, \cdots$ has a convergent subsequence. Without loss of generality, we may assume that the sequence of numbers $\lambda'_{k_1}, \lambda'_{k_2}, \cdots$ itself is convergent to a number $\lambda' \in [0, \lambda]$. Note that

$$f(\boldsymbol{x}_{k_i}) < f(\boldsymbol{x}_{k_i+1}) \leq f(\boldsymbol{x}_{k_{i+1}}).$$

Letting i go to infinity, we obtain

$$f(\boldsymbol{x}^*) \leq \lim_{i \to \infty} f(\boldsymbol{x}_{k_i+1}) \leq f(\boldsymbol{x}^*).$$

Hence

$$\lim_{i \to \infty} f(\boldsymbol{x}_{k_i+1}) = f(\boldsymbol{x}^*).$$

Now, in (3.1), letting k approach to infinity through $k_1, k_2, \cdots$, we obtain

$$0 \geq \lambda \nabla f(\boldsymbol{x}^* + \lambda' \boldsymbol{d}^*)^T \boldsymbol{d}^*$$

where $\boldsymbol{d}^* = \nabla f(\boldsymbol{x}^*)$. Thus,

$$0 \geq \nabla f(\boldsymbol{x}^* + \lambda' \boldsymbol{d}^*)^T \boldsymbol{d}^*$$

Letting λ go to zero, we obtain

$$0 \geq \nabla f(\boldsymbol{x}^*)^T \boldsymbol{d}^* = \|\nabla f(\boldsymbol{x}^*)\|^2$$

that is, $0 = \|\nabla f(\boldsymbol{x}^*)\|$ and hence $\nabla f(\boldsymbol{x}^*) = \mathbf{0}$. Therefore, $\boldsymbol{x}^*$ is a stationary point. □

3.3 Zangwill's Theorem

There is a general tool that helps to prove global convergence of algorithms. It is Zangwill's theorem. To introduce this theorem, let us first explain some concepts.

Let X and Y be two topological spaces. Denote by $\mathcal{P}(X)$ the collection of all subsets of X. A mapping from X to $\mathcal{P}(Y)$ is usually called a *point-to-set mapping*. A point-to-set mapping $A : X \to \mathcal{P}(Y)$ is said to be *closed* at a point $\boldsymbol{x}$ if

$$\left.\begin{array}{l} \boldsymbol{x}_k \to \boldsymbol{x} \\ \boldsymbol{y}_k \to \boldsymbol{y} \\ \boldsymbol{y}_k \in A(\boldsymbol{x}_k) \end{array}\right\} \Longrightarrow \boldsymbol{y} \in A(\boldsymbol{x}). \tag{3.2}$$

A point-to-set mapping is *closed* if it is closed at every point in its definition domain.

For a closed point-to-set mapping $A : X \to \mathcal{P}(Y)$, $A(\boldsymbol{x})$ for every $\boldsymbol{x} \in X$ is a closed set in Y. To see this, suppose a sequence of points $\boldsymbol{y}_1, \boldsymbol{y}_2, \cdots$ in $A(\boldsymbol{x})$ converges to a point $\boldsymbol{y}$. Choose $\boldsymbol{x}_1 = \boldsymbol{x}_2 = \cdots = \boldsymbol{x}$. Then we have $\boldsymbol{x}_k \to \boldsymbol{x}$ and $\boldsymbol{y}_k \in A(\boldsymbol{x}_k)$. By the closedness of mapping A, we see $y \in A(x)$.

Furthermore, if a sequence of points $\boldsymbol{x}_1, \boldsymbol{x}_2, \cdots$ in X converges to a point $\boldsymbol{x}$, then $\cup_{k=1}^{\infty} A(\boldsymbol{x}_k) \cup A(\boldsymbol{x})$ is a closed set. To show this, suppose a sequence of points $\boldsymbol{y}_1, \boldsymbol{y}_2, \cdots$ in $\cup_{k=1}^{\infty} A(\boldsymbol{x}_k) \cup A(\boldsymbol{x})$ converges to a point $\boldsymbol{y}$. Then either there exists a $A(x_{k_i})$ containing an infinite subsequence of the sequence $\{\boldsymbol{y}_k\}$ or there exists a subsequence of points $\boldsymbol{y}_{k_1}, \boldsymbol{y}_{k_2}, \cdots$ such that $\boldsymbol{y}_{k_1} \in A(\boldsymbol{x}_{\ell_1}), \boldsymbol{y}_{k_2} \in A(\boldsymbol{x}_{\ell_2}), \cdots$ with $\ell_1 < \ell_2 < \cdots$. In the former case, we know $y \in A(\boldsymbol{x}_{k_i})$. In the latter case, we know $y \in A(\boldsymbol{x})$.

In the following, we will be interested in the case that $\cup_{k=1}^{\infty} A(\boldsymbol{x}_k) \cup A(\boldsymbol{x})$ is a compact set for every sequence $\{\boldsymbol{x}_k\}$ convergent to $\boldsymbol{x}$.

Consider the following abstract algorithm.

Zangwill's Algorithm: *Let Γ be a subset of a topological space X. Let A be a point-to-set mapping from $X \setminus \Gamma$ to $\mathcal{P}(X)$.*

> Step 0. *Choose an initial point $\boldsymbol{x}_1$.*
>
> Step k. *If $\boldsymbol{x}_k \in \Gamma$, then stop. Otherwise, choose $\boldsymbol{x}_{k+1} \in A(\boldsymbol{x}_k)$.*

The importance of the closedness of point-to-set mappings is included in the following theorem.

Theorem 3.3.1 (Zangwill's Theorem) *Let f be a continuous function on a topological space X and Γ a subset of X. Let A be a closed point-to-set mapping from $X \setminus \Gamma$ to $\mathcal{P}(X)$ such that*

(a) the closure of $\cup_{k=1}^{\infty} A(\boldsymbol{x}_k)$ is compact for every convergent sequence of points $\boldsymbol{x}_1, \boldsymbol{x}_2, \cdots$ in $X \setminus \Gamma$, and

(b) for every $\boldsymbol{x} \in X \setminus \Gamma$ and $\boldsymbol{y} \in A(\boldsymbol{x})$, $f(\boldsymbol{y}) > f(\boldsymbol{x})$.

Then Zangwill's Algorithm either stops at a point in Γ or generates an infinite sequence whose cluster points are all in Γ.

Proof. Clearly, if the algorithm stops at $\boldsymbol{x}_k$, then $\boldsymbol{x}_k \in \Gamma$. Next, suppose that the algorithm generates an infinite sequence $\{\boldsymbol{x}_k\}$. Let $\boldsymbol{x}^*$ be a cluster point of the sequence. Then there exists a subsequence $\{\boldsymbol{x}_k\}_{k \in K}$ such that $\lim_{k \to \infty, k \in K} \boldsymbol{x}_k = \boldsymbol{x}^*$ and $\lim_{k \to \infty, k \in K} \boldsymbol{x}_{k+1} = \boldsymbol{x}^{**}$ for some point $\boldsymbol{x}^{**}$. Suppose to the contrary that $\boldsymbol{x}^* \notin \Gamma$. Then A is closed at $\boldsymbol{x}^*$. So, $\boldsymbol{x}^{**} \in A(\boldsymbol{x}^*)$. It follows that $f(\boldsymbol{x}^{**}) > f(\boldsymbol{x}^*)$. However, since $\{f(\boldsymbol{x}_k)\}$ is increasing, it is easy to see $f(\boldsymbol{x}^*) = \lim_{k\to\infty, k\in K} f(\boldsymbol{x}_k) = \lim_{k\to\infty, k\in K} f(\boldsymbol{x}_{k+1}) = f(\boldsymbol{x}^{**})$, a contradiction. □

Now, we use Blind Man's method as an example to explain how to use Zangwill's Theorem. Let $A(\boldsymbol{x})$ be the set of all points $\boldsymbol{y}$'s satisfying the following conditions:

1. $\boldsymbol{y} = \boldsymbol{x} + \lambda \nabla f(\boldsymbol{x})$ for $\lambda \in [0, \bar{\lambda}]$ and for all $\boldsymbol{z} = \boldsymbol{x} + \zeta \nabla f(\boldsymbol{x})$ with $\zeta \in [0, \lambda]$,
$$\nabla f(\boldsymbol{z})^T \nabla f(\boldsymbol{x}) \geq 0$$
2. either $\lambda = \bar{\lambda}$ or $\nabla f(\boldsymbol{y})^T \nabla f(\boldsymbol{x}) = 0$.

From this definition, it is easy to see that if $\nabla f(\boldsymbol{x}) = \boldsymbol{0}$, then $\boldsymbol{x} \in A(\boldsymbol{x})$.

Mapping A is a closed point-to-set mapping. To show this, consider two sequences $\{\boldsymbol{x}_k\}$ and $\{\boldsymbol{y}_k\}$ such that $\boldsymbol{y}_k \in A(\boldsymbol{x}_k)$, $\boldsymbol{x}_k \to \boldsymbol{x}^*$, and $\boldsymbol{y}_k \to \boldsymbol{y}^*$. By the definition of A,
$$\boldsymbol{y}_k = \boldsymbol{x}_k + \lambda_k \nabla f(\boldsymbol{x}_k)$$
for some $\lambda_k \in [0, \bar{\lambda}]$. If there are infinitely many k's such that $\nabla f(\boldsymbol{x}_k) = \boldsymbol{0}$, then for such k's $\boldsymbol{y}_k = \boldsymbol{x}_k$ and hence $\boldsymbol{y}^* = \boldsymbol{x}^*$ and $\nabla f(\boldsymbol{y}^*) = \boldsymbol{0}$. Hence, $\boldsymbol{y}^* \in A(\boldsymbol{x}^*)$. Therefore, we may assume, without loss of generality, that $\nabla f(\boldsymbol{x}_k) \neq \boldsymbol{0}$ for all k. Since f is continuously differentiable, we have
$$\lim_{k\to\infty} \frac{\|\boldsymbol{y}_k - \boldsymbol{x}_k\|}{\|\nabla f(\boldsymbol{x}_k)\|} = \lambda^*$$
for some $\lambda^* \in [0, \bar{\lambda}]$, and hence
$$\boldsymbol{y}^* = \boldsymbol{x}^* + \lambda^* \nabla f(\boldsymbol{x}^*).$$
If there are infinitely many k's such that $\lambda_k = \bar{\lambda}$, then $\lambda^* = \bar{\lambda}$. Otherwise, there must exist infinitely many k's such that $\nabla f(\boldsymbol{y}_k)^T \nabla f(\boldsymbol{x}_k) = 0$. Letting k go to infinity through such k's, we obtain $\nabla f(\boldsymbol{y}^*)^T \nabla f(\boldsymbol{x}^*) = 0$. Hence, $\boldsymbol{y}^* \in A(\boldsymbol{x}^*)$.

Mapping A satisfies condition (a) in Zangwill's Theorem. In fact, for any convergent sequence $\{\boldsymbol{x}_k\}$, $\cup_{k=1}^{\infty} A(\boldsymbol{x}_k)$ is bounded and hence its closure must be compact.

Define $\Gamma = \{\boldsymbol{x} \mid \nabla f(\boldsymbol{x}) = \boldsymbol{0}\}$. Mapping A satisfies condition (b) in Zangwill's Theorem. In fact, for $\boldsymbol{x} \notin \Gamma$,
$$\left.\frac{df(\boldsymbol{x} + \zeta \nabla f(\boldsymbol{x}))}{d\zeta}\right|_{\zeta=0} = \|\nabla f(\boldsymbol{x})\|^2 > 0.$$

This means that $f(\boldsymbol{x} + \zeta \nabla f(\boldsymbol{x}))$ is strictly increasing at $\zeta = 0$. Moreover, $f(\boldsymbol{x}+\zeta\nabla f(\boldsymbol{x}))$ is nondecreasing with respect to ζ in $[0, \lambda]$. Therefore, $f(\boldsymbol{y}) > f(\boldsymbol{x})$.

Now, we note that although not every sequence generated by Zangwill's Algorithm with defined point-to-set mapping A can be generated by Blind Man's method, every sequence generated by Blind Main's method can be generated by Zangwill's Algorithm with defined point-to-set mapping A. Therefore, by Zangwill's Theorem, we obtain another proof of Theorem 3.2.1.

From the above example, we see the following: To apply Zangwill's Theorem to an iterative algorithm, we may associate the algorithm with a point-to-set mapping such that each iteration can be viewed as taking a new point $\boldsymbol{x}_{k+1}$ from the image of the mapping at the current point $\boldsymbol{x}_k$. However, such a point-to-set mapping may not be closed. In such a case, we may consider a large point-to-set mapping. The following theorem tells us how far we are able to go in this way.

Theorem 3.3.2 *Let f be a continuous function on a compact set X and $\Gamma \subset X$. Let A be a point-to-set mapping from $X \setminus \Gamma$ to $\mathcal{P}(X)$. Then there exists a point-to-set mapping $H : X \setminus \Gamma \to \mathcal{P}(X)$ such that*

(1) *$A(\boldsymbol{x}) \subseteq H(\boldsymbol{x})$ for all $\boldsymbol{x} \in X \setminus \Gamma$,*

(2) *H is closed in $X \setminus \Gamma$, and*

(3) *for every $\boldsymbol{x} \in X \setminus \Gamma$ and $\boldsymbol{y} \in H(\boldsymbol{x})$, $f(\boldsymbol{y}) > f(\boldsymbol{x})$,*

if and only if there exists a lower semi-continuous positive-value function $\epsilon : X\setminus\Gamma \to R_+$ such that for every $\boldsymbol{x} \in X\setminus\Gamma$ and $\boldsymbol{y} \in A(\boldsymbol{x})$, $f(\boldsymbol{y}) \geq f(\boldsymbol{x})+\epsilon(\boldsymbol{x})$.

Proof. For necessity, suppose that there exists a mapping H satisfying (1), (2) and (3). Define

$$\epsilon(\boldsymbol{x}) = \inf\{f(\boldsymbol{y}) \mid \boldsymbol{y} \in H(\boldsymbol{x})\} - f(\boldsymbol{x}).$$

Clearly, for every $\boldsymbol{x} \in X \setminus \Gamma$,

$$\boldsymbol{y} \in A(\boldsymbol{x}) \Longrightarrow \boldsymbol{y} \in H(\boldsymbol{y}) \text{ and } f(\boldsymbol{y}) \geq f(\boldsymbol{x}) + \epsilon(\boldsymbol{x}).$$

Thus, it suffices to verify the lower semi-continuity and positivity of $\epsilon(\boldsymbol{x})$. Consider any $\boldsymbol{x}^* \in X \setminus \Gamma$. Suppose that $\liminf_{\boldsymbol{x}\to\boldsymbol{x}^*} \epsilon(\boldsymbol{x}) = \epsilon^*$. Then there exists a sequence $\{\boldsymbol{x}_k\}$ in $X\setminus\Gamma$ such that $\boldsymbol{x}_k \to \boldsymbol{x}^*$ and $\epsilon(\boldsymbol{x}_k) \to \epsilon^*$ as $k \to \infty$. Since H is closed at $\boldsymbol{x}_k$, it is easy to see that $H(\boldsymbol{x}_k)$ is a closed subset of

X and hence is a compact subset. Therefore, there exists $\boldsymbol{y}_k \in H(\boldsymbol{x}_k)$ such that $f(\boldsymbol{y}_k) = \inf\{f(\boldsymbol{y}) \mid \boldsymbol{y} \in H(\boldsymbol{x}_k)\}$. Note that all $\boldsymbol{y}_k$'s lie in the compact set X. Thus, there is a subsequence $\{\boldsymbol{y}_k\}_{k\in K}$ convergent to a point $\boldsymbol{y}^* \in X$. Furthermore, since H is closed at $\boldsymbol{x}^*$, we have $\boldsymbol{y}^* \in H(\boldsymbol{x}^*)$. Hence

$$\begin{aligned} \epsilon(\boldsymbol{x}^*) &\le f(\boldsymbol{y}^*) - f(\boldsymbol{x}^*) \\ &= \lim_{k\to\infty, k\in K} [f(\boldsymbol{y}_k) - f(\boldsymbol{x}_k)] \\ &= \lim_{k\to\infty, k\in K} \epsilon(\boldsymbol{x}_k) = \epsilon^*. \end{aligned}$$

Thus, $\epsilon(\boldsymbol{x})$ is semi-continuous at $\boldsymbol{x}^*$. Moreover, since H is closed at $\boldsymbol{x}^*$, $H(\boldsymbol{x}^*)$ is a compact set, so that there exists $\boldsymbol{y}^* \in H(\boldsymbol{x}^*)$ such that $f(\boldsymbol{y}^*) = \inf\{f(\boldsymbol{y}) \mid \boldsymbol{y} \in H(\boldsymbol{y}^*)\}$. By (3), $\epsilon(\boldsymbol{x}^*) = f(\boldsymbol{y}^*) - f(\boldsymbol{x}^*) > 0$.

For sufficiency, assume that there is a required function $\epsilon(\boldsymbol{x})$. Define $H : X \to \mathcal{P}(X)$ by

$$H(\boldsymbol{x}) = \{\boldsymbol{y} \in X \mid f(\boldsymbol{y}) \ge f(\boldsymbol{x}) + \epsilon(\boldsymbol{x})\}.$$

Clearly, $A(\boldsymbol{x}) \subseteq H(\boldsymbol{x})$ and for $\boldsymbol{y} \in H(\boldsymbol{x})$, $f(\boldsymbol{y}) \ge f(\boldsymbol{x}) + \epsilon(\boldsymbol{x}) > f(\boldsymbol{x})$. To prove the closeness, suppose that $\boldsymbol{x}_k \to \boldsymbol{x}$, $\boldsymbol{y}_k \to \boldsymbol{y}$ and $\boldsymbol{y}_k \in H(\boldsymbol{x}_k)$ for all k. By the definition of H, $f(\boldsymbol{y}_k) \ge f(\boldsymbol{x}_k) + \epsilon(\boldsymbol{x}_k)$. Letting $k \to \infty$ and by semi-continuity of $\epsilon(\boldsymbol{x})$, we obtain

$$f(\boldsymbol{y}) \ge f(\boldsymbol{x}) + \liminf_{k\to\infty} \epsilon(\boldsymbol{x}_k) \ge f(\boldsymbol{x}) + \epsilon(\boldsymbol{x}).$$

Hence $\boldsymbol{y} \in H(\boldsymbol{x})$, that is, H is closed. □

Exercises

1. Consider problem $\max f(\boldsymbol{x})$ where f is continuously differentiable and there exists $\boldsymbol{x}_0$ such that $\{\boldsymbol{x} \mid f(\boldsymbol{x}) \ge f(\boldsymbol{x}_0)\}$ is a compact set in the n-dimensional space R^n. A general algorithm can be described as follows: Choose an initial point $\boldsymbol{x}_0$ and set $k := 0$. In each iteration, carry out the following steps:

   Step 1 . Compute $\boldsymbol{g}_k = g(\boldsymbol{x}_k)$. If $\boldsymbol{g}_k = 0$, then stop; else, go to Step 2.

   Step 2 . Find a direction $\boldsymbol{d}_k$ such that $\boldsymbol{g}_k^T \boldsymbol{d}_k \ge \rho \|\boldsymbol{g}_k\| \cdot \|\boldsymbol{d}_k\|$ where ρ is a positive constant independent from k.

Step 3 . Find $\lambda_k \geq 0$ such that

$$f(\boldsymbol{x}_k + \lambda_k \boldsymbol{d}_k) = \max\{f(\boldsymbol{x}_k + \lambda \boldsymbol{d}_k) \mid \lambda \geq 0\}.$$

Set $\boldsymbol{x}_{k+1} = \boldsymbol{x}_k + \lambda_k \boldsymbol{d}_k$ and $k := k + 1$.

Show that this algorithm either stops at a stationary point or generates a sequence of points whose cluster points are all stationary points.

2. Let $\phi(\lambda)$ be a twice continuously differentiable function in $[0, b]$. Suppose $\phi'(0) < 0$ and $\lambda^* \in (0, b)$ is a minimum point of $\phi(\lambda)$ over $[0, b]$. Show that $\lambda^* \geq -\phi'(0)/Q$ where Q is a positive upper bound of $\phi''(\lambda)$ over $[0, b]$.

3. Let f be a continuous function on n-dimensional Euclidean space R^n and Γ a subset of R^n. Let A be a point-to-set mapping from R^n to $\mathcal{P}(R^n)$ such that for any point $\boldsymbol{x} \in R^n - \Gamma$, $A(\boldsymbol{x})$ is nonempty and there exists $\varepsilon(\boldsymbol{x}) > 0$ and $\delta(\boldsymbol{x}) > 0$ such that if $\|\boldsymbol{x} - \boldsymbol{x}'\| < \varepsilon(\boldsymbol{x})$ and $\boldsymbol{y}' \in A(\boldsymbol{x}')$, then $f(\boldsymbol{y}') - f(\boldsymbol{x}') \geq \delta(\boldsymbol{x})$.Show that Zhangwill's algorithm either stops at a point in Γ or generates an infinite sequence of points that every cluster point belongs to Γ.

Chapter 4

Hitting Walls

I respect only those who resist me,
but I cannot tolerate them.
CHARLES DE GAULLE

When a blind man hits a wall, what would he do? A popular method is to find a way along the wall. In this chapter, we use this idea to study linearly constrained optimization problems.

4.1 Linearly Constrainted Optimization

Consider linearly constrained optimization problems in the following form:

$$\begin{array}{ll} \text{maximize} & f(\boldsymbol{x}) \\ \text{subject to} & \boldsymbol{a}_j^T \boldsymbol{x} \geq b_j, \quad j = 1, ..., m, \end{array} \tag{4.1}$$

where $\boldsymbol{x} \in R^n$; R^n is the n-dimensional Euclidean space whose points are represented as column vectors. A point is called a *feasible point* if it satisfies the constraints. The set of all feasible points is called the *feasible region.* Without special mentioning, we always assume that the function f is continuously differentiable in a convex set containing the feasible region. The function f is always referred to be an objective function. For simplicity of notations, we denote $\boldsymbol{g}(\boldsymbol{x}) = \nabla f(\boldsymbol{x})$, and especially $\boldsymbol{g}_k = \boldsymbol{g}(\boldsymbol{x}_k)$ and $\boldsymbol{g}^* = \boldsymbol{g}(\boldsymbol{x}^*)$.

A nonzero vector $\boldsymbol{d}$ is called a *feasible direction* at a feasible point $\boldsymbol{x}$ if there exists $\bar{\lambda} > 0$ such that for any $\lambda \in [0, \bar{\lambda}]$, $\boldsymbol{x} + \lambda \boldsymbol{d}$ is feasible.

A constraint is called an *active* constraint at a feasible point $\boldsymbol{x}$ if its equality sign holds at the point. The set of indices of active constraints is called the *active set*, denoted by $J(\boldsymbol{x})$, i.e.

$$J(\boldsymbol{x}) = \{j \mid \boldsymbol{a}_j^T \boldsymbol{x} = b_j\}.$$

In particular, we denote

$$J_k = J(\boldsymbol{x}_k) \text{ and } J^* = J(\boldsymbol{x}^*).$$

For simplicity, we also denote $M = \{1, ..., m\}$. Note that $J \subset J'$ will stand for $J \subseteq J'$ and $J \neq J'$. For a singleton $\{h\}$, we write $J \setminus h$ and $J \cup h$ instead of $J \setminus \{h\}$ and $J \cup \{h\}$, respectively.

For the feasible region in the considered problem (4.1), a direction $\boldsymbol{d}$ is feasible at a feasible point $\boldsymbol{x}$ if and only if $\boldsymbol{a}_j^T \boldsymbol{d} \geq 0$ for $j \in J(\boldsymbol{x})$. In fact, if $\boldsymbol{d}$ is a feasible direction at point $\boldsymbol{x}$, then there exists a number $\lambda > 0$ such that $\boldsymbol{x} + \lambda \boldsymbol{d}$ is a feasible point. Thus, $\boldsymbol{a}_j^T(\boldsymbol{x} + \lambda \boldsymbol{d}) \geq b_j$ for $j \in M$. Since $\boldsymbol{a}_j^T \boldsymbol{x} = b_j$ for $j \in J(\boldsymbol{x})$, we have $\lambda \boldsymbol{a}_j^T \boldsymbol{d} \geq 0$ for $j \in J(\boldsymbol{x})$. Therefore, $\boldsymbol{a}_j^T \boldsymbol{d} \geq 0$ for $j \in J(\boldsymbol{x})$. Conversely, suppose $\boldsymbol{a}_j^T \boldsymbol{d} \geq 0$ for $j \in J(\boldsymbol{x})$. Set

$$\bar{\lambda} = \begin{cases} \min\left\{\frac{b_j - \boldsymbol{a}_j^T \boldsymbol{x}}{\boldsymbol{a}_j^T \boldsymbol{d}} \mid \boldsymbol{a}_j^T \boldsymbol{d} < 0\right\} & \text{if there exists } j \text{ such that } \boldsymbol{a}_j^T \boldsymbol{d} < 0 \\ 1 & \text{otherwise.} \end{cases}$$

Clearly, $\bar{\lambda} > 0$ and for $\lambda \in [0, \bar{\lambda}]$, $\boldsymbol{x} + \lambda \boldsymbol{d}$ is feasible.

Now, define

$$D^1(\boldsymbol{x}) = \{\boldsymbol{d} \mid \boldsymbol{a}_j^T \boldsymbol{d} \geq 0 \text{ for } j \in J(\boldsymbol{x})\}.$$

Then $D^1(\boldsymbol{x})$ is exactly the set of all feasible direction at point $\boldsymbol{x}$.

A feasible point $\boldsymbol{x}^*$ is said to be a *local maximum* of problem (4.1) if there exists a neighborhood of the point $\boldsymbol{x}^*$ such that for any feasible point $\boldsymbol{x}$ in the neighborhood, $f(\boldsymbol{x}) \leq f(\boldsymbol{x}^*)$.

A nonzero vector $\boldsymbol{d}$ is called an *ascendant direction* at a feasible point $\boldsymbol{x}$ if $\boldsymbol{g}(\boldsymbol{x})^T \boldsymbol{d} > 0$. Clearly, if $\boldsymbol{d}$ is a feasible ascendant direction at $\boldsymbol{x}$, then we can find a feasible point $\boldsymbol{x}'$ along the direction $\boldsymbol{d}$ such that $f(\boldsymbol{x}') > f(\boldsymbol{x})$. Therefore, $\boldsymbol{x}$ is a local maximum only if there does not exist a feasible ascendant direction at $\boldsymbol{x}$. That is, a feasible point $\boldsymbol{x}$ is a local maximum only if

$$D^1(\boldsymbol{x}) \cap D^2(\boldsymbol{x}) = \emptyset$$

where

$$D^2(\boldsymbol{x}) = \{\boldsymbol{d} \mid \boldsymbol{g}^T\boldsymbol{d} > 0\}.$$

The following theorem states a necessary and sufficient condition for $D^1(\boldsymbol{x}) \cap D^2(\boldsymbol{x}) = \emptyset$.

Theorem 4.1.1 *Let $\boldsymbol{x}$ be a feasible point. Then $D^1(\boldsymbol{x}) \cap D^2(\boldsymbol{x}) = \emptyset$ if and only if there exist u_j, $j \in J(\boldsymbol{x})$ such that*

$$\boldsymbol{g}(\boldsymbol{x}) = \sum_{j \in J(\boldsymbol{x})} u_j \boldsymbol{a}_j$$

and

$$u_j \leq 0 \;\; \textit{for} \;\; j \in J(\boldsymbol{x}).$$

Proof. Note that $D^1(\boldsymbol{x}) \cap D^2(\boldsymbol{x}) \neq \emptyset$ if and only if $\boldsymbol{d} \in D^1(\boldsymbol{x})$ implies $\boldsymbol{d} \notin D^2(\boldsymbol{x})$. By Farkas Lemma, $\boldsymbol{d} \in D^1(\boldsymbol{x})$ implies $\boldsymbol{d} \notin D^2(\boldsymbol{x})$ if and only if there exist $u_j \leq 0$, $j \in J(\boldsymbol{x})$ such that

$$\boldsymbol{g}(\boldsymbol{x}) = \sum_{j \in J(\boldsymbol{x})} u_j \boldsymbol{a}_j.$$

□

A feasible point $\boldsymbol{x}$ is called a *Kuhn-Tucker point* if there exist u_j, $j \in J(\boldsymbol{x})$ such that

$$\boldsymbol{g}(\boldsymbol{x}) = \sum_{j \in J(\boldsymbol{x})} u_j \boldsymbol{a}_j \tag{4.2}$$

and

$$u_j \leq 0 \;\; \text{for} \;\; j \in J(\boldsymbol{x}). \tag{4.3}$$

Clearly, for a linearly constrained optimization problem, every local maximum is a Kuhn-Tucker point.

A Kuhn-Tucker point may not be a local maximum. However, if the objective function f is concave, then every Kuhn-Tucker point is not only a local maximum but also a global maximum. A strictly concave function has at most one maximum. So, it has at most one Kuhn-Tucker point.

4.2 Projection on Walls

Let us look at the linearly constrained maximization problem as an activity of climbing a mountain within an area surrounded by walls. Suppose a blind man attempts this activity. When he hits a wall, what would he like to do? Try to be away from the wall or go up along the wall? To avoid suffering from hitting walls, the blind man may choose to go up along the wall since this way could reduce the possibility of hitting a new wall (Figure 4.1).

Figure 4.1: Walk along walls.

At a feasible point $\boldsymbol{x}$, all walls passed through $\boldsymbol{x}$ are active constraints. To stay with these walls, i.e. to keep these active constraints active, one has to search within the intersection of these walls. Let us view this intersection, the subspace $D_{J(\boldsymbol{x})} = \{\boldsymbol{d} \mid \boldsymbol{a}_j^T\boldsymbol{d} = 0, j \in J(\boldsymbol{x})\}$, as the tangent plane at the point $\boldsymbol{x}$ although it may have dimension smaller than $n-1$ and hence is not a hyperplane. In fact, the name "tangent plane" could bring us geometric intuition when we talk about it.

What is the steepest ascent within the tangent plane? To find a search direction with the tangent plane, we need to solve this problem. This problem can be formulated into the following quadratic programming:

$$\begin{array}{ll} \text{maximize} & \boldsymbol{g}^T\boldsymbol{d} \\ \text{subject to} & \boldsymbol{a}_j^T\boldsymbol{d} = 0 \quad \text{for } j \in J, \\ & \|\boldsymbol{d}\| \leq 1 \end{array} \tag{4.4}$$

where $\boldsymbol{g} = \boldsymbol{g}(\boldsymbol{x})$ and $J = J(\boldsymbol{x})$.

First of all, we note that D_J is a subspace of n-dimensional Euclidean space R^n. Let $D_J^{\perp}$ denote the complement of D_J. It is well-known from linear algebra that

$$D_J^{\perp} = \{\sum_{j \in J} u_j \boldsymbol{a}_j \mid u_j \in R^1\}.$$

Since R^n is the orthogonal direct sum of subspaces D_J and $D_J^{\perp}$, any vector $\boldsymbol{h}$ can be uniquely decomposed as $\boldsymbol{g} = \boldsymbol{y}+\boldsymbol{z}$ such that $\boldsymbol{y} \in D_J$ and $\boldsymbol{z} \in D_J^{\perp}$. The vector $\boldsymbol{y}$ is the *orthogonal projection* of $\boldsymbol{h}$ on D_J, denoted by P_J. Clearly, P_J is a linear transformation, i.e. an $n \times n$ matrix. So, P_J^2 and P_J^T are well-defined. The following properties are well-known.

Proposition 4.2.1 *$P_J^2 = P_J$ and $P_J^T = P_J$.*

Next, we show that $P_J\boldsymbol{g}/\|P_J\boldsymbol{g}\|$ is an optimal solution of the quadratic programming problem (4.4).

Proposition 4.2.2 *For any $\boldsymbol{d} \in D_J$, $\boldsymbol{g}^T\boldsymbol{d} \leq \|P_J\boldsymbol{g}\| \cdot \|\boldsymbol{d}\|$.*

Proof. By Proposition 4.2.1, for any $\boldsymbol{d} \in D_J$, we have

$$\boldsymbol{g}^T\boldsymbol{d} = \boldsymbol{g}^T P_J \boldsymbol{d} = (P_J\boldsymbol{g})^T\boldsymbol{d} \leq \|P_J\boldsymbol{g}\| \cdot \|\boldsymbol{d}\|.$$

□

Since $\boldsymbol{g}^T(P_J\boldsymbol{g}) = \|P_J(\boldsymbol{x})\boldsymbol{g}\|^2$, by Proposition 4.2.2, we have that for any feasible solution $\boldsymbol{d}$ of (4.4),

$$\boldsymbol{g}^T\boldsymbol{d} \leq \|P_J\boldsymbol{g}\| = \boldsymbol{g}^T P_J\boldsymbol{g}/\|P_J\boldsymbol{g}\|.$$

This means that $P_J\boldsymbol{g}/\|P_J\boldsymbol{g}\|$ is an optimal solution of the quadratic programming problem (4.4). That is, $P_J\boldsymbol{g}$ is the steepest ascent direction in the tangent plane D_J.

If $\boldsymbol{a}_j$, $j \in J$, are linearly independent, then P_J can be represented by a closed form formula. To do so, let A_J denote the matrix consisting of column vectors $\boldsymbol{a}_j$, $j \in J$. For any $\boldsymbol{h} \in R^n$, write

$$\boldsymbol{h} = P_J\boldsymbol{h} + \sum_{j \in J} u_j \boldsymbol{a}_j = P_J\boldsymbol{h} + A_J\boldsymbol{u}$$

where $u = (u_j, j \in J)^T$. Then we have

$$A_J^T\boldsymbol{h} = A_J^T P_J\boldsymbol{h} + A_J^T A_J\boldsymbol{u} = A_J^T A_J\boldsymbol{u}.$$

Since $\boldsymbol{a}_j$, $j \in J$, are linearly independent, $A_J^T A_J$ is a nonsingular square matrix of order $|J|$. Thus

$$\boldsymbol{u} = (A_J^T A_J)^{-1} A_J^T \boldsymbol{h} \tag{4.5}$$

and

$$P_J \boldsymbol{h} = (I - A_J(A_J^T A_J)^{-1} A_J^T)\boldsymbol{h} \tag{4.6}$$

where I is the identity matrix of order n. Therefore

$$P_J = I - A_J(A_J^T A_J)^{-1} A_J^T.$$

A feasible point $\boldsymbol{x}$ is *regular* if all active constraints at $\boldsymbol{x}$ are linearly independent, i.e. $\boldsymbol{a}_j$ for $j \in J(\boldsymbol{x})$ are linearly dependent. The problem (4.1) is *nondegenerate* if its constraints satisfy the following condition:

Regularity Condition. Every feasible point is regular.

In this and the next three chapters, we discuss only nondegenerate problems. Therefore, we always assume the regularity condition holds for linearly constrained optimization problems in these four chapters without special mentioning. Under the regularity condition, the gradient projection on the tangent plane at a.feasible point $\boldsymbol{x}$ can always be represented as

$$P_J \boldsymbol{g} = (I - A_J(A_J^T A_J)^{-1} A_J^T)\boldsymbol{g}$$

where $\boldsymbol{g} = \boldsymbol{g}(\boldsymbol{x})$ and $J = J(\boldsymbol{x})$.

If $P_J \boldsymbol{g} \neq \boldsymbol{0}$, then $P_j \boldsymbol{g}$ can be used as a search direction as discussed with a blind man. Now, the problem is that when $P_J \boldsymbol{g} = \boldsymbol{0}$, what should one do? Of course, if $\boldsymbol{x}$ is a Kuhn-Tucker point, then one can stop. However, it is possible that $\boldsymbol{x}$ is not a Kuhn-Tucker point, but $P_J \boldsymbol{g} = \boldsymbol{0}$. In this case,

$$\begin{aligned} \boldsymbol{g} &= A_J(A_J^T A_J)^{-1} A_J^T \boldsymbol{g} \\ &= \sum_{j \in J} u_j \boldsymbol{a}_j \end{aligned}$$

where

$$(u_j, j \in J)^T = (A_J^T A_J)^{-1} A_J^T \boldsymbol{g}. \tag{4.7}$$

Since $\boldsymbol{x}$ is not a Kuhn-Tucker point, there must exist $h \in J$ such that $u_h > 0$. The following result provides some hints to deal with this case.

Lemma 4.2.3 *For $h \in J$,*

$$\|P_{J\backslash h}\boldsymbol{g}\|^2 = \|P_J\boldsymbol{g}\|^2 + u_h^2\|P_{J\backslash h}\boldsymbol{a}_h\|^2 \tag{4.8}$$

and

$$\boldsymbol{a}_h^T P_{J\backslash i}\boldsymbol{g} = u_h\|P_{J\backslash h}\boldsymbol{a}_h\|^2. \tag{4.9}$$

Proof. Write

$$\boldsymbol{g} = P_J\boldsymbol{g} + \sum_{j\in J} u_j\boldsymbol{a}_j \tag{4.10}$$

$$\boldsymbol{a}_h = P_{J\backslash h}\boldsymbol{a}_h + \sum_{j\in J\backslash h} \alpha_j\boldsymbol{a}_j. \tag{4.11}$$

Substituting (4.11) into (4.10), we obtain

$$\boldsymbol{g} = (P_J\boldsymbol{g} + u_h P_{J\backslash h}\boldsymbol{a}_h) + \sum_{j\in J\backslash h}(u_j + \alpha_j u_h)\boldsymbol{a}_j.$$

Note that the first term is in subspace $D_{J\backslash h}$ and the second term is in its complement subspace $D_{J\backslash h}^{\perp}$. Hence,

$$P_{J\backslash h}\boldsymbol{g} = P_J\boldsymbol{g} + u_h P_{J\backslash h}\boldsymbol{a}_h. \tag{4.12}$$

Furthermore, since $P_J\boldsymbol{g} \in D_J$ and $P_{J\backslash h}\boldsymbol{a}_h \in D_J^{\perp}$, $P_J\boldsymbol{g}$ and $P_{J\backslash h}\boldsymbol{a}_h$ are perpendicular to each other. Thus, we have

$$\|P_{J\backslash h}\boldsymbol{g}\|^2 = \|P_J\boldsymbol{g}\|^2 + u_h^2\|P_{J\backslash h}\boldsymbol{a}_h\|^2.$$

Note that $\boldsymbol{a}_h^T P_J\boldsymbol{g} = 0$. Therefore,

$$\boldsymbol{a}_h^T P_{J\backslash h}\boldsymbol{g} = u_h\boldsymbol{a}_h^T P_{J\backslash h}\boldsymbol{a}_h = u_h\|P_{J\backslash h}\boldsymbol{a}_h\|^2.$$

□

From (4.8) and (4.9), we see that if $u_h > 0$, then $P_{J\backslash h}\boldsymbol{g} \neq \boldsymbol{0}$ and moreover, for all $j \in J$ $\boldsymbol{a}_j^T P_{J\backslash h}\boldsymbol{g} \geq 0$, that is, the gradient projection $P_{J\backslash h}\boldsymbol{g}$ is a feasible ascendant direction. This suggests to us to choose $h \in J$ such that

$$u_h = \max_{j\in J} u_j.$$

With this choice, u_h would have an interesting property.

Lemma 4.2.4 *A feasible point $\boldsymbol{x}$ is a Kuhn-Tucker point if and only if*

$$P_J\boldsymbol{g} = \boldsymbol{0} \text{ and } u_h \le 0$$

where $J = J(\boldsymbol{x})$, $\boldsymbol{g} = \boldsymbol{g}(\boldsymbol{x})$, and

$$\begin{aligned} (u_j, j \in J) &= (A_J^T A_J)^{-1} A^T \boldsymbol{g} \\ u_h &= \max_{j \in J} u_j. \end{aligned}$$

Lemma 4.2.4 suggests to choose the search direction as follows:

$$\boldsymbol{d} = \begin{cases} P_J\boldsymbol{g} & \text{if } \|P_J\boldsymbol{g}\| > c \cdot u_h, \\ P_{J\setminus h}\boldsymbol{g} & \textit{otherwise}, \end{cases}$$

where $0 \le c \le +\infty$. The following theorem states the property of this choice.

Theorem 4.2.5 *Let $0 < c \le +\infty$. For $J = J(\boldsymbol{x})$, define*

$$\boldsymbol{d} = \begin{cases} P_J\boldsymbol{g} & \textit{if } \|P_J\boldsymbol{g}\| > c \cdot u_h, \\ P_{J\setminus h}\boldsymbol{g} & \textit{otherwise}, \end{cases}$$

where $u_h = \max_{j\in J} u_j$. Then $\boldsymbol{d} = \boldsymbol{0}$ if and only if $\boldsymbol{x}$ is a Kuhn-Tucker point. Furthermore, if $\boldsymbol{d} \ne \boldsymbol{0}$, then $\boldsymbol{d}$ is an ascendant feasible direction.

Proof. Suppose $\boldsymbol{x}$ is a Kuhn-Tucker point. Then $P_J\boldsymbol{g} = \boldsymbol{0}$ and $u_h \le 0$. If $u_h < 0$, then $\|P_J\boldsymbol{g}\| > c \cdot u_h$ and hence $\boldsymbol{d} = P_J\boldsymbol{g} = \boldsymbol{0}$. If $u_h = 0$, then by (4.8) $P_{J\setminus h}\boldsymbol{g} = 0$ and hence $\boldsymbol{d} = \boldsymbol{0}$. Conversely, suppose that $\boldsymbol{d} = \boldsymbol{0}$. Then either $P_J\boldsymbol{g} = \boldsymbol{0}$ and $\|P_J\boldsymbol{g}\| > c \cdot u_h$ or $P_{J\setminus h}\boldsymbol{g} = \boldsymbol{0}$. In the former case, we must have $c > 0$, so $u_h \le 0$ and hence x is a Kuhn-Tucker point. In the latter case, by (4.8), we have $P_J\boldsymbol{g} = \boldsymbol{0}$ and $u_h = 0$, so $\boldsymbol{x}$ is a Kuhn-Tucker point.

Now, suppose that $\boldsymbol{d} \ne \boldsymbol{0}$. If $\boldsymbol{d} = P_J\boldsymbol{g}$, then obviously, $\boldsymbol{d}$ is a feasible direction. If $\boldsymbol{d} = P_{J\setminus h}\boldsymbol{g}$, then $\|P_J\boldsymbol{g}\| \le c \cdot u_h$. We claim that $u_h \ge 0$. In fact, $u_h < 0$ implies that $c = 0$ and $P_J\boldsymbol{g} = \boldsymbol{0}$. Thus, $\boldsymbol{x}$ is a Kuhn-Tucker point, contradicting $\boldsymbol{d} \ne \boldsymbol{0}$. Hence, our claim holds. By (4.9), we have $\boldsymbol{a}_h^T\boldsymbol{d} = \boldsymbol{a}_h^T P_{J\setminus h}\boldsymbol{g} \ge 0$. It follows that $\boldsymbol{d}$ is a feasible direction. Finally, since $\boldsymbol{g}^T\boldsymbol{d} = \|\boldsymbol{d}\|^2 > 0$, $\boldsymbol{d}$ is an ascendant feasible direction. □

4.3 Rosen's Gradient Projection Method

Theorem 4.2.5 suggests the following algorithm:

Rosen's Method: *Initially, choose a feasible point* $\boldsymbol{x}_1$. *At each iteration* $k = 1, 2, \cdots$, *the algorithm carries out the following steps.*

1. *Choose a positive number* c_k. *Compute a search direction by the following formula:*

$$\boldsymbol{d}_k = \begin{cases} P_{J_k}\boldsymbol{g}_k & \text{if } \|P_{J_k}\boldsymbol{g}_k\| > c_k u_{kh_k}, \\ P_{J_k \setminus h_k}\boldsymbol{g}_k & \text{otherwise,} \end{cases}$$

where

$$(u_{kj}, j \in J_k)^T = (A_{J_k}^T A_{J_k})^{-1} A_{J_k}^T \boldsymbol{g}_k$$

and

$$u_{kh_k} = \max\{u_{kj} \mid j \in J_k\}.$$

2. *If* $\boldsymbol{d}_k = \boldsymbol{0}$, *then stop;* $\boldsymbol{x}_k$ *is a Kuhn-Tucker point. If* $\boldsymbol{d}_k \neq \boldsymbol{0}$, *then compute*

$$\bar{\lambda}_k = \begin{cases} 1 & \text{if } \boldsymbol{a}_j^T \boldsymbol{d}_k \geq 0 \text{ for all } j \notin J_k, \\ \min\left\{ \frac{b_j - \boldsymbol{a}_j^T \boldsymbol{x}_k}{\boldsymbol{a}_j^T \boldsymbol{d}_k} \;\middle|\; \boldsymbol{a}_j^T \boldsymbol{d}_k < 0 \right\} & \text{otherwise,} \end{cases}$$

and find a new point $\boldsymbol{x}_{k+1} = \boldsymbol{x}_k + \lambda_k \boldsymbol{d}_k$ $(0 < \lambda_k \leq \bar{\lambda}_k)$ *by a line search procedure.*

In the above algorithm, when the line search meets a new constraint, an index is added to the active index set. In this case, the projection matrix needs to be computed again. To simplify the computation, the following recursive formula is helpful:

$$P_J = P_{J\setminus h} - \frac{P_{J\setminus h}\boldsymbol{a}_h \boldsymbol{a}_h^T P_{J\setminus h}}{\boldsymbol{a}_h^T P_{J\setminus h} \boldsymbol{a}_h}. \tag{4.13}$$

There are two different ways to compute search directions with gradient projection:

$$\boldsymbol{d}_k = \begin{cases} P_{J_k}\boldsymbol{g}_k & \text{if } \|P_{J_k}\boldsymbol{g}_k\| > 0 \text{ or } u_{kh_k} \leq 0, \\ P_{J_k \setminus h_k}\boldsymbol{g}_k & \text{otherwise,} \end{cases}$$

and

$$d_k = \begin{cases} P_{J_k} g_k & \text{if } u_{kh_k} \le 0, \\ P_{J_k \setminus h_k} g_k & \text{otherwise.} \end{cases}$$

They correspond to the case $c_k = 0$ and the case $c_k = +\infty$ respectively. Clearly, they give simpler formulas for d_k. This is probably the reason why most textbooks introduce Rosen's method in these two forms. Unfortunately, in these two cases, Rosen's method is not globally convergent. In fact, the algorithm is not globally convergent even if the c_k's are positive and converge to zero or to infinity. In the following, we explain it by presenting two counterexamples. Note that we have to choose a specific line search procedure to perform the computation described in the example. Here, we use Armijo's rule to do line search.

Armijo's Rule: *This procedure runs as follows:*

begin
$\lambda^+ = \bar{\lambda}_k$;
while

$$\frac{1}{2}\lambda^+ g_k^T d_k < f(x^+) - f(x_k)$$

where

$$x^+ = x_k + \lambda^+ d_k$$

do $\lambda^+ := \lambda^+/2$;
set $x_{k+1} = x_k + \lambda^+ d_k$;
end

The reader may follow the same idea to find similar examples for other line search procedures.

Counterexample 1 Consider the following problem:

$$\begin{array}{ll} \text{maximize} & f(x) = -x_1^2 - 3x_2^2 - (x_3 - 1)^2 \\ \text{subject to} & x_1 \ge 0, \; x_3 \ge 0. \end{array}$$

Since $f(x)$ is strictly concave, the problem has only one Kuhn-Tucker point, $x^* = (0, 0, 1)^T$. Choose an initial point $x_1 = (3/2, -1/2, 0)^T$ and parameters $0 \le c_k \le 1/2^k$. Then Rosen's method generates a sequence of points $x_k = (3/2^k, (-1/2)^k, 0)^T$. In fact, $x_3 \ge 0$ is the only active constraint for all k.

So, the projection matrix is always:

$$P_{J_k} = \begin{bmatrix} 1 & 0 & 0 \\ 0 & 1 & 0 \\ 0 & 0 & 0 \end{bmatrix}.$$

Thus,

$$P_{J_k}\boldsymbol{g}_k = \begin{pmatrix} -6/2^k \\ -6(-1/2)^k \\ 0 \end{pmatrix}$$

and

$$\|P_{J_k}\boldsymbol{g}_k\| = 6\sqrt{2}/2^k > c_k u_{kh_k} = 2c_k.$$

Hence $\boldsymbol{d}_k = P_{J_k}\boldsymbol{g}_k$. Then,

$$f(\boldsymbol{x}_k + \lambda\boldsymbol{d}_k) = -\frac{3}{2^{2k}}\left[48(\lambda - \frac{1}{4})^2 + 1\right] - 1$$

and

$$\bar{\lambda}_k = \frac{1}{2}.$$

Since

$$f(\boldsymbol{x}_k + \bar{\lambda}_k\boldsymbol{d}_k) - f(\boldsymbol{x}_k) = 0 < \frac{1}{2}\bar{\lambda}_k\boldsymbol{g}_k^T\boldsymbol{d}_k = \frac{18}{2^{2k}}$$

and

$$f(\boldsymbol{x}_k + \frac{1}{2}\bar{\lambda}_k\boldsymbol{d}_k) - f(\boldsymbol{x}_k) = \frac{9}{2^{2k}} = \frac{1}{4}\bar{\lambda}_k\boldsymbol{g}_k^T\boldsymbol{d}_k,$$

we have $\lambda_k = \frac{1}{2}\bar{\lambda}_k = \frac{1}{4}$. It follows that

$$\boldsymbol{x}_{k+1} = (3/2^{k+1}, (-1/2)^{k+1}, 0)^T.$$

As k goes to infinity, this sequence approaches $(0,0,0)^T$ which is not a Kuhn-Tucker point.

Counterexample 2. Consider the following problem:

$$\begin{array}{ll} \text{maximize} & f(\boldsymbol{x}) = -\sqrt[3]{3/2}(x_1^2 + 9x_2^2)^{2/3} + x_3 \\ \text{subject to} & x_1 + 3x_2 \geq 0, \\ & x_1 - 3x_2 \geq 0, \\ & x_3 \leq 10. \end{array}$$

Note that $f(\boldsymbol{x})$ is concave and that the maximum value is achieved at a unique point $\boldsymbol{x}^* = (0,0,10)^T$. So, the problem has only one Kuhn-Tucker point $\boldsymbol{x}^*$. Choose an initial point $\boldsymbol{x}_1 = (3/2, 1/2, 0)^T$ and parameters $15 \cdot 2^{k/3} \leq c_k \leq +\infty$. We show that Rosen's method generates a sequence of points $\boldsymbol{x}_k = (3/2^k, (-1/2)^k, \frac{3}{2}\sum_{i=1}^{k-1} 2^{-2i/3})^T$.

There is only one active constraint at $\boldsymbol{x}_k$, which can be represented as $x_1 + (-1)^{k+1}3x_2 \geq 0$. Thus

$$P_{J_k}\boldsymbol{g}_k = (-1.2 \cdot 2^{-k/3}, 0.4 \cdot (-1)^{k-1}2^{-k/3}, 1)^T.$$

Since $\|P_{J_k}\boldsymbol{g}_k\| = (1 + 1.6 \cdot 2^{-2k/3})^{1/2} < c_k u_{kh_k} = 0.2 \cdot 2^{-2k/3}c_k$, we have

$$\boldsymbol{d}_k = P_{J_k \setminus h_k}\boldsymbol{g}_k = \boldsymbol{g}_k = (-2^{-k/3}, -(-1)^k 2^{-k/3}, 1)^T$$

and

$$\bar{\lambda}_k = \frac{3}{2} \cdot 2^{-2k/3}.$$

Note that

$$\begin{aligned} f(\boldsymbol{x}_k + \bar{\lambda}_k\boldsymbol{d}_k) - f(\boldsymbol{x}_k) &> 3 \cdot 2^{-4k/3} + \bar{\lambda}_k \\ &> \frac{1}{2}\bar{\lambda}_k\boldsymbol{g}_k^T\boldsymbol{d}_k = \frac{1}{2}(1 + 1.6 \cdot 2^{-2k/3})\bar{\lambda}_k. \end{aligned}$$

Thus

$$\boldsymbol{x}_{k+1} = \boldsymbol{x}_k + \bar{\lambda}_k\boldsymbol{d}_k = (\frac{3}{2^{k+1}}, (-\frac{1}{2})^{k+1}, \frac{3}{2}\sum_{i=1}^{k} 2^{-2i/3})^T.$$

As $k \to \infty$, this sequence goes to $(0, 0, \frac{3}{2(\sqrt[3]{4}-1)})^T \neq \boldsymbol{x}^*$.

From the above counterexamples, we see that in order to have global convergence, the parameters c_k have to be bounded away from zero and infinity. An easy way is to choose c_k to be a positive constant $c > 0$. With this choice, Rosen's method is indeed globally convergent. However, the proof is not so easy. We have to develop some new tools.

In fact, the global convergence of Rosen's method cannot be proved by Zangwill's theorem. To see this, we consider a compact linearly constrained feasible region Ω. Suppose that Γ consists of all Kuhn-Tucker points in Ω. Define $A : \Omega \setminus \Gamma \to \mathcal{P}(\Omega)$ by setting $A(\boldsymbol{x})$ to be the set of possible $\boldsymbol{x}_{k+1}$'s when set $\boldsymbol{x}_k = \boldsymbol{x}$. If the convergence of the algorithm is provable by Zangwill's theorem, then there must exist a point-to-set mapping H satisfying conditions (1), (2) and (3) in Theorem 3.3.2. By Theorem 3.3.2, there exists a lower semi-continuous, positive-value function $\epsilon(\boldsymbol{x})$ such that $\boldsymbol{y} \in A(\boldsymbol{x})$ implies $f(\boldsymbol{y}) \geq f(\boldsymbol{x}) + \epsilon(\boldsymbol{x})$. Thus, we have

$$\left.\begin{array}{l} \boldsymbol{x}_k \to \boldsymbol{x} \in \Omega \setminus \Gamma \\ \boldsymbol{y}_k \to \boldsymbol{y} \\ \boldsymbol{y}_k \in A(\boldsymbol{x}_k) \end{array}\right\} \Longrightarrow f(\boldsymbol{y}) > f(\boldsymbol{x}). \tag{4.14}$$

This means that for our purpose, it suffices to find a counterexample for (4.14).

Consider the following instance:

$$\begin{array}{ll} \text{maximize} & 2x_1 - x_2 \\ \text{subject to} & 2x_1 + x_2 \leq 6, \\ & x_1 \geq 0,\ x_2 \geq 0. \end{array}$$

Let $\boldsymbol{x}_k = (2, 2 - \frac{1}{k})^T$ and $\boldsymbol{x} = \lim_{k\to\infty} \boldsymbol{x}_k = (2,2)^T$. For each $\boldsymbol{x}_k$, there is no active constraint. Thus, $\boldsymbol{d}_k = \boldsymbol{g}_k = (2,-1)^T$. Searching along $\boldsymbol{d}_k$ by Armijo's rule, we find

$$A(\boldsymbol{x}_k) = \{(2 + \frac{2}{3k}, 2 - \frac{4}{3k})^T\}.$$

Let $\boldsymbol{y}_k = (2 + \frac{2}{3k}, 2 - \frac{4}{3k})^T$. Then $\boldsymbol{y} = \lim_{k\to\infty} \boldsymbol{y}_k = (2,2)^T$. Therefore, $f(\boldsymbol{x}) = f(\boldsymbol{y})$, contradicting (4.14).

Exercises

1. Prove that if we choose

$$\boldsymbol{d} = \begin{cases} P_J\boldsymbol{g} & \text{if } \|P_J\boldsymbol{g}\| > 0 \text{ or } u_h \leq 0, \\ P_{J\setminus h}\boldsymbol{g} & \textit{otherwise}, \end{cases}$$

then the conclusion of Theorem 4.2.5 is still true.

2. Prove

$$P_J = P_{J\setminus h} + \frac{P_{J\setminus h}\boldsymbol{a}_h\boldsymbol{a}_h^T P_{J\setminus h}}{\boldsymbol{a}_h^T P_{J\setminus h}\boldsymbol{a}_h}.$$

3. Prove that if we choose $c = +\infty$ in Theorem 4.2.5, then $\boldsymbol{d}$ is a continuous function of $\boldsymbol{x}$.

Chapter 5

Slope and Path Length

Men trip not on mountains,
they stumble on stones.
HINDUSTANI PROVERBS

When you passed by a mountain, you may have noticed that the road somehow looks like a snake in s-shape. Why was the road built in such a way? The answer is easy, increase the length to decrease the slope. The longer its length is, the smaller its slope is. However, did you think that as its length approaches infinity its slope approaches zero (Figure 5.1)? In this chapter, we will use this idea to develop a tool to establish the global convergence of some algorithms.

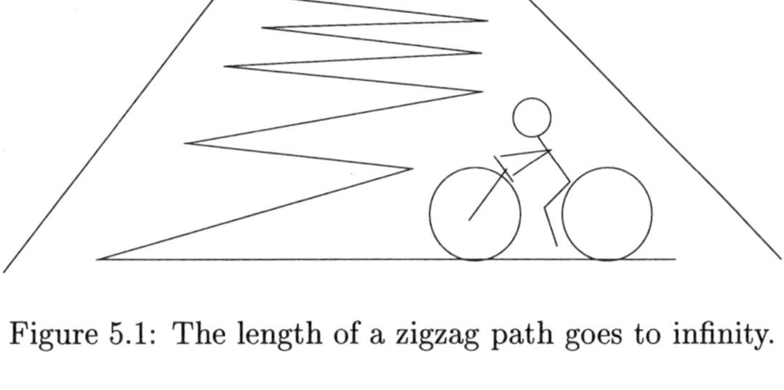

Figure 5.1: The length of a zigzag path goes to infinity.

5.1 The First Slope Lemma

Let $f : R^n \to R$ be a continuously differentiable function. Consider a convergent nondecreasing sequence $f(\boldsymbol{x}_1)$, $f(\boldsymbol{x}_2)$, $\cdots$. A path going up to a top of a mountain is formed by

$$(\boldsymbol{x}_1, f(\boldsymbol{x}_1)) \to (\boldsymbol{x}_2, f(\boldsymbol{x}_2)) \to \cdots \to (\boldsymbol{x}_k, f(\boldsymbol{x}_k)) \to \cdots.$$

At each piece $[\boldsymbol{x}_k, \boldsymbol{x}_{k+1}]$, the slope is computed by

$$\frac{f(\boldsymbol{x}_{k+1}) - f(\boldsymbol{x}_k)}{\|\boldsymbol{x}_{k+1} - \boldsymbol{x}_k\|}.$$

We claim that if $\sum_{k=1}^{\infty} \|\boldsymbol{x}_{k+1} - \boldsymbol{x}_k\| = +\infty$, then there exists a subsequence $\{\boldsymbol{x}_k\}_{k \in K}$ such that

$$\lim_{k \to \infty, k \in K} \frac{f(\boldsymbol{x}_{k+1}) - f(\boldsymbol{x}_k)}{\|\boldsymbol{x}_{k+1} - \boldsymbol{x}_k\|} = 0.$$

By contradition, suppose such a subsequence does not exist. Then there exists a positive number ζ such that

$$\frac{f(\boldsymbol{x}_{k+1}) - f(\boldsymbol{x}_k)}{\|\boldsymbol{x}_{k+1} - \boldsymbol{x}_k\|} \geq \zeta$$

for all $k = 1, 2, \cdots$. Thus,

$$\sum_{k=1}^{\infty} (f(\boldsymbol{x}_{k+1}) - f(\boldsymbol{x}_k)) \geq \zeta \sum_{k=1}^{\infty} \|\boldsymbol{x}_{k+1} - \boldsymbol{x}_k\| = +\infty$$

contradicting the convergence of the sequence $\{f(\boldsymbol{x}_k)\}$.

The following lemma is motivated from the above observation.

Lemma 5.1.1 (The First Slope Lemma) *Let $\{\boldsymbol{x}_k\}$ be a sequence of feasible points such that $f(\boldsymbol{x}_k) < f(\boldsymbol{x}_{k+1})$ for $k = 1, 2, \cdots$. Let $\boldsymbol{x}^*$ be a cluster point of the sequence such that for any subsequence $\{\boldsymbol{x}_k\}_{k \in K}$ converging to $\boldsymbol{x}^*$, $\boldsymbol{x}_{k+1} - \boldsymbol{x}_k \to \mathbf{0}$ as $k \to \infty$, $k \in K$. If $\{\boldsymbol{x}_k\}$ does not converge to $\boldsymbol{x}^*$, then there exists a subsequence $\{\boldsymbol{x}_k\}_{k \in K}$ such that $\boldsymbol{x}_k \to \boldsymbol{x}^*$ as $k \to \infty$, $k \in K$, and*

$$\lim_{k \to \infty, k \in K} \frac{\boldsymbol{g}_k^T (\boldsymbol{x}_{k+1} - \boldsymbol{x}_k)}{\|\boldsymbol{x}_{k+1} - \boldsymbol{x}_k\|} = 0. \tag{5.1}$$

Proof. First, we show that there exists a subsequence $\{\boldsymbol{x}_k\}_{k\in K}$ convergent to $\boldsymbol{x}^*$ such that

$$\lim_{k\to\infty, k\in K} \frac{f(\boldsymbol{x}_{k+1}) - f(\boldsymbol{x}_k)}{\|\boldsymbol{x}_{k+1} - \boldsymbol{x}_k\|} = 0. \tag{5.2}$$

Note that compared with the subsequence obtained at the beginning of this section, we have an additional request on this subsequence that this subsequence needs to converge to the given cluster point $\boldsymbol{x}^*$. Thus, we need a new proof.

By contradiction, suppose such a subsequence does not exist. Then there exists a positive number ε and a neighborhood of $\boldsymbol{x}^*$, $B(\delta) = \{\boldsymbol{x} \mid \|\boldsymbol{x}-\boldsymbol{x}^*\| \leq \delta\}$ $(\delta > 0)$, such that for every $\boldsymbol{x}_k \in B(\delta)$,

$$\frac{f(\boldsymbol{x}_{k+1}) - f(\boldsymbol{x}_k)}{\|\boldsymbol{x}_{k+1} - \boldsymbol{x}_k\|} \geq \varepsilon.$$

Since $\{\boldsymbol{x}_k\}$ does not converge to $\boldsymbol{x}^*$, there exists a neighborhood $B(\delta')$ of $\boldsymbol{x}^*$ such that there are infinitely many $\boldsymbol{x}_k$'s not in $B(\delta')$. Choose $\zeta = \min(\delta, \delta') > 0$. Choose $\boldsymbol{x}_{k_1} \in B(\zeta/2)$ and $\boldsymbol{x}_{k_1'} \notin B(\zeta)$, $k_1 < k_1'$, such that all $\boldsymbol{x}_{k_1+1}, \boldsymbol{x}_{k_1+2}, \cdots, \boldsymbol{x}_{k_1'-1}$ belong to $B(\zeta)$ (Figure 5.2). Choose $\boldsymbol{x}_{k_2} \in B(\zeta/2)$

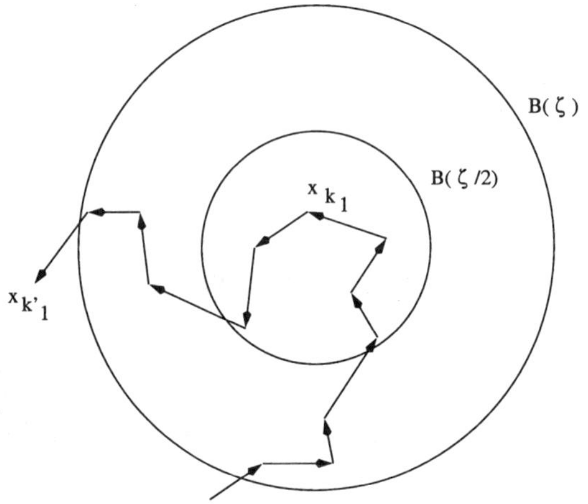

Figure 5.2: The proof of the first slope lemma.

and $\boldsymbol{x}_{k_2'} \notin B(\zeta)$, $k_1' < k_2 < k_2'$, such that all $\boldsymbol{x}_{k_2+1}, \boldsymbol{x}_{k_2+2}, \cdots, \boldsymbol{x}_{k_2'-1}$ belong to $B(\zeta)$, and so on. We would obtain a subsequence $\{k_1 < k_1' < k_2 < k_2' < \cdots\}$ such that $\boldsymbol{x}_{k_i} \in B(\zeta/2)$ for every i and after point $\boldsymbol{x}_{k_i}$ $\boldsymbol{x}_{k_i'}$ is the first point not in $B(\zeta)$. Now, we have

$$\begin{aligned} f(\boldsymbol{x}^*) - f(\boldsymbol{x}_{k_i}) &\geq \sum_{i=1}^{\infty} \sum_{j=k_i}^{k_i'-1} (f(\boldsymbol{x}_{j+1}) - f(\boldsymbol{x}_j)) \\ &\geq \sum_{i=1}^{\infty} \sum_{j=k_i}^{k_i'-1} \frac{\varepsilon}{2} \|\boldsymbol{x}_{j+1} - \boldsymbol{x}_j\| \qquad \text{by (2.2)} \\ &\geq \sum_{i=1}^{\infty} \frac{\varepsilon}{2} \|\boldsymbol{x}_{k_i'} - \boldsymbol{x}_{k_i}\| \\ &\geq \sum_{i=1}^{\infty} \frac{\varepsilon}{2} \cdot \frac{\zeta}{2} = \infty, \end{aligned}$$

a contradiction. Therefore, there exists a subsequence $\{\boldsymbol{x}_k\}_{k \in K}$ convergent to $\boldsymbol{x}^*$ such that (5.2) holds.

Next, write $f(\boldsymbol{x}_{k+1}) - f(\boldsymbol{x}_k) = \boldsymbol{g}(\boldsymbol{y}_k)^T(\boldsymbol{x}_{k+1} - \boldsymbol{x}_k)$. Since $\boldsymbol{g}$ is continuous and $\|\boldsymbol{x}_{k+1} - \boldsymbol{x}_k\| \to 0$ as k goes to infinity through $k \in K$, we have

$$\lim_{k \to \infty} \frac{(\boldsymbol{g}(\boldsymbol{y}_k) - \boldsymbol{g}_k)^T(\boldsymbol{x}_{k+1} - \boldsymbol{x}_k)}{\|\boldsymbol{x}_{k+1} - \boldsymbol{x}_k\|} = 0.$$

Thus,

$$\liminf_{k \to \infty, k \in K} \frac{\boldsymbol{g}_k^T(\boldsymbol{x}_{k+1} - \boldsymbol{x}_k)}{\|\boldsymbol{x}_{k+1} - \boldsymbol{x}_k\|} = 0.$$

□

5.2 A Property of Line Searches

In the first slope lemma, there is a condition that for every subsequence $\{\boldsymbol{x}_k\}_{k \in K}$ convergent to $\boldsymbol{x}^*$, $\boldsymbol{x}_{k+1} - \boldsymbol{x}_k \to 0$ as $k \to \infty$ through $k \in K$. In this section, we are going to indicate that this condition could be satisfied if the line search procedure is chosen properly.

The line search is a basic part in many algorithms for nonlinear optimization. During the past thirty years, many line search procedures have been discovered. Most of them share an important property, which will meet the condition in the first slope lemma.

Consider an iterative algorithm. At each iteration, the algorithm first finds a feasible ascendant direction and then computes a new point along this direction by a line search procedure. Let Ω denote a feasible region. Given a feasible point $\boldsymbol{x}$, a feasible ascendant direction $\boldsymbol{d}$ and an appropriate number $\bar{\lambda} > 0$ such that for $0 \leq \lambda \leq \bar{\lambda}$, $\boldsymbol{x} + \lambda \boldsymbol{d}$ lies in the feasible region Ω, the following procedures are used very often for finding a new point $\boldsymbol{x}^+$.

Exact Line Search: *The procedure computes a solution λ^+ for the problem of maximizing $f(\boldsymbol{x} + \lambda \boldsymbol{d})$ subject to $0 \leq \lambda \leq \bar{\lambda}$, and set $\boldsymbol{x}^+ = \boldsymbol{x} + \lambda^+ \boldsymbol{d}$.*

Curry Test: *The procedure finds the first local maximum λ^+ for the function $f(\boldsymbol{x} + \lambda \boldsymbol{d})$ with respect to λ over $[0, \bar{\lambda}]$, and set $\boldsymbol{x}^+ = \boldsymbol{x} + \lambda^+ \boldsymbol{d}$.*

Goldstein Test: *Select a constant α from $(0, 1/2)$. The procedure computes a new feasible point $\boldsymbol{x}^+ = \boldsymbol{x} + \lambda^+ \boldsymbol{d}$ $(0 < \lambda^+ \leq \bar{\lambda})$ such that*

$$\alpha \lambda^+ \boldsymbol{g}(\boldsymbol{x})^T \boldsymbol{d} \leq f(\boldsymbol{x}^+) - f(\boldsymbol{x})$$

and either $\lambda^+ = \bar{\lambda}$ or

$$(1 - \alpha) \lambda^+ \boldsymbol{g}(\boldsymbol{x})^T \boldsymbol{d} \geq f(\boldsymbol{x}^+) - f(\boldsymbol{x}).$$

Wolfe Test: *Select a constant α from $(0, 1/2)$. The procedure computes a feasible point $\boldsymbol{x}^+ = \boldsymbol{x} + \lambda^+ \boldsymbol{d}$ $(0 < \lambda^+ \leq \bar{\lambda})$ such that*

$$\alpha \lambda^+ \boldsymbol{g}(\boldsymbol{x})^T \boldsymbol{d} \leq f(\boldsymbol{x}^+) - f(\boldsymbol{x})$$

and either $\lambda^+ = \bar{\lambda}$ or

$$(1 - \alpha) \boldsymbol{g}(\boldsymbol{x})^T \boldsymbol{d} \geq \boldsymbol{g}(\boldsymbol{x}^+)^T \boldsymbol{d}.$$

Armijo's Rule: *The procedure finds a point $\boldsymbol{x}^+ = \boldsymbol{x} + \lambda^+ \boldsymbol{d}$ $(0 < \lambda^+ \leq \bar{\lambda})$ such that*

$$\frac{1}{2} \lambda^+ \boldsymbol{g}(\boldsymbol{x})^T \boldsymbol{d} \leq f(\boldsymbol{x}^+) - f(\boldsymbol{x})$$

and either $\lambda^+ = \bar{\lambda}$ or

$$\lambda^+ \boldsymbol{g}(\boldsymbol{x})^T \boldsymbol{d} > f(\boldsymbol{x} + 2\lambda^+ \boldsymbol{d}) - f(\boldsymbol{x}).$$

Any line search procedure can be seen as a point-to-set mapping from the point $(\boldsymbol{x}, \boldsymbol{d}, \bar{\lambda})$ to the set of all possible $\boldsymbol{x}^+$'s. Usually, a line search procedure T should have the following properties:

(a). $\forall \boldsymbol{x}^+ \in T(\boldsymbol{x}, \boldsymbol{d}, \bar{\lambda}) : f(\boldsymbol{x}^+) > f(\boldsymbol{x})$.

(b). If $\boldsymbol{g}(\boldsymbol{x})^T \boldsymbol{d} > 0$ and $0 < \bar{\lambda} < \infty$, then $T(\boldsymbol{x}, \boldsymbol{d}, \bar{\lambda}) \neq \emptyset$.

A line search procedure T is *normal* if in addition, it satisfies

$$\left.\begin{array}{l} \boldsymbol{x}_k \to \boldsymbol{x}^* \\ \boldsymbol{d}_k \to \boldsymbol{d}^* \\ \boldsymbol{x}_k^+ \in T(\boldsymbol{x}_k, \boldsymbol{d}_k, \bar{\lambda}_k) \\ f(\boldsymbol{x}_k^+) \to f(\boldsymbol{x}^*) \end{array}\right\} \Longrightarrow \text{ either } \lambda_k^+ \to 0 \text{ or } (\boldsymbol{g}^*)^T \boldsymbol{d}^* \leq 0 \tag{5.3}$$

and

$$\left.\begin{array}{l} \boldsymbol{x}_k \to \boldsymbol{x}^* \\ \boldsymbol{d}_k \to \boldsymbol{d}^* \\ \boldsymbol{x}_k^+ \in T(\boldsymbol{x}_k, \boldsymbol{d}_k, \bar{\lambda}_k) \\ f(\boldsymbol{x}_k^+) \to f(\boldsymbol{x}^*) \end{array}\right\} \Longrightarrow \begin{array}{l} \text{either } (\boldsymbol{g}^*)^T \boldsymbol{d}^* \leq 0 \\ \text{or } \exists\ k_0\ \forall\, k \geq k_0 : \lambda_k^+ = \bar{\lambda}_k \end{array} \tag{5.4}$$

where $\boldsymbol{x}_k^+ = \boldsymbol{x}_k + \lambda_k^+ \boldsymbol{d}_k$.

Remark: For unconstrained optimization, we always have $\bar{\lambda}_k = +\infty$. If an algorithm with normal line search generates an infinite sequence, then we must have $\lim_{k\to\infty} \boldsymbol{g}_k^T \boldsymbol{d}_k = 0$.

Clearly, the above line search procedures have properties (a) and (b). In the following, we show that they are all normal.

Proposition 5.2.1 *Exact line search and Curry test are normal.*

Proof. Suppose that there is a positive number a and an infinite set K such that for $k \in K$, $\lambda_k^+ > a$. Note that

$$\begin{aligned} f(\boldsymbol{x}_k^+) - f(\boldsymbol{x}_k) &\geq f(\boldsymbol{x}_k + \lambda \boldsymbol{d}_k) - f(\boldsymbol{x}_k) \quad (\lambda \in (0, a)) \\ &= \lambda \boldsymbol{g}(\boldsymbol{x}_k + \theta_k(\lambda) \boldsymbol{d}_k) \end{aligned}$$

where $\theta_k(\lambda) \in (0, \lambda)$. Choosing a convergent subsequence $\{\theta_k(\lambda)\}_{k\in K'}$ ($K' \subset K$) and letting $k \to \infty$, $k \in K'$, we obtain $0 \geq \lambda \boldsymbol{g}(\boldsymbol{x}^* + \theta(\lambda)\boldsymbol{d}^*)^T \boldsymbol{d}^*$ for some $\theta(\lambda) \in (0, \lambda)$. Thus, $\boldsymbol{g}(\boldsymbol{x}^* + \theta(\lambda)\boldsymbol{d}^*)^T \boldsymbol{d}^* \leq 0$. Furthermore, letting $\lambda \to 0$, we obtain $(\boldsymbol{g}^*)^T \boldsymbol{d}^* \leq 0$. Thus, (5.3) holds.

To show (5.4), suppose that there is an infinite set K such that for $k \in K$, $\lambda_k^+ < \bar{\lambda}_k$. Then for $k \in K$, $\boldsymbol{g}_k^T \boldsymbol{d}_k = 0$. Letting $k \to \infty$, we obtain $(\boldsymbol{g}^*)^T \boldsymbol{d} = 0$. $\square$

Proposition 5.2.2 *Armijo's rule, Goldstein test and Wolfe test are normal.*

Proof. We only give a proof for Armijo's rule since proofs for the other two procedures are similar. For (5.3), suppose that $(\boldsymbol{g}^*)^T \boldsymbol{d}^* > 0$. Note that

$$0 \leq \lambda_k^+ \leq 2(f(\boldsymbol{x}_k^+) - f(\boldsymbol{x}_k))/\boldsymbol{g}_k^T \boldsymbol{d}_k.$$

Letting $k \to \infty$, we obtain $\lambda_k^+ \to 0$. For (5.4), suppose that $(\boldsymbol{g}^*)^T \boldsymbol{d}^* > 0$. By (5.3), $\lambda_k^+ \to 0$ as $k \to \infty$. Suppose to the contrary that there are infinitely many k such that $\lambda_k^+ < \bar{\lambda}_k$. Then for such k,

$$\begin{aligned} \lambda_k^+ \boldsymbol{g}_k^T \boldsymbol{d}_k &> f(\boldsymbol{x}_k + 2\lambda_k^+ \boldsymbol{d}_k) - f(\boldsymbol{x}_k) \\ &= 2\lambda_k^+ \boldsymbol{g}(\boldsymbol{x}_k + \lambda_k' \boldsymbol{d}_k)^T \boldsymbol{d}_k \end{aligned}$$

where $\lambda_k' \in (0, \lambda_k^+)$. Eliminating λ_k^+ and letting $k \to \infty$, we obtain $(\boldsymbol{g}^*)^T \boldsymbol{d}^* \geq 2(\boldsymbol{g}^*)^T \boldsymbol{d}^*$, contradicting $(\boldsymbol{g}^*)^T \boldsymbol{d}^* > 0$. □

Throughout this book, we study mainly iterative algorithms each generating a sequence of points increasing in the value of the objective function $f(\boldsymbol{x})$. When $\{\boldsymbol{x}_k\}$ is a sequence generated by such an algorithm, we denote $\boldsymbol{x}_{k+1} = \boldsymbol{x}_k^+$ and $\lambda_k = \lambda_k^+$. Since $f(\boldsymbol{x}_k) < f(\boldsymbol{x}_{k+1})$, we have $f(\boldsymbol{x}_k) \to f(\boldsymbol{x}^*)$ for any cluster point $\boldsymbol{x}^*$ of $\{\boldsymbol{x}_k\}$.

Theorem 5.2.3 *Let $\{\boldsymbol{x}_k\}$ be an infinite sequence of feasible points generated by an algorithm with a normal line search procedure. Let $\boldsymbol{d}_k$ be the search direction used at $\boldsymbol{x}_k$, i.e. $\boldsymbol{d}_k$ is an ascendant feasible direction at $\boldsymbol{x}_k$ such that $\boldsymbol{x}_{k+1} = \boldsymbol{x}_k + \lambda_k \boldsymbol{d}_k$. Suppose that $\{(\boldsymbol{x}_k, \boldsymbol{d}_k)\}$ has a cluster point $(\boldsymbol{x}^*, \boldsymbol{d}^*)$ and $\{(\boldsymbol{x}_k, \boldsymbol{d}_k)\}_{k \in K}$ is a subsequence converging to $(\boldsymbol{x}^*, \boldsymbol{d}^*)$. Then either $(\boldsymbol{g}^*)^T \boldsymbol{d}^* = 0$ or the following both hold.*
(1) $\boldsymbol{x}_{k+1} - \boldsymbol{x}_k \to \boldsymbol{0}$ *as* $k \to \infty$, $k \in K$.
(2) $\exists\, k_0 \ \forall\, k > k_0$ *and* $k \in K$, $\lambda_k = \bar{\lambda}_k$.

Proof. Since $\boldsymbol{g}_k^T \boldsymbol{d}_k > 0$ for all k, we have $(\boldsymbol{g}^*)^T \boldsymbol{d}^* \geq 0$. If $(\boldsymbol{g}^*)^T \boldsymbol{d}^* \neq 0$, then $(\boldsymbol{g}^*)^T \boldsymbol{d}^* > 0$. Since the line search procedure is normal, (2) holds and $\lambda_k \to 0$ as $k \to \infty$, $k \in K$. Note that $\{\boldsymbol{d}_k\}_{k \in K}$ is bounded. Thus, $\boldsymbol{x}_{k+1} - \boldsymbol{x}_k = \lambda_k \boldsymbol{d}_k \to 0$ as $k \to \infty$, $k \in K$. □

In nonlinearly constrained optimization problems, given a feasible direction $\boldsymbol{d}$ at a feasible point $\boldsymbol{x}$, how do we find a $\bar{\lambda}$ such that for all $\lambda \in [0, \bar{\lambda}]$ $\boldsymbol{x} + \lambda \boldsymbol{d}$ is in the feasible region Ω? This is a practical problem. For a convex feasible region Ω, if $\boldsymbol{x} + \bar{\lambda} \in \Omega$, then $\boldsymbol{x} + \lambda \boldsymbol{d} \in \Omega$ for all $\lambda \in [0, \bar{\lambda}]$. So, $\bar{\lambda}$ can be chosen easily. However, when Ω is not convex, it is hard to check if $\boldsymbol{x} + \lambda \boldsymbol{d} \in \Omega$ for all $\lambda \in [0, \bar{\lambda}]$. In this case, it is very difficult to satisfy such a strong condition for $\bar{\lambda}$. In fact, for a normal line search procedure, we can usually find a practical condition for $\bar{\lambda}$ such that (5.3) and (5.4) hold. For example, for Amijo's rule, we can ask $\bar{\lambda}$ to satisfy the following:

1. $\bar{\lambda}$ is bigger than a certain number or $\boldsymbol{x} + 2\bar{\lambda}\boldsymbol{d} \notin \Omega$.
2. There exists a natural number k such that for $i = 0, 1, \cdots, k$, $\boldsymbol{x} + \frac{\bar{\lambda}}{2^i}\boldsymbol{d} \in \Omega$ and
$$\frac{1}{2^{k+1}}\bar{\lambda}\boldsymbol{g}(\boldsymbol{x})^T\boldsymbol{d} \leq f(\boldsymbol{x} + \frac{1}{2^k}\bar{\lambda}\boldsymbol{d}) - f(\boldsymbol{x}).$$

Such a $\bar{\lambda}$ together with λ^+ can be found by the following procedure:

> **begin**
> $\bar{\lambda} := 1$; $\lambda^+ := 1$;
> **repeat until** $\boldsymbol{x} + \lambda^+\boldsymbol{d} \in \Omega$ and $\frac{1}{2}\boldsymbol{g}^T\boldsymbol{d} \leq f(\boldsymbol{x} + \lambda^+\boldsymbol{d}) - f(\boldsymbol{x})$ **do**
>
> > **begin**
> > **if** $\boldsymbol{x} + \lambda^+\boldsymbol{d} \notin \Omega$ **then** $\bar{\lambda} := \lambda^+/2$;
> > $\lambda^+ := \lambda^+/2$;
> > **end**;
>
> **end**.

Since no extra technique is needed to prove the global convergence property for such a practical $\bar{\lambda}$, for simplicity, we will always choose $\bar{\lambda}$ to satisfy $\boldsymbol{x} + \lambda\boldsymbol{d} \in \Omega$ for all $\lambda \in [0, \bar{\lambda}]$, although this choice for a nonlinearly constrained feasible region is sometimes impractical.

5.3 Consequences of the First Slope Lemma

We discuss consequences of the first slope lemma in this section. To make our results in a general form, we consider nonlinearly constrained optimization problems as follows.

$$\begin{array}{lll} \text{maximize} & f(\boldsymbol{x}) & \\ \text{subject to} & h_j(\boldsymbol{x}) = 0, & j = 1, ..., m', \\ & h_j(\boldsymbol{x}) \geq 0, & j = m' + 1, ..., m, \end{array} \tag{5.5}$$

where $\boldsymbol{x} \in R^n$; R^n is the n-dimensional Euclidean space whose points are represented as column vectors.

Define $J(\boldsymbol{x}) = \{j \mid h_j(\boldsymbol{x}) = 0\}$ and $D^1(\boldsymbol{x}) = \{\boldsymbol{d} \mid \nabla h_j(\boldsymbol{x})^T\boldsymbol{d} = 0, j \in M'$ and $\nabla h_j(\boldsymbol{x}) \geq 0, j \in J(\boldsymbol{x}) \setminus M'\}$. We claim that any feasible direction $\boldsymbol{d}$ at $\boldsymbol{x}$ belongs to $D^1(\boldsymbol{x})$. In fact, for contradiction, suppose $\boldsymbol{d} \notin D^1(\boldsymbol{x})$. Then either $\nabla h_j(\boldsymbol{x})^T\boldsymbol{d} \neq 0$ for some $j \in M'$ or $\nabla h_j(\boldsymbol{x})^T\boldsymbol{d} < 0$ for some $j \in J(\boldsymbol{x}) \setminus M'$. If the former occurs, then noticing that

$$h_j(\boldsymbol{x} + \lambda\boldsymbol{d}) = \lambda\nabla h_j(\boldsymbol{x})^T\boldsymbol{d} + o(\lambda), \tag{5.6}$$

we have $h(\boldsymbol{x} + \lambda\boldsymbol{d}) \neq 0$ for sufficiently small $\lambda > 0$, contradicting the feasibility of $\boldsymbol{d}$. If the latter occurs then we have $h(\boldsymbol{x} + \lambda\boldsymbol{d}) < 0$ for sufficiently small $\lambda > 0$, also contradicting the feasibility of $\boldsymbol{d}$.

In general, not all vectors in $D^1(\boldsymbol{x})$ are feasible directions. For example, consider the feasible region $\{(x, y) \mid 0 \leq y \leq x^3\}$ and the point $\boldsymbol{x}_0 = (0, 0)$. Clearly, $D^1(\boldsymbol{x}_0) = \{(d_1, 0) \mid d_1 \neq 0\}$. However, $(-1, 0) \in D^1(\boldsymbol{x}_0)$ is not a feasible direction.

A feasible point $\boldsymbol{x}$ is called a *Kuhn-Tucker point* if

$$\boldsymbol{g}(\boldsymbol{x}) = \sum_{j \in J(\boldsymbol{x})} u_j \nabla h_j(\boldsymbol{x})$$

and

$$u_j \leq 0 \quad \text{for} \quad j \in J(\boldsymbol{x}) \setminus M'.$$

A feasible point $\boldsymbol{x}$ is called a *Fritz John point* if there exist $u_0 \geq 0$ and u_j, $j \in J(\boldsymbol{x})$ such that

$$u_0\boldsymbol{g}(\boldsymbol{x}) = \sum_{j \in J(\boldsymbol{x})} u_j \nabla h_j(\boldsymbol{x}), \tag{5.7}$$

$$u_j \leq 0 \quad \text{for} \quad j \in J(\boldsymbol{x}) \setminus M', \tag{5.8}$$

and

$$u_0 - \sum_{j \in J(\boldsymbol{x})} u_j > 0.$$

A Kuhn-Tucker point must be a Fritz John point. The converse does not hold. However, a regular Fritz John point must be a Kuhn-Tucker point. The following theorem is due to Fritz John.

Theorem 5.3.1 *A local maximum must be a Fritz John point.*

Usually, by the *global convergence* of an iterative algorithm, we mean that the algorithm either stops at a Fritz John point or generates an infinite sequence whose cluster points are Fritz John points.

Now, we consider an algorithm in which each iteration can be divided into two steps. At the first step, an ascendant feasible direction $\boldsymbol{d}_k$ at the current feasible point $\boldsymbol{x}_k$ is computed. At the second step, a new feasible point is searched along the search direction by a line search procedure. From the first slope lemma, we can derive the following theorem for such algorithms.

Theorem 5.3.2 *Let $\{\boldsymbol{x}_k\}$ be a sequence of feasible points generated by an algorithm with a normal line search procedure, satisfying that $f(\boldsymbol{x}_{k+1}) > f(\boldsymbol{x}_k)$ for all k. Let $\boldsymbol{d}_k$ be the search direction at $\boldsymbol{x}_k$, i.e. an ascendant feasible direction at $\boldsymbol{x}_k$. Consider a cluster point $\boldsymbol{x}^*$ of the sequence $\{\boldsymbol{x}_k\}$. Suppose that for every subsequence $\{\boldsymbol{x}_k\}_{k\in K}$ converging to $\boldsymbol{x}^*$, $\{\boldsymbol{d}_k\}_{k\in K}$ is bounded. If $\{\boldsymbol{x}_k\}$ does not converge to $\boldsymbol{x}^*$, then there exists a subsequence $\{\boldsymbol{x}_k\}_{k\in K}$ such that $\lim_{k\to\infty,k\in K} \boldsymbol{d}_k = \boldsymbol{d}^*$ and $(\boldsymbol{g}^*)^T\boldsymbol{d}^* = 0$.*

Proof. Consider two cases:

Case 1. There is a subsequence $\{\boldsymbol{x}_k\}_{k\in K}$ converging to $\boldsymbol{x}^*$ such that $\|\boldsymbol{x}_{k+1} - \boldsymbol{x}_k\| > \varepsilon$ for all $k \in K$ and a positive number ε. Since $\{\boldsymbol{d}_k\}_{k\in K}$ is bounded, there is a subsequence $\{\boldsymbol{x}_k\}_{k\in K'}$ $(K' \subset K)$ such that $\{\boldsymbol{d}_k\}_{k\in K'}$ converges to a vector $\boldsymbol{d}^*$. Since $\boldsymbol{d}_k$ is an ascendant direction at $\boldsymbol{x}_k$, we have $\boldsymbol{g}_k^T\boldsymbol{d}_k > 0$ for every k. Thus, $(\boldsymbol{g}^*)^T\boldsymbol{d}^* \geq 0$. By Theorem 5.2.3, we can conclude that $(\boldsymbol{g}^*)^T\boldsymbol{d}^* = 0$.

Case 2. For every subsequence $\{\boldsymbol{x}_k\}_{k\in K}$ converging to $\boldsymbol{x}^*$, $\boldsymbol{x}_{k+1} - \boldsymbol{x}_k \to \boldsymbol{0}$ as $k \to \infty$, $k \in K$. By the first slope lemma, there exists a subsequence $\{\boldsymbol{x}_k\}_{k\in K}$ such that

$$\lim_{k\to\infty,k\in K} \frac{\boldsymbol{g}_k^T(\boldsymbol{x}_{k+1} - \boldsymbol{x}_k)}{\|\boldsymbol{x}_{k+1} - \boldsymbol{x}_k\|} = 0.$$

Let $\boldsymbol{d}^*$ be a cluster point of $\{\boldsymbol{d}_k\}_{k\in K}$. Then

$$\frac{(\boldsymbol{g}^*)^T\boldsymbol{d}^*}{\|\boldsymbol{d}^*\|} = \lim_{k\to\infty,k\in K} \frac{\boldsymbol{g}_k^T\boldsymbol{d}_k}{\|\boldsymbol{d}_k\|} = 0.$$

Therefore, $(\boldsymbol{g}^*)^T\boldsymbol{d}^* = 0$. □

Many feasible direction algorithms which are not of Newton type generate bounded search directions. Thus, for them, in order to prove the global convergence in the case that the generated sequence is not convergent, it suffices to prove that for any subsequence $\{\boldsymbol{x}_k\}_{k\in K}$ of a generated sequence, the following holds:

$$\left.\begin{array}{l} \boldsymbol{x}_k \to \boldsymbol{x}^* \\ \boldsymbol{d}_k \to \boldsymbol{d}^* \\ (\boldsymbol{g}^*)^T \boldsymbol{d}^* = 0 \end{array}\right\} \Longrightarrow \begin{array}{l} \boldsymbol{x}^* \text{ is a Kuhn-Tucker point} \\ \text{(or a Fritz John point)} \end{array} \tag{5.9}$$

where $\boldsymbol{d}_k$ is the search direction at $\boldsymbol{x}_k$ generated by the algorithm. There are many algorithms satisfying condition (5.9). In particular, Rosen's gradient projection method satisfies (5.9). This is one of the key points for proving the global convergence of Rosen's method. This proof will be presented in the next chapter. Before doing so, we first give an example which is a well-known algorithm having a counterexample for its global convergence.

Consider problem (5.5) with $M' = \emptyset$. The following is a popular algorithm cited in the earlier literature.

Zoutendijk's Method: *Initially, choose a starting feasible point* $\boldsymbol{x}_1$. *At each iteration* $k = 1, 2, \cdots$, *solve the following linear programming*

$$\begin{array}{lll} \textit{maximize} & z & \\ \textit{subject to} & \boldsymbol{g}_k^T \boldsymbol{d} - z \geq 0, & \\ & \nabla h_j(\boldsymbol{x}_k)^T \boldsymbol{d} - z \geq 0 & \textit{for } j \in J_k, \\ & -1 \leq d_i \leq 1 & \textit{for } i = 1, \cdots, n, \end{array} \tag{5.10}$$

where d_i *is the* i*th component of* $\boldsymbol{d}$ *and* $J_k = J(\boldsymbol{x}_k)$. *Let* $(z_k, \boldsymbol{d}_k)$ *be an optimal solution of the linear programming. If* $z_k = 0$, *then stop;* $\boldsymbol{x}_k$ *is a Fritz John point. If* $z_k > 0$, *then choose an appropriate* $\bar{\lambda}_k$ *such that for* $\lambda \in [0, \bar{\lambda}_k]$, $\boldsymbol{x}_k + \lambda \boldsymbol{d}_k$ *is feasible, and find a new feasible point* $\boldsymbol{x}_{k+1} = \boldsymbol{x}_k + \lambda_k \boldsymbol{d}_k$ $(0 < \lambda_k \leq \bar{\lambda}_k)$ *by a line search procedure.*

Zoutendijk's method has been shown by a counterexample that it is not globally convergent. In fact, it may generate a sequence of points convergent to a point which is not a Fritz John point. But, an interesting point is the fact that with the first slope lemma, we can show that if Zoutendijk's method generates a sequence which is not convergent, then every cluster point of the sequence is a Fritz John point.

Lemma 5.3.3 *Zoutendijk's method satisfies* (5.9).

Proof. Let $\boldsymbol{x}_k \to \boldsymbol{x}^*$ and $\boldsymbol{d}_k \to \boldsymbol{d}^*$ such that $(\boldsymbol{g}^*)^T\boldsymbol{d}^* = 0$ where $\boldsymbol{d}_k$ is determined at $\boldsymbol{x}_k$ by Zountendijk's method, that is, there is a number z_k such that $(z_k, \boldsymbol{d}_k)$ is an optimal solution of the linear programming (5.10). Denote $z^* = \limsup_{k\to\infty, k\in K} z_k$. Since $\boldsymbol{g}_k^T\boldsymbol{d}_k - z_k \geq 0$, we have $-z^* \geq 0$. However, since $(z, d) = (0, \boldsymbol{0})$ is always a feasible solution of (5.10), we have $z_k \geq 0$ for all k. Thus, $z^* = 0$. Without loss of generality, we assume that for every k, $J_k = J$ for a certain index set J. Next, we show that z^* is the optimal value of the following problem:

$$\begin{array}{lll} \text{maximize} & z & \\ \text{subject to} & (\boldsymbol{g}^*)^T\boldsymbol{d} - z \geq 0, & \\ & \nabla h_j(\boldsymbol{x}^*)^T\boldsymbol{d} - z \geq 0 & \text{for } j \in J, \\ & -1 \leq d_i \leq 1 & \text{for } i = 1, \cdots, n. \end{array} \tag{5.11}$$

In fact, if z^* is not the optimal value, then there exists a feasible solution of (5.11), $(z', \boldsymbol{d}')$ with $z' > 0$. It follows that for sufficiently large k, $(z'/2, \boldsymbol{d}')$ is a feasible solution of (5.10). So, $z_k \geq z'/2$ for such k's. Letting $k \to \infty$, we have $0 = z^* \geq z'/2$, a contradiction. Therefore, $z^* = 0$ is the optimal value of (5.11). It implies that the following system of inequalities has no solution.

$$\begin{aligned} z &> 0, \\ (\boldsymbol{g}^*)^T\boldsymbol{d} - z &\geq 0, \\ \nabla h_j(\boldsymbol{x}^*)^T\boldsymbol{d} - z &\geq 0, \text{ for } j \in J. \end{aligned}$$

(If the system has a solution $(z, \boldsymbol{d})$, we can obtain a feasible solution with $z > 0$ for (5.11) by normalizing $\boldsymbol{d}$.) Hence, there exist $u > 0$, $v > 0$ and $w_j \geq 0$ for $j \in J$ such that

$$\begin{aligned} v\boldsymbol{g}^* + \sum_{j\in J} w_j \nabla h_j(\boldsymbol{x}^*) &= \boldsymbol{0}, \\ u - v - \sum_{j\in J} w_j &= 0. \end{aligned}$$

The second equality is equivalent to $v + \sum_{j\in J} w_j \neq 0$. Therefore, x^* is a Fritz John point. □

Clearly, the search direction determined by Zoutendijk's method is bounded. Therefore, by Theorem 5.3.2 and Lemma 5.3.3, we have:

Theorem 5.3.4 *If an infinite sequence of points generated by Zoutendijk's method is not convergent, then every cluster point of the sequence is a Fritz John point.*

For algorithms considered before, each iteration consists of two parts. In the first part, a search direction is computed depending on an index set. In the second part, a line search procedure is performed. This structure motivates us to abstract them in the following way.

Let Ω be a feasible region and M the index set of all constraints. Let D be a compact set. Suppose that Γ is a solution set, i.e. the set of Kuhn-Tucker points or the set of Fritz John points. Let $\{F_J\}_{J \in \mathcal{P}(M)}$ be a family of point-to-set mappings from $\Omega \setminus \Gamma$ to $\mathcal{P}(D)$ such that

$$\left.\begin{array}{l} J \subseteq J(\boldsymbol{x}) \\ \boldsymbol{x} \in \Omega \setminus \Gamma \\ \boldsymbol{d} \in F_J(\boldsymbol{x}) \end{array}\right\} \Longrightarrow \boldsymbol{d} \text{ is an ascendant feasible direction at } \boldsymbol{x}. \qquad (5.12)$$

Consider the following algorithm.

Algorithm A: Given an initial feasible point $\boldsymbol{x}_1$, at iteration k, carry out the following two steps.

1. Choose a search direction $\boldsymbol{d}_k$ from $F_J(\boldsymbol{x}_k)$ where $J = J(\boldsymbol{x}_k)$. If $\boldsymbol{d}_k$ is not an ascending feasible direction, then stop. Otherwise, go to the next step.

2. Compute a $\bar{\lambda}_k$ such that $\boldsymbol{x}_k + \lambda \boldsymbol{d}_k$ for all $\lambda \in [0, \bar{\lambda}_k]$ lies in Ω. Find a new point $\boldsymbol{x}_{k+1} = \boldsymbol{x}_k + \lambda_k \boldsymbol{d}_k$ $(0 < \lambda_k \leq \bar{\lambda}_k)$ by a normal line search procedure.

The following theorem is a consequence of the first slope lemma.

Theorem 5.3.5 *Suppose that for $J \subseteq J(\boldsymbol{x})$, $F_J(\boldsymbol{x}) \neq \emptyset$, F_J is closed at $\boldsymbol{x}$ and (5.12) holds. Let $\{\boldsymbol{x}_k\}$ be an infinite sequence generated by Algorithm A. If $\{\boldsymbol{x}_k\}$ is not convergent, then its cluster points all belong to Γ.*

Proof. It is sufficient to prove that for any subsequence $\{\boldsymbol{x}_k\}_{k \in K}$, the following holds

$$\left.\begin{array}{l} \boldsymbol{x}_k \to \boldsymbol{x}^* \\ \boldsymbol{d}_k \to \boldsymbol{d}^* \\ (\boldsymbol{g}^*)^T \boldsymbol{d}^* = 0 \end{array}\right\} \Longrightarrow \boldsymbol{x}^* \in \Gamma. \qquad (5.13)$$

Since the number of index sets is finite, there exists an infinite subset K' of K such that for $k \in K'$, $J_k = J$. Clearly, $J \subseteq J(\boldsymbol{x}^*)$. If $\boldsymbol{x}^* \notin \Gamma$, then F_J is closed at $\boldsymbol{x}^*$. Thus, $\boldsymbol{d}^* \in F_J(\boldsymbol{x}^*)$. Hence $\boldsymbol{d}^*$ is an ascending direction, contradicting $(\boldsymbol{g}^*)^T \boldsymbol{d}^* = 0$. □

The closeness of F_J is not a strong condition. Since J is fixed, this condition does not suffer from the changing of the active set. In fact, for many methods of choosing search directions, we can find a possibly larger mapping satisfying this condition. For example, for Rosen's method, we may define $F_J(\boldsymbol{x}) = \emptyset$ for $J \not\subseteq J(\boldsymbol{x})$ and

$$F_J(\boldsymbol{x}) = \begin{cases} \{P_J \boldsymbol{g}\} & \text{if } \|P_J \boldsymbol{g}\| > u_h, \\ \{P_J \boldsymbol{g}, P_{J \setminus h} \boldsymbol{g}\} & \text{if } \|P_J \boldsymbol{g}\| = u_h, \\ \{P_{J \setminus h} \boldsymbol{g}\} & \text{if } \|P_J \boldsymbol{g}\| < u_h \end{cases}$$

for $J \subseteq J(\boldsymbol{x})$.

Instead of a point-to-set mapping, using a family of point-to-set mappings with an index set as its parameter, it is probably a better way to give abstract models for algorithms, because more exact, natural and easily-verified convergence conditions could be given to such a kind of models. For one more example, we will give such an abstract algorithm in section 6.2 for problem (5.5) with $M' = \emptyset$.

Exercises

1. Select a positive number c, $0 < c < 1$. The Percentage test computes a new point $\boldsymbol{x}^+ = \boldsymbol{x} + \lambda^+ \boldsymbol{d}$ $(0 < \lambda^+ \leq \bar{\lambda})$ satisfying $|\lambda^+ - \lambda^*| < c\lambda^+$ for some local maximum λ^*. Find a counterexample to show that the Percentage Test is not normal.

2. Let $f(x)$ be a function on $[a, b]$. The Golden section search computes a local maximum of $f(x)$ in the following way: Choose $x_1 = a + \frac{3-\sqrt{5}}{2}(b-a)$ and $x_2 = a + \frac{\sqrt{5}-1}{2}(b-a)$ and compare $f(x_1)$ and $f(x_2)$. If $f(x_1) < f(x_2)$, then delete $[a, x_1)$. Otherwise, delete $(x_2, b]$. Repeat the above computation on the remainder of the interval. Prove that the Golden Section Search with the Percentage Test as its termination rule is normal for convex objective functions.

3. Find a linearly constrained optimization problem which has a Fritz John point not being a Kuhn-Tucker point.

Chapter 6

Average Slope

The creative person is
both more primitive and more cultivated, ...,
than the average person.
FRANK BARRON

Consider a path going up on a mountain. The average slope is the ratio of the height increment over the length of the projection of the path onto the level plane. In this chapter, we study the properties of the average slope and develop some tools based on these properties.

6.1 The Second Slope Lemma

Consider a path

$$(\boldsymbol{x}_1, f(\boldsymbol{x}_1)) \to (\boldsymbol{x}_2, f(\boldsymbol{x}_2)) \to \cdots \to (\boldsymbol{x}_k, f(\boldsymbol{x}_k)) \to \cdots$$

satisfying

$$f(\boldsymbol{x}_1) < f(\boldsymbol{x}_2) < \cdots < f(\boldsymbol{x}_k) < \cdots$$

and

$$\boldsymbol{x}_k \to \boldsymbol{x}^* \text{ as } k \to \infty.$$

The average slope from $(\boldsymbol{x}_k, f(\boldsymbol{x}_k))$ to $(\boldsymbol{x}^*, f(\boldsymbol{x}^*))$ is defined by

$$\frac{f(\boldsymbol{x}^*) - f(\boldsymbol{x}_k)}{\|\boldsymbol{x}_k - \boldsymbol{x}_{k+1}\| + \|\boldsymbol{x}_{k+1} - \boldsymbol{x}_{k+2}\| + \cdots}.$$

There are two properties about the average slope:

The first one is that the average slope of any path from $(\boldsymbol{x}_k, f(\boldsymbol{x}_k))$ to $(\boldsymbol{x}^*, f(\boldsymbol{x}^*))$ cannot exceed

$$\frac{f(\boldsymbol{x}^*) - f(\boldsymbol{x}_k)}{\|\boldsymbol{x}^* - \boldsymbol{x}_k\|}.$$

Mathematically, this property is trivial since

$$\sum_{i=k}^{\infty} \|\boldsymbol{x}_i - \boldsymbol{x}_{i+1}\| \geq \|\boldsymbol{x}_k - \boldsymbol{x}^*\|.$$

Intuitively, this property means that the average slope of any path cannot exceed the slope of the direct path connecting the same pair of locations (Figure 6.1).

Figure 6.1: The steepest path is the direct path.

The second property of the average slope is that

$$\inf_{i \geq k} \frac{f(\boldsymbol{x}_{i+1}) - f(\boldsymbol{x}_i)}{\|\boldsymbol{x}_{i+1} - \boldsymbol{x}_i\|} \leq \frac{f(\boldsymbol{x}^*) - f(\boldsymbol{x}_k)}{\sum_{i=k}^{\infty} \|\boldsymbol{x}_{i+1} - \boldsymbol{x}_i\|} \leq \sup_{i \geq k} \frac{f(\boldsymbol{x}_{i+1}) - f(\boldsymbol{x}_i)}{\|\boldsymbol{x}_{i+1} - \boldsymbol{x}_i\|}.$$

Let us prove only one side of the above inequality, since the other side can be proved in a similar way. Denote

$$\zeta = \inf_{i \geq k} \frac{f(\boldsymbol{x}_{i+1}) - f(\boldsymbol{x}_i)}{\|\boldsymbol{x}_{i+1} - \boldsymbol{x}_i\|}.$$

Then we have

$$f(\boldsymbol{x}_{i+1}) - f(\boldsymbol{x}_i) \geq \zeta \|\boldsymbol{x}_{i+1} - \boldsymbol{x}_i\|.$$

for $i \geq k$. Taking the sum of these inequalities over $i \geq k$, we obtain

$$f(\boldsymbol{x}^*) - f(\boldsymbol{x}_k) \geq \zeta \sum_{i \geq k} \|\boldsymbol{x}_{i+1} - \boldsymbol{x}_i\|.$$

It follows from the above two properties that

$$\inf_{i \geq k} \frac{f(\boldsymbol{x}_{i+1}) - f(\boldsymbol{x}_i)}{\|\boldsymbol{x}_{i+1} - \boldsymbol{x}_i\|} \leq \frac{f(\boldsymbol{x}^*) - f(\boldsymbol{x}_k)}{\|\boldsymbol{x}^* - \boldsymbol{x}_k\|}.$$

Therefore,

$$\lim_{k \to \infty} \inf_{i \geq k} \frac{f(\boldsymbol{x}_{i+1}) - f(\boldsymbol{x}_i)}{\|\boldsymbol{x}_{i+1} - \boldsymbol{x}_i\|} \leq \lim_{k \to \infty} \inf \frac{f(\boldsymbol{x}^*) - f(\boldsymbol{x}_k)}{\|\boldsymbol{x}^* - \boldsymbol{x}_k\|}.$$

This means that there exists a subsequence $\{\boldsymbol{x}_k\}_{k \in K}$ such that the

$$\lim_{k \to \infty, k \in K} \frac{f(\boldsymbol{x}_{i+1}) - f(\boldsymbol{x}_i)}{\|\boldsymbol{x}_{i+1} - \boldsymbol{x}_i\|}$$

exists, and

$$0 \leq \lim_{k \to \infty, k \in K} \frac{f(\boldsymbol{x}_{i+1}) - f(\boldsymbol{x}_i)}{\|\boldsymbol{x}_{i+1} - \boldsymbol{x}_i\|} \leq \lim_{k \to \infty} \inf \frac{f(\boldsymbol{x}^*) - f(\boldsymbol{x}_k)}{\|\boldsymbol{x}^* - \boldsymbol{x}_k\|}. \tag{6.1}$$

Now, we consider the linearly constrained maximization problem (4.1) stated again here for convenience:

$$\begin{array}{ll} \text{maximize} & f(\boldsymbol{x}) \\ \text{subject to} & \boldsymbol{a}_j^T \boldsymbol{x} \geq b_j, \quad j \in M = \{1, 2, \cdots, m\}. \end{array}$$

Recall that for any $J \subseteq M$, P_J is the orthogonal projection operator for the subspace $D_J = \{\boldsymbol{d} \mid \boldsymbol{a}_j^T \boldsymbol{d} = 0, j \in J\}$. Suppose that $\boldsymbol{x}_1$, $\boldsymbol{x}_2$, $\cdots$ are all feasible points. Let J be an index set such that $J = J(\boldsymbol{x}_k)$ for infinitely many k. Then, $\boldsymbol{x}_k \to \boldsymbol{x}^*$ implies $J \subseteq J(\boldsymbol{x}^*)$. Thus, for those k with $J = J(\boldsymbol{x}_k)$, we have $P_J(\boldsymbol{x}^* - \boldsymbol{x}_k) = \boldsymbol{x}^* - \boldsymbol{x}_k$ and hence, by the mean value theorem,

$$\begin{aligned} \frac{f(\boldsymbol{x}^*) - f(\boldsymbol{x}_k)}{\|\boldsymbol{x}^* - \boldsymbol{x}_k\|} &= \frac{\boldsymbol{g}(\boldsymbol{y}_k)^T (\boldsymbol{x}^* - \boldsymbol{x}_k)}{\|\boldsymbol{x}^* - \boldsymbol{x}_k\|} \\ &= \frac{\boldsymbol{g}(\boldsymbol{y}_k)^T P_J (\boldsymbol{x}^* - \boldsymbol{x}_k)}{\|\boldsymbol{x}^* - \boldsymbol{x}_k\|} \\ &= \frac{(P_J \boldsymbol{g}(\boldsymbol{y}_k))^T (\boldsymbol{x}^* - \boldsymbol{x}_k)}{\|\boldsymbol{x}^* - \boldsymbol{x}_k\|} \\ &\leq \|P_J \boldsymbol{g}(\boldsymbol{y}_k)\| \end{aligned}$$

where $\boldsymbol{y}_k \in [\boldsymbol{x}_k, \boldsymbol{x}^*]$, i.e. $\boldsymbol{y}_k$ is on the straight line segment between $\boldsymbol{x}_k$ and $\boldsymbol{x}^*$. Since $\boldsymbol{x}_k \to \infty$ as $k \to \infty$, we have that as k goes to infinity through such k's, $\boldsymbol{y}_k$ converges to $\boldsymbol{x}^*$ and $\|P_J\boldsymbol{g}(\boldsymbol{y}_k)\|$ converges to $\|P_J\boldsymbol{g}(\boldsymbol{x}^*)$. Therefore,

$$\liminf_{k\to\infty} \frac{f(\boldsymbol{x}^*) - f(\boldsymbol{x}_k)}{\|\boldsymbol{x}^* - \boldsymbol{x}_k\|} \le \|P_J\boldsymbol{g}(\boldsymbol{x}^*)\|. \tag{6.2}$$

Moreover, by the mean value theorem,

$$\frac{f(\boldsymbol{x}_{k+1}) - f(\boldsymbol{x}_k)}{\|\boldsymbol{x}_{k+1} - \boldsymbol{x}_k\|} = \frac{\boldsymbol{g}(\boldsymbol{z}_k)^T(\boldsymbol{x}_{k+1} - \boldsymbol{x}_k)}{\|\boldsymbol{x}_{k+1} - \boldsymbol{x}_k\|} \tag{6.3}$$

for some $\boldsymbol{z}_k \in [\boldsymbol{x}_k, \boldsymbol{x}_{k+1}]$, and

$$\lim_{k\to\infty} \frac{\boldsymbol{g}(\boldsymbol{x}_k) - \boldsymbol{g}(\boldsymbol{z}_k)}{\|\boldsymbol{x}_{k+1} - \boldsymbol{x}_k\|} = 0. \tag{6.4}$$

Putting (6.1), (6.2), (6.3), and (6.4) together, we have

$$0 \le \lim_{k\to\infty, k\in K} \frac{\boldsymbol{g}_k^T(\boldsymbol{x}_{k+1} - \boldsymbol{x}_k)}{\|\boldsymbol{x}_{k+1} - \boldsymbol{x}_k\|} \le \|P_J\boldsymbol{g}(\boldsymbol{x}^*)\|.$$

The following lemma summarizes the above result.

Lemma 6.1.1 (The Second Slope Lemma) *Let $\{\boldsymbol{x}_k\}$ be a sequence of feasible points, in a linearly constrained region, convergent to $\boldsymbol{x}^*$. Suppose $f(\boldsymbol{x}_k) < f(\boldsymbol{x}_{k+1})$ for all k. Then there exists a subsequence $\{\boldsymbol{x}_k\}_{k\in K}$ such that for every J with $J = J_k$ for infinitely many k,*

$$0 \le \lim_{k\to\infty, k\in K} \frac{\boldsymbol{g}_k^T(\boldsymbol{x}_{k+1} - \boldsymbol{x}_k)}{\|\boldsymbol{x}_{k+1} - \boldsymbol{x}_k\|} \le \|P_J\boldsymbol{g}(\boldsymbol{x}^*)\|.$$

Corollary 6.1.2 *Under the conditions in Lemma 6.1.1, there exists a subsequence $\{\boldsymbol{x}_k\}_{k\in K}$ such that for $k \in K$, $J_k = J$, and*

$$0 \le \lim_{k\to\infty, k\in K} \frac{\boldsymbol{g}_k^T(\boldsymbol{x}_{k+1} - \boldsymbol{x}_k)}{\|\boldsymbol{x}_{k+1} - \boldsymbol{x}_k\|} \le \|P_J\,\boldsymbol{g}^*\|.$$

Proof. Let J' be an index with the least value of $\|P_{J'}\,\boldsymbol{g}^*\|$ for which $J_k = J'$ for infinitely many k. Let $\{\boldsymbol{x}_k\}_{k\in K}$ be a sequence in Lemma 6.1.1. Since the number of index sets is finite, there is an infinite subset K' of K such that for all $k \in K'$, $J_k = J$ for a certain index set J. Since $\|P_{J'}\,\boldsymbol{g}^*\| \le \|P_J\,\boldsymbol{g}^*\|$, the corollary is proved. □

6.2 The Global Convergence of Rosen's Method

An immediate application of the first and the second slope lemmas is to show the global convergence of Rosen's method as follows.

Theorem 6.2.1 *Let α and β be two positive numbers with $\alpha < \beta$. Suppose that $\alpha \le c_k \le \beta$ for all k. Then Rosen's method with a normal line search procedure either stops at a Kuhn-Tucker point or generates an infinite sequence whose cluster points are all Kuhn-Tucker points.*

To prove it, let us first prove a lemma.

Lemma 6.2.2 *Under the conditions in Theorem 6.2.1, suppose Rosen's method with a normal line search procedure generates an infinite sequence of points, $\{\boldsymbol{x}_k\}$. Then every convergent subsequence $\{\boldsymbol{x}_k\}_{k\in K}$ satisfies*

$$\left.\begin{array}{l} \boldsymbol{x}_k \to \boldsymbol{x}^* \\ \boldsymbol{d}_k \to \boldsymbol{d}^* \\ (\boldsymbol{g}^*)^T\boldsymbol{d}^* \le 0 \end{array}\right\} \Longrightarrow \boldsymbol{x}^* \text{ is a Kuhn-Tucker point}$$

where $\boldsymbol{d}_k$ is the search direction generated by Rosen's method in the kth iteration.

Proof. Let $\{\boldsymbol{x}_k\}_{k\in K}$ be a subsequence of a generated sequence such that $\boldsymbol{x} \to \boldsymbol{x}^*$, $\boldsymbol{d}_k \to \boldsymbol{d}^*$ and $(\boldsymbol{g}^*)^T\boldsymbol{d}^* = 0$. Since the number of index sets is finite, we can find an infinite subset K' of K such that for all $k \in K'$, $J_k = J$ and $h_k = h$ for a certain index set J and a certain index h. In addition, either

$$\boldsymbol{d}_k = P_J\boldsymbol{g}_k \text{ for all } k \in K' \tag{6.5}$$

or

$$\boldsymbol{d}_k = P_{J\setminus h}\boldsymbol{g}_k \text{ for all } k \in K'. \tag{6.6}$$

If (6.5) holds, then $\boldsymbol{d}^* = P_J\boldsymbol{g}^*$ and $\|P_J\boldsymbol{g}^*\| \ge \alpha u_h^*$ where

$$(u_j^*, j \in J)^T = (A_J^T A_J)^{-1} A_J^T \boldsymbol{g}^* \tag{6.7}$$

and

$$u_h^* = \max\{u_j^* \mid j \in J \setminus M'\}. \tag{6.8}$$

Since $0 \geq (\boldsymbol{g}^*)^T \boldsymbol{d}^* = \|P_J \boldsymbol{g}^*\|^2$, we have $P_J \boldsymbol{g}^* = \mathbf{0}$ and $0 \geq u_h^*$. Note that $\boldsymbol{x}_k \to \boldsymbol{x}^*$ implies $J \subseteq J^*$. Hence, $\boldsymbol{x}^*$ is a Kuhn-Tucker point. If (6.6) holds, then $\boldsymbol{d}^* = P_{J \setminus h} \boldsymbol{g}^*$. Hence, $0 \geq (\boldsymbol{g}^*)^T \boldsymbol{d}^* = \|P_{J \setminus h} \boldsymbol{g}^*\|^2$, By (4.8), we obtain $P_J \boldsymbol{g}^* = \mathbf{0}$ and $u_h^* = 0$. Therefore, $\boldsymbol{x}^*$ is a Kuhn-Tucker point. □

Proof of Theorem 6.2.1. If the algorithm terminates at a point $\boldsymbol{x}_k$, then by Theorem 4.2.5, $\boldsymbol{x}_k$ is a Kuhn-Tucker point. Next, we consider an infinite sequence $\{\boldsymbol{x}_k\}$ generated by Rosen's method. Let $\boldsymbol{x}^*$ be a cluster point of the sequence $\{\boldsymbol{x}_k\}$. There are two cases, (a) the sequence $\{\boldsymbol{x}_k\}$ is not convergent and (b) the sequence $\{\boldsymbol{x}_k\}$ is convergent. We will let the first slope lemma deal with the first case and the second slope lemma deal with the second case. To do so, we need to note that

$$\frac{\boldsymbol{g}_k^T(\boldsymbol{x}_{k+1} - \boldsymbol{x}_k)}{\|\boldsymbol{x}_{k+1} - \boldsymbol{x}_k\|} = \frac{\boldsymbol{g}_k^T \boldsymbol{d}_k}{\|\boldsymbol{d}_k\|} = \frac{\boldsymbol{g}_k^T P_J \boldsymbol{g}_k}{\|P_J \boldsymbol{g}_k\|} = \|P_J \boldsymbol{g}_k\| = (\boldsymbol{g}_k^T \boldsymbol{d}_k)^{1/2}$$

where $J = J_k$ or $J_k \setminus h_k$.

First, assume that $\{\boldsymbol{x}_k\}$ is not convergent. Suppose that the subsequence $\{\boldsymbol{x}_k\}_{k \in K}$ is convergent to $\boldsymbol{x}^*$. Note that for any k, $\|\boldsymbol{d}_k\| \leq \|\boldsymbol{g}_k\|$. Thus, the set of corresponding search directions $\{\boldsymbol{d}_k \mid k \in K\}$ is bounded. We may further assume, without loss of generality, that the corresponding subsequence $\{\boldsymbol{d}_k\}_{k \in K}$ is convergent to $\boldsymbol{d}^*$. Since the line search procedure is normal, we have either $(\boldsymbol{g}^*)^T \boldsymbol{d}^* \leq 0$ or $\lambda_k \to 0$ as $k \to \infty$ through $k \in K$. If the former occurs, then by Lemma 6.2.2, $\boldsymbol{x}^*$ is a Kuhn-Tucker point. If the latter occurs, what should we do? We may assume

$$\lambda_k \to 0 \text{ as } k \to \infty \text{ through } k \in K$$

for every subsequence $\{\boldsymbol{x}_k\}_{k \in K}$ convergent to $\boldsymbol{x}^*$ with the convergent corresponding subsequence of search directions, $\{\boldsymbol{d}_k\}_{k \in K}$ (since one of such subsequence in the former case would imply that $\boldsymbol{x}^*$ is a Kuhn-Tucker point). Now, we further claim that for every subsequence $\{\boldsymbol{x}_k\}_{k \in K}$ convergent to $\boldsymbol{x}^*$,

$$\lambda_k \to 0 \text{ as } k \to \infty \text{ through } k \in K$$

that is,

$$\|\boldsymbol{x}_{k+1} - \boldsymbol{x}_k\| \to 0 \text{ as } k \to \infty \text{ through } k \in K.$$

In fact, for contradiction, suppose this is not true, i.e. there exists a subsequence $\{\boldsymbol{x}_k\}$ convergent to $\boldsymbol{x}^*$ such that $\{\lambda_k\}_{k \in K}$ is not convergent to 0. Then there exists a positive number ζ such that $\lambda_k \geq \zeta$ for infinitely many k

in K. From those k's, we can further find a convergent subsequence $\{\boldsymbol{d}_k\}_{k\in K'}$ where $K' \subset K$. Thus, $\{\boldsymbol{x}_k\}_{k\in K'}$ is convergent to $\boldsymbol{x}^*$ with convergent $\{\boldsymbol{d}_k\}$, but $\lambda_k \geq \zeta$ for $k \in K'$, a contradiction.

Now, all conditions in the first slope lemma hold. By the first slope lemma, there exists a subsequence $\{\boldsymbol{x}_k\}_{k\in K}$ convergent to $\boldsymbol{x}^*$ such that

$$\lim_{k\to\infty, k\in K} \frac{\boldsymbol{g}_k^T(\boldsymbol{x}_{k+1} - \boldsymbol{x}_k)}{\|\boldsymbol{x}_{k+1} - \boldsymbol{x}_k\|} = 0$$

that is,

$$\lim_{k\to\infty, k\in K} \boldsymbol{g}_k^T \boldsymbol{d}_k = 0.$$

Choose $K' \subset K$ such that $\{\boldsymbol{d}_k\}_{k\in K'}$ is convergent, say to $\boldsymbol{d}^*$. Then $(\boldsymbol{g}^*)^T\boldsymbol{d}^* = 0$. Hence, $\boldsymbol{x}^*$ is a Kuhn-Tucker point.

Next, assume that $\{\boldsymbol{x}_k\}$ is convergent to $\boldsymbol{x}^*$. It follows that $\{\boldsymbol{d}_k\}$ is bounded. If there are infinitely many k such that $\lambda_k \neq \bar{\lambda}_k$, then the corresponding $\boldsymbol{d}_k$'s must have a cluster point $\boldsymbol{d}^*$. Since the line search procedure is normal, we have $(\boldsymbol{g}^*)^T\boldsymbol{d}^* = 0$. By Lemma 6.2.2, $\boldsymbol{x}^*$ is a Kuhn-Tucker point. Therefore, we may assume that there exists k' such that for $k > k'$, $\lambda_k = \bar{\lambda}_k$. For a similar reason, we may assume that $\lambda_k \to 0$ as $k \to \infty$. Thus, there exists $k'' > k'$ such that for $k > k''$, $J_{k+1} \setminus J_k \neq \emptyset$, i.e. at least one inactive constraint becomes active at iteration k. Note that at each iteration, at most one active constraint becomes inactive. Since the number of constraints is finite, there exists k^* such that for $k > k^*$, exactly one active constraint changes to an inactive constraint and exactly one inactive constraint changes to an active constraint at iteration k, that is,

$$|J_k \setminus J_{k+1}| = |J_{k+1} \setminus J_k| = 1. \tag{6.9}$$

By Corollary 6.1.2, there is a subsequence $\{\boldsymbol{x}_k\}_{k\in K}$ such that $J_k = J$ for all $k \in K$ and a certain J, and

$$\lim_{k\to\infty, k\in K} \boldsymbol{g}_k^T \boldsymbol{d}_k / \|\boldsymbol{d}_k\| \leq \|P_J \boldsymbol{g}^*\|. \tag{6.10}$$

Furthermore, we may assume without loss of generality that $h_k = h$ for all $k \in K$ and $\boldsymbol{d}_k \to \boldsymbol{d}^*$ as $k \to \infty$, $k \in K$. By (6.9), $\boldsymbol{d}_k = P_{J\setminus h}\boldsymbol{g}_k$. Thus $\boldsymbol{d}^* = P_{J\setminus h}\boldsymbol{g}^*$. By (6.10), $\|P_{J\setminus h}\boldsymbol{g}^*\| \leq \|P_J\boldsymbol{g}^*\|$. It follows from (4.8) that $u_h^* = 0$ where u_h^* is defined by (6.7) and (6.8). However, (6.9) implies $\beta u_{kh} \geq \|P_J\boldsymbol{g}_k\|$ for all $k \in K$. Thus, $0 = \beta u_h^* \geq \|P_J\boldsymbol{g}^*\|$, and herefore, $\boldsymbol{x}^*$ is a Kuhn-Tucker point. □

From the above proof, it is easy to see the following.

Corollary 6.2.3 *Suppose that the conditions in Theorem 6.2.1 hold and* $\boldsymbol{x}_k \to \boldsymbol{x}^*$ *as* $k \to \infty$. *Then* $\mathbf{0}$ *is a cluster point of* $\{\boldsymbol{d}_k\}$.

The normality of the line search procedure is crucial in Theorem 6.2.1. The following is a counterexample which shows that Rosen's method may not be globally convergent if the used line search procedure is not normal.

Counterexample. Consider the following problem:

$$\begin{array}{ll} \text{maximize} & f(x,y) = y^2 - x^2 \\ \text{subject to} & x \geq 0,\ y \geq 0, \\ & x - y \leq 1. \end{array}$$

We use Rosen's method with $c_k = 1/3$ for all k and a line search procedure which finds $\boldsymbol{x}_{k+1} = \boldsymbol{x}_k + \lambda_k \boldsymbol{d}_k$ $(0 < \lambda_k \leq \bar{\lambda}_k)$ such that $f(\boldsymbol{x}_{k+1}) > f(\boldsymbol{x}_k)$ and $f(\boldsymbol{x}_{k+1}) \geq f(\boldsymbol{x}_k + \lambda \boldsymbol{d}_k) - \varepsilon_k$ for all $\lambda \in [0, \bar{\lambda}_k]$ where $\varepsilon_k = 2^{1-k}$. Choose $\boldsymbol{x}_1 = (3/2, 1/2)^T$ as the initial point. We show that the sequence $\{\boldsymbol{x}_k = (1 + 2^{-k}, 2^{-k})^T\}$ can be generated by the algorithm. In fact, all $\boldsymbol{x}_k$'s lie on the boundary $x - y = 1$. Thus, $P_{J_k} \boldsymbol{g}_k = -(1,1)^T$ and $u_{kh} = 1 + 2^{1-k}$. Since $\|P_{J_K} \boldsymbol{g}_k\| = \sqrt{2} > (1+2^{1-k})/3 = c_k u_{kh}$, we have $\boldsymbol{d}_k = -(1,1)^T$ and $\bar{\lambda}_k = 2^{-k}$. Note that $f(\boldsymbol{x}_k + \lambda \boldsymbol{d}_k)$ gets its maximum value at $\lambda = \bar{\lambda}_k$. Moreover,

$$f(\boldsymbol{x}_{k+1}) = -(1 + 2^{-k}) > f(\boldsymbol{x}_k + \bar{\lambda}_k \boldsymbol{d}_k) - \varepsilon_k = -(1 + 2^{1-k}).$$

Thus, $\{\boldsymbol{x}_k\}$ can be generated by the algorithm. However, $(0,0)^T$ is the unique Kuhn-Tucker point and $\lim_{k \to \infty} \boldsymbol{x}_k = (1,0)^T$. Therefore, Rosen's method with this line search procedure is not globally convergent. Since Rosen's method with a normal line search procedure is globally convergent, it also follows that the considered line search procedure is not normal. In fact, this line search procedure satisfies (5.3) but does not satisfy (5.4).

Proposition 6.2.4 *The line search procedure in the above counterexample satisfies (5.3).*

Proof. Suppose that $\bar{\lambda}_k$ for $k \in K$ does not converge to 0. Then there exists a positive number μ such that for infinitely many $k \in K$, $\bar{\lambda}_k \geq \mu$. For such k, we have

$$\begin{aligned} f(\boldsymbol{x}_{k+1}) - f(\boldsymbol{x}_k) &\geq f(\boldsymbol{x}_k + \lambda \boldsymbol{d}_k) - f(\boldsymbol{x}_k) - \varepsilon_k \\ &= \lambda \boldsymbol{g}(\boldsymbol{x}_k + \mu_k \boldsymbol{d}_k)^T \boldsymbol{d}_k - \varepsilon_k \end{aligned}$$

where $\mu_k \in [0, \lambda]$ and $\lambda \in [0, \mu]$. Choose a subset K' of K such that μ_k converges to a number μ' as k goes to infinity through K'. Fixing λ and letting $k \to \infty$, we obtain

$$0 \geq \lambda \boldsymbol{g}(\boldsymbol{x}^* + \mu' \boldsymbol{d}^*)^T \boldsymbol{d}^*.$$

Thus,

$$0 \geq \boldsymbol{g}(\boldsymbol{x}^* + \mu' \boldsymbol{d}^*)^T \boldsymbol{d}^*.$$

Letting $\lambda \to 0$, we obtain $(\boldsymbol{g}^*)^T \boldsymbol{d}^* \leq 0$. □

Zangwill's theorem cannot prove the convergence of Rosen's method and slope lemmas can. This means that the "proof power" of the slope lemmas is stronger than Zangwill's theorem.

In the literature, there is another popular proof technique for the global convergence of feasible direction methods, called *Bazaraa-Shetty's condition* which is as follows: Any sequence generated by a feasible direction method with the exact line search procedure cannot have a subsequence $\{\boldsymbol{x}_k\}_{k \in K}$ satisfying the following four conditions:

1. $\lim_{k \to \infty, k \in K} \boldsymbol{x}_k = \boldsymbol{x}^*$
2. There exists a positive number μ such that $\bar{\lambda}_k \geq \mu$ for $k \in K$.
3. Let $\boldsymbol{d}_k$ be the search direction at $\boldsymbol{x}_k$. Then $\lim_{k \to \infty, k \in K} \boldsymbol{d}_k = \boldsymbol{d}^*$.
4. $(\boldsymbol{g}^*)^T \boldsymbol{d}^* > 0$.

Bazaraa-Shetty's condition is actually a property of the exact line search. This property can also be stated as follows: For any subsequence $\{\boldsymbol{x}_k\}_{k \in K}$, the following holds

$$\left.\begin{array}{l} \boldsymbol{x}_k \to \boldsymbol{x}^* \\ \boldsymbol{d}_k \to \boldsymbol{d}^* \end{array}\right\} \Longrightarrow \begin{array}{l} \text{either } (\boldsymbol{g}^*)^T \boldsymbol{d}^* \leq 0 \\ \text{or } \bar{\lambda}_k \to 0. \end{array} \tag{6.11}$$

From Theorem 5.2.3, it is easy to see that the normality implies (6.11). However, (6.11) does not imply the normality. In fact, the line search procedure in the above counterexample satisfies (6.11) but is not normal. The counterexample shows that Bazaraa-Shetty's condition does not give enough information on the line search in order to guarantee the global convergence of Rosen's method and hence is not strong enough to show the global convergence of Rosen's method. This means that in some sense, the new approach gives more proof power than Bazaraa-Shetty's condition does.

The continuous differentiability of the objective function in the entire feasible region is also a necessary condition for Theorem 6.2.1. The following is a counterexample to show this:

$$\begin{array}{ll} \text{maximize} & f(\boldsymbol{x}) = -\sqrt{x_1^2 + x_2^2 + x_3} \\ \text{subject to} & x_1 + 3x_2 \geq 0, \\ & x_1 - 3x_2 \geq 0, \\ & 0 \leq x_3 \leq 10. \end{array}$$

The problem has the unique Kuhn-Tucker point $\boldsymbol{x}^* = (0,0,10)^T$. Starting from $\boldsymbol{x}_1 = (3/2, -1/2, \sqrt{3})^T$ and choosing $c_k = 10$, Rosen's method with Armijo's rule generates a sequence $\{\boldsymbol{x}_k = (3/2^k, (-1/2)^k, \sum_{i=0}^{k-1} \sqrt{3}/2^i)^T\}$. However, $\boldsymbol{x}_k \to (0,0,2\sqrt{3})^T$ which is not a Kuhn-Tucker point. Why can't Theorem 6.2.1 apply to this example? Consider

$$\boldsymbol{g}(\boldsymbol{x}) = \left(\frac{-x_1}{\sqrt{x_1^2 + 3x_2^2}}, \frac{-3x_2}{\sqrt{x_1^2 + 3x_2^2}}, 1 \right)^T.$$

Clearly, $\boldsymbol{g}(\boldsymbol{x})$ does not exist on the segment $\{(0,0,x_3) \,|\, 0 \leq x_3 \leq 10\}$. Hence, the continuous differentiability of $f(\boldsymbol{x})$ required by Theorem 6.2.1 does not hold.

When the regularity condition is not assumed, we may encounter a linearly dependent set of active constraints and the multipliers u_{kj} are non-uniquely determined. If the multipliers are chosen arbitrarily a dropped constraint may cause a step of length zero to be taken, and since the dropped constraint is a function of $\boldsymbol{x}_k$, the algorithm cycles at the current iteration. In the following, we introduce a method to deal with this degenerative case.

Theorem 6.2.5 *Let Jm be the index set of a maximal independent subset of $\{\boldsymbol{a}_j \mid j \in J(\boldsymbol{x})\}$. The following statements are equivalent.*

(1) *$\boldsymbol{x}$ is a Kuhn-Tucker point.*

(2) *For any Jm,*

  (a) *$P_{Jm}\boldsymbol{g}(\boldsymbol{x}) = \boldsymbol{0}$, and*

  (b) *$\{\boldsymbol{y} \mid A_{Jm}^T A_{J(\boldsymbol{x})}\boldsymbol{y} = -A_{Jm}^T \boldsymbol{g}(\boldsymbol{x})$ and $y_j \geq 0, j \in J(\boldsymbol{x}) \setminus M'\} \neq \emptyset$.*

(3) *There exists Jm such that (a) and (b) hold.*

Proof. It is obvious that (1) $\Longrightarrow$ (2) $\Longrightarrow$ (3). Next, we prove (3) $\Longrightarrow$ (1). Suppose that (a) and (b) hold for Jm. Note that from (b), we can obtain

$$\boldsymbol{y}_{Jm} = (y_j, j \in Jm)^T = -(A_{Jm}^T A_{Jm})^{-1} A_{Jm}^T \boldsymbol{g}(\boldsymbol{x}).$$

Thus

$$g(x) = -\sum_{j \in Jm} y_j a_j \text{ and } y_j \geq 0 \text{ for } j \in Jm \setminus M'.$$

This implies that x is a Kuhn-Tucker point. □

It is easy to check if (a) holds. Next, we present a method to check if (b) holds. Consider the following linear programming problem:

$$\begin{array}{ll} \text{maximize} & Z = \sum_{j \in J(x) \setminus Jm} y_j \\ \text{subject to} & A_{Jm}^T A_{J(x)} y = A_{Jm}^T g(x), \\ & y_j \geq 0 \text{ for } j \in J(x) \setminus M'. \end{array} \tag{6.12}$$

Let J be a feasible basis of (6.12). Then $A_{Jm}^T A_J$ is a nonsingular square matrix. So, $|J| = |Jm|$ and J is the index set of a maximal independent subset of $\{a_j \mid j \in J(x)\}$. Therefore, there exists a nonsingular square matrix Q of order $|J|$ such that $A_{Jm} = A_J Q$, and we have

$$\begin{aligned} y_J &= (y_j, j \in J)^T = -(A_{Jm}^T A_J)^{-1} A_{Jm}^T g(x) \\ &= -((A_J Q)^T A_J)^{-1} (A_J Q)^T g(x) = (A_J^T A_J)^{-1} A_J^T g(x). \end{aligned}$$

It follows that the simplex tableau corresponding to J is as follows.

$$\begin{aligned} (c_J (A_J^T A_J)^{-1} A_J^T A_{J(x)} - c_{J(x)}) y &= -Z + c_J (A_J^T A_J)^{-1} A_J^T g(x), \\ (A_J^T A_J)^{-1} A_J^T A_{J(x)} y &= (A_J^T A_J)^{-1} A_J^T g(x), \end{aligned}$$

where

$$c_{J(x)} = (c_j, j \in J(x)), \quad c_J = (c_j, j \in J)$$

and

$$c_j = \begin{cases} 0 & j \in J, \\ 1 & j \in J(x) \setminus J. \end{cases}$$

Let

$$\begin{aligned} (\alpha_{0j}, j \in J(x)) &= c_J (A_J^T A_J)^{-1} A_J^T A_{J(x)} - c_{J(x)}, \\ (\alpha_{ij}, i \in J, j \in J(x)) &= (A_J^T A_J)^{-1} A_J^T A_{J(x)}, \\ (u_j, j \in J)^T &= (A_J^T A_J)^{-1} A_J^T g(x). \end{aligned}$$

To test if (b) holds, we could use the following dual simplex procedure.

Step 1. If $\max\{u_j \mid j \in J \setminus M'\} < 0$, then stop ((b) holds); otherwise, determine h such that $u_h = \max\{u_j \mid j \in J \setminus M'\}$.

Step 2. If $\alpha_{hj} \geq 0$ for all $j \in J(\boldsymbol{x})$, then stop ((b) does not hold); otherwise, choose $\alpha_{ht} < 0$ such that according to the lexicographic ordering $\boldsymbol{\alpha}_t/\alpha_{ht}$ is the minimum in

$$\{\boldsymbol{\alpha}_j/\alpha_{hj} \mid \alpha_{hj} < 0, j \in J(\boldsymbol{x})\}$$

where $\boldsymbol{\alpha}_j = (\alpha_{0j}, \alpha_{ij}, j \in J)^T$ is the column vector whose index is j in the simplex tableau.

Step 3. Let $J := (J \setminus h) \cup t$ and let α_{ht} be the pivot element. Perform the pivoting operation in the simplex tableau. Go to Step 1.

Let $J_*(\boldsymbol{x})$ denote the index set of the basis obtained when the above computation stops. Obviously, it has the following property:

Theorem 6.2.6 *(b) holds iff $u_j \geq 0$ for $j \notin M'$, where*

$$(u_j, j \in J_*(\boldsymbol{x}))^T = (A^T_{J_*(\boldsymbol{x})} A_{J_*(\boldsymbol{x})})^{-1} A^T_{J_*(\boldsymbol{x})} \boldsymbol{g}(\boldsymbol{x}).$$

If (b) does not hold, then there exists an index $h \in J_*(\boldsymbol{x})$ such that

(1) $u_h = \max\{u_j \mid j \in J_*(\boldsymbol{x})\} > 0$, and

(2) for each $\boldsymbol{a}_j$, when it is expressed as the linear combination of those $\boldsymbol{a}_i$, $i \in J_*(\boldsymbol{x}) : \boldsymbol{a}_j = \sum_{i \in J_*(\boldsymbol{x})} \alpha_{ij} \boldsymbol{a}_i$, we have $\alpha_{hj} \geq 0$.

In the following, we use h_* to denote such an h.

Theorem 6.2.7 *Let $0 \leq c \leq +\infty$. Denote $J = J_*(\boldsymbol{x})$ and $\boldsymbol{g} = \boldsymbol{g}(\boldsymbol{x})$. Define*

$$\mathbf{d} = \begin{cases} P_J \boldsymbol{g} & \text{if } \|P_J \boldsymbol{g}\| > c u_{h_*}, \\ P_{J \setminus h_*} \boldsymbol{g} & \text{otherwise.} \end{cases}$$

Then $\mathbf{d} = \mathbf{0}$ iff $\boldsymbol{x}$ is a Kuhn-Tucker point. Furthermore, if $\mathbf{d} \neq \mathbf{0}$, then it is an ascending feasible direction.

The proof of Theorem 6.2.7 is similar to that of Theorem 4.2.5. In Rosen's method, replacing J_k and h by $J_*(\boldsymbol{x}_k)$ and h_*, we obtain an algorithm without assuming the regularity condition. For such an algorithm, we can also prove a global convergence theorem similar to Theorem 6.2.1. Finally, it is worth mentioning that Bland's anticycling rule can also be used here instead of the lexicographic simplex procedure.

6.3 The Third Slope Lemma

In this section we only consider the linearly constrained problem (4.1). First, we strengthen the results obtained by the first and the second slope lemmas.

Lemma 6.3.1 (The Third Slope Lemma) *Let $\{\boldsymbol{x}_k\}$ be a sequence of feasible points such that for all k, $f(\boldsymbol{x}_k) < f(\boldsymbol{x}_{k+1})$. Let $\boldsymbol{x}^*$ be a cluster point of the sequence. Suppose that for any subsequence $\{\boldsymbol{x}_k\}_{k\in K}$ converging to $\boldsymbol{x}^*$, $\lim_{k\to\infty,k\in K}(\boldsymbol{x}_{k+1}-\boldsymbol{x}_k) = \mathbf{0}$. If there exists a positive number μ such that for all k,*

$$\frac{\boldsymbol{g}_k^T(\boldsymbol{x}_{k+1}-\boldsymbol{x}_k)}{\|\boldsymbol{x}_{k+1}-\boldsymbol{x}_k\|} \geq \mu \max\{\|P_{J_k}\boldsymbol{g}_k\|, \|P_{J_{k+1}}\boldsymbol{g}_k\|\}, \tag{6.13}$$

then the following two statements are equivalent:

(a) There exists a subsequence $\{\boldsymbol{x}_k\}_{k\in K}$ converging to $\boldsymbol{x}^*$ such that

$$\lim_{k\to\infty,k\in K} \frac{\boldsymbol{g}_k^T(\boldsymbol{x}_{k+1}-\boldsymbol{x}_k)}{\|\boldsymbol{x}_{k+1}-\boldsymbol{x}_k\|} = 0. \tag{6.14}$$

(b) For every subsequence $\{\boldsymbol{x}_k\}_{k\in K}$ converging to $\boldsymbol{x}^*$, (6.14) holds.

Proof. It suffices to prove (a) $\Longrightarrow$ (b). First, we notice that there exists $\delta_1 > 0$ such that $\|\boldsymbol{x}_k - \boldsymbol{x}^*\| \leq \delta_1$ implies $J_k \subseteq J(\boldsymbol{x}^*)$. Define

$$\alpha = \begin{cases} 1 & \text{if } P_J\boldsymbol{g}^* = \mathbf{0} \text{ for all } J \subseteq J(\boldsymbol{x}^*), \\ \min\{\|P_J\boldsymbol{g}\| \mid J \subseteq J(\boldsymbol{x}^*) \ \text{ and } P_J\boldsymbol{g}^* \neq \mathbf{0}\} & \text{otherwise.} \end{cases}$$

We claim that there exists $\delta_2 > 0$ such that $\|\boldsymbol{x}_k - \boldsymbol{x}^*\| < \delta_2$ and $f(\boldsymbol{x}_{k+1}) - f(\boldsymbol{x}_k) < \frac{1}{2}\alpha\mu\|\boldsymbol{x}_{k+1}-\boldsymbol{x}_k\|$ imply $J_k \subseteq J(\boldsymbol{x}^*)$, $P_{J_k}\boldsymbol{g}^* = \mathbf{0}$ and $P_{J_{k+1}}\boldsymbol{g}^* = \mathbf{0}$. For otherwise, suppose that such a positive number δ_2 does not exist. Then we can find a subsequence $\{\boldsymbol{x}_k\}_{k\in K}$ satisfying the following conditions:

(1) $\boldsymbol{x}_k \to \boldsymbol{x}^*$ as $k \to \infty$, $k \in K$.

(2) $J_k = J$ and $J_{k+1} = J'$ for all $k \in K$.

(3) $P_J\boldsymbol{g}^* \neq \mathbf{0}$ or $P_{J'}\boldsymbol{g}^* \neq \mathbf{0}$.

(4) $f(\boldsymbol{x}_{k+1}) - f(\boldsymbol{x}_k) < \frac{1}{2}\mu\alpha\|\boldsymbol{x}_{k+1}-\boldsymbol{x}_k\|$ for all $k \in K$.

Since $\lim_{k\to\infty,k\in K}(\boldsymbol{x}_{k+1}-\boldsymbol{x}_k)=\mathbf{0}$, it is easy to see that

$$\begin{aligned}\liminf_{k\to\infty,k\in K}\frac{f(\boldsymbol{x}_{k+1})-f(\boldsymbol{x}_k)}{\|\boldsymbol{x}_{k+1}-\boldsymbol{x}_k\|} &= \liminf_{k\to\infty,k\in K}\frac{\boldsymbol{g}_k^T(\boldsymbol{x}_{k+1}-\boldsymbol{x}_k)}{\|\boldsymbol{x}_{k+1}-\boldsymbol{x}\|}\\ &\geq \liminf_{k\to\infty,k\in K}\mu\max\{\|P_J\boldsymbol{g}_k\|,\|P_{J'}\boldsymbol{g}_k\|\}\\ &= \mu\max\{\|P_J\boldsymbol{g}^*\|,\|P_{J'}\boldsymbol{g}^*\|\}\geq\mu\alpha.\end{aligned}$$

Thus, for sufficiently large $k\in K$,

$$f(\boldsymbol{x}_{k+1})-f(\boldsymbol{x}_k)\geq\frac{1}{2}\alpha\mu\|\boldsymbol{x}_{k+1}-\boldsymbol{x}_k\|$$

contradicting (4). So, the number δ_2 exists. Now, suppose to the contrary that (a) holds but (b) does not hold. Then there exist $\varepsilon\in(0,\frac{1}{2}\alpha\mu)$ and a subsequence $\{\boldsymbol{x}_k\}_{k\in K}$ converging to $\boldsymbol{x}^*$ such that

$$\lim_{k\to\infty,k\in K}\frac{f(\boldsymbol{x}_{k+1})-f(\boldsymbol{x}_k)}{\|\boldsymbol{x}_{k+1}-\boldsymbol{x}_k\|}=\left|\lim_{k\to\infty,k\in K}\frac{\boldsymbol{g}_k^T(\boldsymbol{x}_{k+1}-\boldsymbol{x}_k)}{\|\boldsymbol{x}_{k+1}-\boldsymbol{x}_k\|}\right|>\varepsilon.$$

Therefore, there exists $\delta_3>0$ such that $k\in K$ and $\|\boldsymbol{x}_k-\boldsymbol{x}^*\|\leq\delta_3$ imply $f(\boldsymbol{x}_{k+1})-f(\boldsymbol{x}_k)\geq\frac{1}{2}\varepsilon\|\boldsymbol{x}_{k+1}-\boldsymbol{x}_k\|$. Define $\mathcal{J}=\{J\mid J\subseteq J(\boldsymbol{x}^*)$ and $P_J\boldsymbol{g}^*=\mathbf{0}\}$. Let $L(J,J')$ denote the linear subspace generated by $D_J\cup D_{J'}$. Denote by $P_{L(J,J')}$ the operator of the orthogonal projection onto the subspace $L(J,J')$. Clearly, for any $J,J'\in\mathcal{J}$, $P_{L(J,J')}\boldsymbol{g}^*=\mathbf{0}$. Therefore, there exists $\delta_4>0$ such that $\|\boldsymbol{x}_k-\boldsymbol{x}^*\|\leq\delta_4$ implies $\|P_{L(J,J')}\boldsymbol{g}(\boldsymbol{x})\|<\frac{1}{2}\varepsilon$ for any $J,J'\in\mathcal{J}$. Let $\delta=\min\{\delta_1,\delta_2,\delta_3,\delta_4\}$. Consider the index set

$$K'=\{k\mid f(\boldsymbol{x}_{k+1})-f(\boldsymbol{x}_k)\geq\frac{1}{2}\varepsilon\|\boldsymbol{x}_{k+1}-\boldsymbol{x}_k\|\}.$$

K' is an infinite set containing all but finitely many elements of K. By (a), there exists a subsequence $\{\boldsymbol{x}_k\}_{k\in K^*}$, converging to $\boldsymbol{x}^*$, such that

$$\lim_{k\to\infty,k\in K^*}\frac{f(\boldsymbol{x}_{k+1})-f(\boldsymbol{x}_k)}{\|\boldsymbol{x}_{k+1}-\boldsymbol{x}_k\|}=0.$$

Thus, there are infinitely many k not in K'. This implies the existence of infinitely many consecutive chains $\{k',k'+1,\cdots,k''\}$ satisfying

$$k'-1\notin K',k'\in K',k'+1\in K',\cdots,k''\in K',k''+1\notin K'.$$

Next, we consider two cases:

Case 1. There exists a consecutive chain $\{k', k'+1, \cdots, k''\}$ such that $\boldsymbol{x}_{k'}$ and $\boldsymbol{x}_{k''+1}$ lie in the ball $B(\delta) = \{\boldsymbol{x} \mid \|\boldsymbol{x} - \boldsymbol{x}^*\| \le \delta\}$. Since $k'-1, k''+1 \notin K'$, we have

$$\begin{aligned} f(\boldsymbol{x}_{k'}) - f(\boldsymbol{x}_{k'-1}) &\le \frac{\varepsilon}{2}\|\boldsymbol{x}_{k'} - \boldsymbol{x}_{k'-1}\|, \\ f(\boldsymbol{x}_{k''+2}) - f(\boldsymbol{x}_{k''+1}) &\le \frac{\varepsilon}{2}\|\boldsymbol{x}_{k''+2} - \boldsymbol{x}_{k''+1}\|. \end{aligned}$$

By the choice of δ_1 and δ_2, we obtain $J_{k'} \in \mathcal{J}$ and $J_{k''+1} \in \mathcal{J}$. Furthermore, by the choice of δ_4 and Proposition 4.2.2, we have

$$\frac{f(\boldsymbol{x}_{k''+1}) - f(\boldsymbol{x}_{k'})}{\|\boldsymbol{x}_{k''+1} - \boldsymbol{x}_{k'}\|} = \frac{\boldsymbol{g}(\boldsymbol{y})^T(\boldsymbol{x}_{k''+1} - \boldsymbol{x}_{k'})}{\|\boldsymbol{x}_{k''+1} - \boldsymbol{x}_{k'}\|} \le \|P_{L(J_{k''+1}, J_{k'})}\boldsymbol{g}(\boldsymbol{y})\| < \frac{1}{2}\varepsilon$$

where $\boldsymbol{y}$ lies on segment $[\boldsymbol{x}_{k''+1}, \boldsymbol{x}_{k'}]$. But this contradicts the following:

$$\begin{aligned} f(\boldsymbol{x}_{k''+1}) - f(\boldsymbol{x}_{k'}) &= \sum_{k=k'}^{k''} (f(\boldsymbol{x}_{k+1}) - f(\boldsymbol{x}_k)) \\ &> \frac{\varepsilon}{2} \sum_{k=k'}^{k''} \|\boldsymbol{x}_{k+1} - \boldsymbol{x}_k\| \ge \frac{\varepsilon}{2}\|\boldsymbol{x}_{k''+1} - x_{k'}\|. \end{aligned}$$

Case 2. For any consecutive chain $\{k', k'+1, \cdots, k''\}$, either $\|\boldsymbol{x}_{k'} - \boldsymbol{x}^*\| > \delta$ or $\|\boldsymbol{x}_{k''+1} - \boldsymbol{x}^*\| > \delta$. Since $\{\boldsymbol{x}_k\}_{k \in K}$ converges to $\boldsymbol{x}^*$ and $K \cap K'$ is infinite, there exist infinitely many consecutive chains $\{k', k'+1, \cdots, k''\}$ such that $k' > k^*$ and $\|\boldsymbol{x}_k - \boldsymbol{x}^*\| < \delta/2$ for some $k \in \{k', k'+1, \cdots, k''\}$. For each link of such a chain, we have

$$f(\boldsymbol{x}_{k''+1}) - f(\boldsymbol{x}_{k'}) = \sum_{k=k'}^{k''} (f(\boldsymbol{x}_{k+1}) - f(\boldsymbol{x}_k)) \ge \frac{\varepsilon}{2} \sum_{k=k'}^{k''} \|\boldsymbol{x}_{k+1} - \boldsymbol{x}_k\| > \frac{\varepsilon\delta}{4}.$$

Therefore, $f(\boldsymbol{x}^*) - f(\boldsymbol{x}_1) = \infty$, a contradiction. □

An application of the third slope lemma is as follows.

Theorem 6.3.2 *Let α and β be two positive numbers. Suppose that in Rosen's method, c_k's are selected to satisfy $c_k \ge \alpha$ whenever $J_k \subseteq J_{k-1}$ and $c_k \le \beta$ whenever $J_k \not\subseteq J_{k-1}$, and $\bar{\lambda}_k$'s are computed by the following formula:*

$$\bar{\lambda}_k = \begin{cases} 1 & \text{if } \boldsymbol{a}_j^T(\boldsymbol{x}_k + \boldsymbol{d}_k) \ge b_j \text{ for all } j \notin J_k \\ \min\left\{ \frac{b_j - \boldsymbol{a}_j^T \boldsymbol{x}_k}{\boldsymbol{a}_k^T \boldsymbol{d}_k} \;\middle|\; \boldsymbol{a}_j^T(\boldsymbol{x}_k + \boldsymbol{d}_k) < b_j \right\} & \text{otherwise.} \end{cases} \quad (6.15)$$

Then Rosen's method with a normal line search procedure either stops at a Kuhn-Tucker point or generates an infinite sequence whose cluster points are all Kuhn-Tucker points.

To prove this theorem, let us first prove two lemmas.

Lemma 6.3.3 *Under the conditions in Theorem 6.3.2, for any convergent subsequence* $\{\boldsymbol{x}_k\}_{k\in K}$, $\lim_{k\to\infty,k\in K}(\boldsymbol{x}_{k+1} - \boldsymbol{x}_k) = \boldsymbol{0}$.

Proof. Clearly, for any convergent subsequence $\{\boldsymbol{x}_k\}_{k\in K}$, $\{\boldsymbol{d}_k\}_{k\in K}$ is a bounded set. Note that for every k, $\bar{\lambda}_k \leq 1$. So, $\{\boldsymbol{x}_{k+1} - \boldsymbol{x}_k\}_{k\in K}$ is bounded. Thus, to prove $\lim_{k\to\infty,k\in K}(\boldsymbol{x}_{k+1} - \boldsymbol{x}_k) = \boldsymbol{0}$, it suffices to prove that every cluster point of $\{\boldsymbol{x}_{k+1} - \boldsymbol{x}_k\}_{k\in K}$ is $\boldsymbol{0}$. Suppose that $\boldsymbol{y}$ is a cluster point. Then there exists an infinite subset K' of K such that $\lim_{k\to\infty,k\in K'} \boldsymbol{d}_k = \boldsymbol{d}^*$, $\lim_{k\to\infty,k\in K'} \lambda_k = \lambda^*$ and $\boldsymbol{y} = \lambda^*\boldsymbol{d}^*$. Note that $(\boldsymbol{g}^*)^T\boldsymbol{d}^* = \|\boldsymbol{d}^*\|^2$. If $(\boldsymbol{g}^*)^T\boldsymbol{d}^* = 0$, then $\boldsymbol{d}^* = \boldsymbol{0}$, and hence $\boldsymbol{y} = \boldsymbol{0}$. If $(\boldsymbol{g}^*)^T\boldsymbol{d}^* \neq 0$, then by Theorem 1.4.3, $\boldsymbol{y} = \boldsymbol{0}$. □

Lemma 6.3.4 *Under the conditions in Theorem 6.3.2, (5.9) holds.*

Proof. Let $\boldsymbol{x}^*$ be a cluster point of an infinite generated sequence $\{\boldsymbol{x}_k\}$. Suppose that there exists a subsequence of $\{\boldsymbol{x}_k\}$ converging to $\boldsymbol{x}^*$ such that the corresponding subsequence of $\{\boldsymbol{g}_k^T\boldsymbol{d}_k\}$ converges to 0. Our aim is to prove that $\boldsymbol{x}^*$ is a Kuhn-Tucker point. First, by Lemma 6.3.3 and the third slope lemma, we can conclude that for every subsequence converging to $\boldsymbol{x}^*$, the corresponding subsequence of $\{\boldsymbol{g}_k^T\boldsymbol{d}_k\}$ converges to 0. Since $\boldsymbol{g}_k^T\boldsymbol{d}_k = \|\boldsymbol{d}_k\|^2$, the corresponding subsequence of $\{\boldsymbol{d}_k\}$ converges to $\boldsymbol{0}$. Next, we consider two cases.

Case 1. There exists a subsequence $\{\boldsymbol{x}_k\}_{k\in K}$ converging to $\boldsymbol{x}^*$ such that for all $k \in K$, $\boldsymbol{d}_k = P_{J_k\setminus h_k}\boldsymbol{g}_k$. Note that $\{\boldsymbol{d}_k\}_{k\in K}$ converges to $\boldsymbol{0}$. It follows from Lemma 4.2.3 that $\lim_{k\to\infty,k\in K} P_{J_k}\boldsymbol{g}_k = \boldsymbol{0}$ and $\lim_{k\to\infty,k\in K} u_{kh_k} = 0$. Choose an infinite subset K' of K such that $J_k = J$ and $h_k = h$ for $k \in K'$. Then $P_J\boldsymbol{g}^* = 0$ and $\max\{u_j^* \mid j \in J \setminus M'\} = 0$ where $(u_j^*, j \in J)^T = (A_J^TA_J)^{-1}A_J^T\boldsymbol{g}^*$. Therefore, $\boldsymbol{x}^*$ is a Kuhn-Tucker point.

Case 2. There exists a subsequence $\{\boldsymbol{x}_k\}_{k\in K}$ converging to $\boldsymbol{x}^*$ such that for all $k \in K$, $c_k \geq \alpha$ and $\boldsymbol{d}_k = P_{J_k}\boldsymbol{g}_k$. In this case, $\|P_{J_k}\boldsymbol{g}_k\| \geq \alpha u_{kh_k}$. Thus, $\lim_{k\to\infty,k\in K} \boldsymbol{d}_k = \boldsymbol{0}$ implies $\lim_{k\to\infty,k\in K} P_{J_k}\boldsymbol{g}_k = \boldsymbol{0}$ and $\lim_{k\to\infty,k\in K} u_{kh_k} = 0$. Hence, $\boldsymbol{x}^*$ is a Kuhn-Tucker point.

In the remaining part, we prove that either Case 1 or Case 2 occurs. Suppose to the contrary that both Case 1 and Case 2 do not occur. Then there exists a positive number δ such that $\|\boldsymbol{x}_k - \boldsymbol{x}^*\| < \delta$ implies $c_k < \alpha$ and $\boldsymbol{d}_k = P_{J_k}\boldsymbol{g}_k$, that is, $J_k \not\subseteq J_{k-1}$ and $J_k \subseteq J_{k+1}$. Consider an integer $t > m$. Let $\{\boldsymbol{x}_k\}_{k\in K}$ be a subsequence converging to $\boldsymbol{x}^*$. By Lemma 6.3.3, for any fixed i, $\boldsymbol{x}_{k+i} \to \boldsymbol{x}^*$ as $k \to \infty$ and $k \in K$. Thus, there exists $k \in K$ such that $\boldsymbol{x}_k, \cdots, \boldsymbol{x}_{k+t}$ all lie in the ball $B(\delta) = \{\boldsymbol{x} \mid \|\boldsymbol{x} - \boldsymbol{x}^*\| < \delta\}$. Therefore, $J_k \subset J_{k+1} \subset \cdots \subset J_{k+t}$. It follows that J_{k+t} contains at least t elements, which is impossible. □

Now, we prove the theorem.

Proof of Theorem 6.3.2. By Theorem 4.2.5, if the algorithm stops at $\boldsymbol{x}_k$, then $\boldsymbol{x}_k$ is a Kuhn-Tucker point. By Lemma 6.3.4 and the first slope lemma, if the algorithm generates an infinite sequence not convergent, then every cluster point is a Kuhn-Tucker point. Thus, we need only to consider a generated sequence $\{\boldsymbol{x}_k\}$ converging to $\boldsymbol{x}^*$. In this case, $\{\boldsymbol{d}_k\}$ is bounded. If there exist infinitely many k such that $\bar{\lambda}_k = \lambda_k$, then by Theorem 5.2.3, we can find a subsequence of $\{\boldsymbol{x}_k\}$ such that the corresponding subsequence of $\{\boldsymbol{g}_k^T\boldsymbol{d}_k\}$ converges to 0. By Lemma 6.3.4, $\boldsymbol{x}^*$ is a Kuhn-Tucker point. If there exist only finitely many k such that $\bar{\lambda}_k = \lambda_k$, then there exists $k' > 0$ such that for $k > k'$, $|J_{k+1} \setminus J_k| = |J_k \setminus J_{k+1}| = 1$. It follows that for $k > k'$, $c_k \le \beta$ and $\boldsymbol{d}_k = P_{J_k\setminus h_k}\boldsymbol{g}_k$. By Corollary 6.1.2, there is a subsequence $\{\boldsymbol{x}_k\}_{k\in K}$ such that for all $k \in K$, $J_k = J$ for a certain J and

$$\lim_{k\to\infty, k\in K} \boldsymbol{g}_k^T\boldsymbol{d}_k/\|\boldsymbol{d}_k\| \le \|P_J\boldsymbol{g}^*\|. \tag{6.16}$$

Furthermore, we can find an infinite subset K' of K such that for $k \in K'$, $h_k = h$ and $\lim_{k\to\infty,k\in K'} \boldsymbol{d}_k = \boldsymbol{d}^* = P_{J\setminus h}\boldsymbol{g}^*$. By (6.16), $\|P_{J\setminus h}\boldsymbol{g}^*\| \le \|P_J\boldsymbol{g}^*\|$. It follows from (4.8) that $u_h^* = 0$ where $u^* = \max\{u_j^* \mid j \in J \setminus M'\}$ and $(u_j^*, j \in J)^T = (A_J^TA_J)^{-1}A_J^T\boldsymbol{g}^*$. However, $J_k \setminus J_{k+1} \ne \emptyset$ implies that $\beta u_{kh} \ge \|P_J\boldsymbol{g}^*\|$. Letting $k \to \infty$, we obtain $0 = \beta u_h^* \ge \|P_J\boldsymbol{g}^*\|$. Therefore, $\boldsymbol{x}^*$ is a Kuhn-Tucker point. □

Does Theorem 6.3.2 still hold when the $\bar{\lambda}_k$'s are determined by the formula in section 3.1? It is an open question. We believe that the answer is negative. In the other words, the boundedness of $\{\lambda_k\}$ is necessary for Theorem 6.3.2 to hold.

Exercises

1. Suppose that problem (4.1) satisfies the strong regularity condition, i.e. for any feasible point $\boldsymbol{x}$, $J \subset J' \subseteq J(\boldsymbol{x})$ implies $\|P_J \boldsymbol{g}(\boldsymbol{x})\| > \|P_{J'} \boldsymbol{g}(\boldsymbol{x})\|$. Prove that under such a condition, Rosen's method with $c_k = \infty$ for all k and a normal line search, either stops at a Kuhn-Tucker point, or generates an infinite sequence whose cluster points are all Kuhn-Tucker points.

2. Prove the global convergence theorem for the above-mentioned algorithm.

3. Let β be a positive number. Consider Rosen's method with a normal line search procedure. Suppose that c_k's are selected such that $c_k \leq \beta$ whenever $J_k \not\subseteq J_{k-1}$ and $\bar{\lambda}_k$'s are determined by (6.15). Prove the following:

   (a) If the algorithm generates an infinite sequence lying in a compact region, then $\lim_{k\to\infty} \boldsymbol{d}_k = \boldsymbol{0}$.

   (b) If the algorithm generates an infinite sequence converging to a point which is not a Kuhn-Tucker point, then $\lim_{k\to\infty} c_k = 0$ and for sufficiently large k, $J_k = J_{k+1}$.

Chapter 7

Inexact Active Constraints

If you want work well done,
select a busy man: the other kind has no time.
ELBERT HUBBARD

When an approximate value is computed from each computation, one may have difficulty to decide whether a constraint is active or not. One strategy, called ε-active set strategy, is used to overcome this obstacle by treating near-active constraints just like active constraints. We study this topic in this chapter.

7.1 Gradient Projection

Before discussing how to do gradient projection with the ε-active set strategy, let us show some property of a linearly constrained feasible region.

A linearly constrained set is also called a *polyhedron*. In this section, we prove a classical result about decomposition of a polyhedron. This result will be used to establish a property of ϵ-active sets in the next section.

Since an equality can be replaced by two inequalities, any polyhedron Q can be represented in the form $\{\boldsymbol{x} \mid A^T\boldsymbol{x} \geq \boldsymbol{b}\}$ where A is an $n \times m$ matrix and $\boldsymbol{b}$ an m-dimensional vector. Let $\boldsymbol{a}_i$ denote the ith column vector of A and b_i the ith component of $\boldsymbol{b}$. If $\emptyset \neq Q \cap \{\boldsymbol{x} \mid \boldsymbol{a}_i^T\boldsymbol{x} = b_i\} \neq Q$, then $Q \cap \{\boldsymbol{x} \mid \boldsymbol{a}_i^T\boldsymbol{x} = b_i\}$ is called a *proper face* of Q. A polyhedron is *primitive* if it cannot be represented as a convex hull of its proper faces.

Lemma 7.1.1 *Every polyhedron Q is the convex hull of finitely many primitive polyhedrons.*

Proof. If Q is primitive, then it is done. If Q is not primitive, then Q is a convex hull of some of its proper faces, $Q_1, \cdots, Q_k$. If every Q_i is primitive, then it is done. If Q_i is not primitive, then Q_i can also be represented as a convex hull of some of its proper faces Q_{ij}'s. This argument must halt in finitely many steps since the chain $Q \supset Q_i \supset Q_{ij} \supset \cdots$ cannot be longer than m. Thus, the lemma holds. □

A (linear) *manifold* is a polyhedron without a proper face. It is easy to see that any manifold can be represented in the form $\{\boldsymbol{x} \mid A^T\boldsymbol{x} = \boldsymbol{b}\}$. A (linear) *halfmanifold* is an intersection of a manifold and a halfspace $\{\boldsymbol{x} \mid \boldsymbol{a}^T\boldsymbol{x} \geq b\}$.

Lemma 7.1.2 *Any primitive polyhedron Q is either a manifold or a halfmanifold.*

Proof. If Q has no proper face, then Q is a manifold. In the following, we suppose that Q has at least one proper face. Let F be a maximal element among proper faces under the partial ordering $\subset$. We first show that every proper face is a subset of F. For contradiction, suppose that there is a proper face F' such that $F' \setminus F \neq \emptyset$. Since F is maximal, we also have $F' \setminus F \neq \emptyset$. Choose two points $\boldsymbol{v} \in F \setminus F'$ and $\boldsymbol{v}' \in F' \setminus F$. For any point $\boldsymbol{x} \in Q$, we consider two points $\boldsymbol{y} = \boldsymbol{x} + \alpha(\boldsymbol{v} - \boldsymbol{v}')$ and $\boldsymbol{y}' = \boldsymbol{x} + \beta(\boldsymbol{v}' - \boldsymbol{v})$ where

$$\alpha = \min\left\{ \frac{\boldsymbol{a}_i^T\boldsymbol{x} - b_i}{\boldsymbol{a}_i^T(\boldsymbol{v}' - \boldsymbol{v})} \;\middle|\; \boldsymbol{a}_i^T(\boldsymbol{v}' - \boldsymbol{v}) > 0 \right\}, \tag{7.1}$$

$$\beta = \min\left\{ \frac{\boldsymbol{a}_i^T\boldsymbol{x} - b_i}{\boldsymbol{a}_i^T(\boldsymbol{v} - \boldsymbol{v}')} \;\middle|\; \boldsymbol{a}_i^T(\boldsymbol{v} - \boldsymbol{v}') > 0 \right\}.$$

Clearly, $\boldsymbol{y}$ and $\boldsymbol{y}'$ belong to Q and for some index i_0, $\alpha = (\boldsymbol{a}_{i_0}^T\boldsymbol{x} - b_{i_0})/\boldsymbol{a}_{i_0}^T(\boldsymbol{v}' - \boldsymbol{v})$ and $\boldsymbol{a}_{i_0}^T(\boldsymbol{v}' - \boldsymbol{v}) > 0$. We claim that $Q \not\subseteq \{\boldsymbol{x} \mid \boldsymbol{a}_{i_0}^T\boldsymbol{x} = b_{i_0}\}$. Suppose the opposite, i.e. $Q \subseteq \{\boldsymbol{x} \mid \boldsymbol{a}_{i_0}^T\boldsymbol{x} = b_{i_0}\}$. Then $\boldsymbol{a}_{i_0}^T(\boldsymbol{v}' - \boldsymbol{v}) = b_{i_0} - b_{i_0} = 0$, a contradiction. It follows that $Q \cap \{\boldsymbol{x} \mid \boldsymbol{a}_{i_0}^T\boldsymbol{x} = b_{i_0}\}$ is a proper face. Thus, $\boldsymbol{y}$ belongs to a proper face. Similarly, $\boldsymbol{y}'$ belongs to a proper face. If $\alpha = 0$, then $\boldsymbol{x} = \boldsymbol{y}$. If $\alpha > 0$, then $\alpha + \beta > 0$ and

$$\boldsymbol{x} = \frac{\alpha\boldsymbol{y} + \beta\boldsymbol{y}'}{\alpha + \beta}.$$

Therefore, Q is the convex hull of all its proper faces, contradicting the primitivity of Q.

Now, suppose $F = Q \cap \{\boldsymbol{x} \mid \boldsymbol{a}_1^T \boldsymbol{x} = b_1\}$. Denote $H = \{\boldsymbol{x} \mid \boldsymbol{a}_1^T \boldsymbol{x} \geq b_1\}$. Let G be the intersection of all hyperplanes $\{\boldsymbol{x} \mid \boldsymbol{a}_i^T \boldsymbol{x} = b_i\}$ containing Q. Clearly, G is a manifold and $Q \subseteq (H \cap G)$. Next, we show $(H \cap G) \subseteq Q$. By contradiction, suppose that there exists $\boldsymbol{v} \in (H \cap G) \setminus Q$. Then for some j, $\boldsymbol{a}_j^T \boldsymbol{v} < b_j$ and $Q \not\subseteq \{\boldsymbol{x} \mid \boldsymbol{a}_j^T \boldsymbol{x} = b_j\}$. It follows that $Q \cap \{\boldsymbol{x} \mid \boldsymbol{a}_j^T \boldsymbol{x} = b_j\}$ is either an empty set or a proper face and hence is a subset of F. Choose $\boldsymbol{v}' \in Q \setminus F$. Then $\boldsymbol{a}_j^T \boldsymbol{v}' > b_j$. So, $\boldsymbol{a}_j^T(\boldsymbol{v}' - \boldsymbol{v}) > 0$. Define α by substituting $\boldsymbol{x} = \boldsymbol{v}'$ into (7.3). Consider $\boldsymbol{y} = \boldsymbol{v}' + \alpha(\boldsymbol{v} - \boldsymbol{v}')$. By an argument similar to the above, we can prove that $\boldsymbol{y}$ belongs to a proper face. Hence, $\boldsymbol{y} \in F$ and $\boldsymbol{a}_1^T \boldsymbol{y} = b_1$. However, $\boldsymbol{a}_1^T \boldsymbol{v}' > b_1$ implies

$$\boldsymbol{a}_1^T \boldsymbol{y} = \boldsymbol{a}_1^T((1-\alpha)\boldsymbol{v} - \alpha \boldsymbol{v}') > b_1,$$

a contradiction. Therefore, $Q = H \cap G$, i.e. Q is a halfmanifold. □

Theorem 7.1.3 *For any polyhedron Q, there exist finitely many points $\boldsymbol{x}_1$, $\cdots$, $\boldsymbol{x}_u$ in Q and finitely many vectors $\boldsymbol{d}_1, \cdots, \boldsymbol{d}_v$ such that $\boldsymbol{x} \in Q$ if and only if*

$$\boldsymbol{x} = \sum_{i=1}^{u} \lambda_i \boldsymbol{x}_i + \sum_{j=1}^{v} \mu_j \boldsymbol{d}_j$$

where $\lambda_i \geq 0, \mu_j \geq 0$ and $\sum_{i=1}^{u} \lambda_i = 1$.

Proof. By Lemmas 7.1.1 and 7.1.2, any polyhedron Q is a convex hull of finitely many manifolds and halfmanifolds. For each manifold or halfmanifold, the theorem holds obviously, which implies the truth of the theorem for any polyhedron. □

Consider the linearly constrained set $\Omega = \{\boldsymbol{x} \mid \boldsymbol{a}_j^T \boldsymbol{x} = b_j, j \in M'$ and $\boldsymbol{a}_j^T \boldsymbol{x} \geq b_j, j \in M \setminus M'\}$. Denote $J_\varepsilon(\boldsymbol{x}) = \{j \mid \boldsymbol{a}_j^T \boldsymbol{x} - b_j \leq \varepsilon\}$.

Theorem 7.1.4 *Suppose Ω satisfies the regularity condition. Then there exists $\varepsilon > 0$ such that for any $\boldsymbol{x} \in \Omega$, $\boldsymbol{a}_j$, $j \in J_\varepsilon(\boldsymbol{x})$, are linearly independent.*

Proof. By Theorem 7.1.3, there exist $\boldsymbol{x}_1$, $\cdots$, $\boldsymbol{x}_u$ in Ω and $\boldsymbol{d}_1$, $\cdots$, $\boldsymbol{d}_v$ in R^n such that $\boldsymbol{x} \in \Omega$ if and only if $\boldsymbol{x}$ can be represented as

$$\boldsymbol{x} = \sum_{k=1}^{u} \lambda_k \boldsymbol{x}_k + \sum_{\ell=1}^{v} \mu_\ell \boldsymbol{d}_\ell$$

where $\lambda_k \geq 0$, $\mu_\ell \geq 0$ and $\sum_{k=1}^u \lambda_k = 1$. First, we show that for every $\boldsymbol{d}_\ell$, $\boldsymbol{a}_j^T \boldsymbol{d}_\ell = 0$ for $j \in M'$ and $\boldsymbol{a}_j^T \boldsymbol{d}_\ell \geq 0$ for $j \in M \setminus M'$. Denote $\boldsymbol{x}(\mu) = \boldsymbol{x}_1 + \mu \boldsymbol{d}_\ell$ for $\mu > 0$. Then $\boldsymbol{x}(\mu) \in \Omega$. Thus, for $j \in M'$,

$$\boldsymbol{a}_j^T \boldsymbol{d}_\ell = \boldsymbol{a}_j^T (\boldsymbol{x}(\mu) - \boldsymbol{x}_1)/\mu = 0,$$

and for $j \in M \setminus M'$,

$$\boldsymbol{a}_j^T \boldsymbol{d}_\ell = \boldsymbol{a}_j^T (\boldsymbol{x}(\mu) - \boldsymbol{x}_1)/\mu \geq (b_j - \boldsymbol{a}_j^T \boldsymbol{x}_1)/\mu.$$

Letting $\mu \to \infty$, we obtain $\boldsymbol{a}_j^T \boldsymbol{d}_\ell \geq 0$. Now, define

$$\varepsilon = \frac{1}{u+1} \min\{\boldsymbol{a}_j^T \boldsymbol{x}_k - b_j \mid \boldsymbol{a}_j^T \boldsymbol{x}_k > b_j, k = 1, \cdots, u, \text{ and } j \in M\}.$$

We show that ε meets our requirement. Since $\lambda_k \geq 0$ and $\sum_{k=1}^u \lambda_k = 1$, there exists k_0 such that $\lambda_{k_0} \geq 1/u$. So, for any $j \in M$, we have

$$\begin{aligned} \boldsymbol{a}_j^T \boldsymbol{x} - b_j &= \sum_{k=1}^u \lambda_k (\boldsymbol{a}_j^T \boldsymbol{x}_k - b_j) + \sum_{\ell=1}^v \mu_\ell \boldsymbol{a}_j^T \boldsymbol{d}_\ell \\ &\geq \lambda_{k_0} (\boldsymbol{a}_j^T \boldsymbol{x}_{k_0} - b_j). \end{aligned}$$

Thus, for $j \notin J(\boldsymbol{x}_{k_0})$, $\boldsymbol{a}_j^T \boldsymbol{x} - b_j \geq \lambda_{k_0}(u+1)\varepsilon > \varepsilon$. This means that $J_\varepsilon(\boldsymbol{x}) \subseteq J(\boldsymbol{x}_{k_0})$. By the regularity condition, $\boldsymbol{a}_j$, $j \in J_\varepsilon(\boldsymbol{x})$ are linearly independent. □

Suppose that an algorithm, using the ε-active set strategy, generates an infinite sequence $\{\boldsymbol{x}_k\}$ and at point $\boldsymbol{x}_k$, the search direction is found depending on $J_{\varepsilon_k}(\boldsymbol{x}_k)$. Usually, one expects to have a positive number ε such that $\varepsilon_k > \varepsilon$ for all k. In fact, it is a key point for the ε-active set strategy to work. In order to have such a positive number ε, Theorem 7.1.4 suggests us to choose ε_k recursively in the following way:

ϵ-Procedure L: *Suppose that ε_{k-1} has been chosen. Then choose ε_k as follows.*

begin
$\varepsilon := \varepsilon_{k-1}$; $J := J_\varepsilon(\boldsymbol{x}_k)$;
while *rank*$(A_J) < |J|$ **do** $\varepsilon := \varepsilon/2$ *and* $J := J_\varepsilon(\boldsymbol{x}_k)$;
$\varepsilon_k = \varepsilon$;
end.

Corollary 7.1.5 *Suppose Ω satisfies the regularity condition. If $\{\varepsilon_k\}$ is generated by the ϵ-procedure L for a sequence of feasible points, then there exists $\varepsilon > 0$ such that $\varepsilon_k > \varepsilon$ for all k.*

Proof. By Theorem 7.1.4, there exists k' such that for $k > k'$, $\varepsilon_k = \varepsilon_{k-1}$. □

Usually, with the ε-active set strategy, the algorithm has a good convergence property and the proof is also simple. To see how it works, let us look at an example.

RD Method: *Choose an initial point $\boldsymbol{x}_1$ and a positive number ε_0. At each iteration $k = 1, 2, \cdots$, we carry out the following steps:*

1. *Compute the ε_k by ϵ-procedure L and set $J := J_{\varepsilon_k}(\boldsymbol{x}_k)$, $\boldsymbol{g} := \boldsymbol{g}_k$ and $\boldsymbol{x} := \boldsymbol{x}_k$. Compute*

$$(u_j, j \in J)^T = (A_J^T A_J)^{-1} A_J^T \boldsymbol{g}, \tag{7.2}$$

$$u_s = \max\{u_j \mid j \in J \setminus M'\}, \tag{7.3}$$

$$u_t(b_t - \boldsymbol{a}_t^T) = \max\{u_j(b_j - \boldsymbol{a}_j^T \boldsymbol{x}) \mid j \in J \setminus M'\}, \tag{7.4}$$

$$h = \begin{cases} s & \text{if } u_t(b_t - \boldsymbol{a}_t^T \boldsymbol{x}) < u_s, \\ t & \text{otherwise,} \end{cases} \tag{7.5}$$

and

$$\boldsymbol{d}_k = \begin{cases} P_J \boldsymbol{g} & \text{if } \|P_J \boldsymbol{g}\| \geq u_t(b_t - \boldsymbol{a}_t^T \boldsymbol{x}) \geq u_s, \\ P_{J \setminus h} \boldsymbol{g} & \text{otherwise.} \end{cases}$$

If $\boldsymbol{d}_k = \boldsymbol{0}$, then stop; otherwise, go to the next step.

2. *Compute*

$$\bar{\lambda}_k = \begin{cases} 1 & \text{if } \boldsymbol{a}_j^T \boldsymbol{d}_k \geq 0 \text{ for all } j \notin J_k, \\ \min\left\{ \frac{b_j - \boldsymbol{a}_j^T \boldsymbol{x}_k}{\boldsymbol{a}_j^T \boldsymbol{d}_k} \,\middle|\, \boldsymbol{a}_j^T \boldsymbol{d}_k < 0 \right\} & \text{otherwise.} \end{cases}$$

Find a new point $\boldsymbol{x}_{k+1} = \boldsymbol{x}_k + \lambda_k \boldsymbol{d}_k$ $(0 < \lambda_k \leq \bar{\lambda}_k)$ by a normal line search procedure.

Lemma 7.1.6 *Let $\boldsymbol{x}$ be a feasible point and $J \supseteq J(\boldsymbol{x})$. Denote $\boldsymbol{g} = \boldsymbol{g}(\boldsymbol{x})$. Suppose that u_s and $u_t(b_t - \boldsymbol{a}_t^T \boldsymbol{x})$ are determined by the equations in step 1 above. Then $\boldsymbol{x}$ is a Kuhn-Tucker point if one of the following occurs:*

(1) $0 = \|P_J \boldsymbol{g}\| \geq u_t(b_t - \boldsymbol{a}_t^T \boldsymbol{x}) \geq u_s$.
(2) $P_{J \setminus s} \boldsymbol{g} = \boldsymbol{0}$ *and* $u_t(b_t - \boldsymbol{a}_t^T \boldsymbol{x}) \leq u_s$.
(3) $P_{J \setminus t} \boldsymbol{g} = \boldsymbol{0}$ *and* $u_t(b_t - \boldsymbol{a}_t^T \boldsymbol{x}) \geq u_s$.

Proof. Note that in every case, we can obtain $P_J \boldsymbol{g} = \boldsymbol{0}$, $u_s \leq 0$ and $u_t(b_t - \boldsymbol{a}_t^T \boldsymbol{x}) \leq 0$. Thus, $\boldsymbol{g} = \sum_{j \in J} u_j \boldsymbol{a}_j$ and $u_j \leq 0$ for $j \in J \setminus M'$. Moreover, $u_t(b_t - \boldsymbol{a}_t^T \boldsymbol{x}) \leq 0$ implies $u_j \geq 0$ for $j \in J \setminus J(\boldsymbol{x})$. Therefore, $u_j = 0$ for $j \in J \setminus J(\boldsymbol{x})$. Hence, $\boldsymbol{x}$ is a Kuhn-Tucker point. □

Lemma 7.1.7 *If $\boldsymbol{d}_k = \boldsymbol{0}$, then $\boldsymbol{x}_k$ is a Kuhn-Tucker point. Moreover, the RD method satisfies*

$$\left.\begin{array}{l} \boldsymbol{x}_k \to \boldsymbol{x}^* \\ \boldsymbol{d}_k \to \boldsymbol{d}^* \\ (\boldsymbol{g}^*)^T \boldsymbol{d}^* \leq 0 \end{array}\right\} \Longrightarrow \boldsymbol{x}^* \text{ is a Kuhn-Tucker point.}$$

Proof. The first part follows immediately from Lemma 7.1.6. For the second part, suppose that $\{\boldsymbol{x}_k\}_{k \in K}$ is a subsequence of the generated sequence such that $\boldsymbol{x}_k \to \boldsymbol{x}^*$, $\boldsymbol{d}_k \to \boldsymbol{d}^*$ and $(\boldsymbol{g}^*)^T \boldsymbol{d}^* = 0$. Without loss of generality, we can assume that at every iteration $k \in K$, the same index set J and the same indices s and t are obtained. By Corollary 7.1.5, it is easy to see that $J \supseteq J(\boldsymbol{x}^*)$. By taking limit, it is also easily seen that one of the conditions in Lemma 7.1.6 must hold for $\boldsymbol{x}^*$. Thus, $\boldsymbol{x}^*$ is a Kuhn-Tucker point. □

Theorem 7.1.8 *Under the regularity condition, the RD method either stops at a Kuhn-Tucker point or generates an infinite sequence whose cluster points are all Kuhn-Tucker points.*

Proof. If the algorithm stops at iteration k, then by Lemma 7.1.7, $\boldsymbol{x}_k$ is a Kuhn-Tucker point. Next, suppose that the algorithm generates an infinite sequence $\{\boldsymbol{x}_k\}$ with a cluster point $\boldsymbol{x}^*$. If the sequence does not converge to $\boldsymbol{x}^*$, then by Lemma 7.1.7 and Theorem 5.3.2, $\boldsymbol{x}^*$ is a Kuhn-Tucker point. Thus, we need to consider only the case where the sequence converges to $\boldsymbol{x}^*$. In this case, $\lim_{k \to \infty} \lambda_k = 0$ and $\{\boldsymbol{d}_k\}$ is bounded. If $\lambda_k \neq \bar{\lambda}_k$ for infinitely many k, then the corresponding $\boldsymbol{d}_k$ must have a cluster point $\boldsymbol{d}^*$. Since the line search procedure is normal, we have $(\boldsymbol{g}^*)^T \boldsymbol{d}^* = 0$. By Lemma 7.1.7, $\boldsymbol{x}^*$ is a Kuhn-Tucker point. Thus, we need only to consider the case where for sufficiently large k, $\bar{\lambda}_k = \lambda_k < 1$. Next, we show that this case is impossible. Note that for such k, at least one index gets in the active set. So, for sufficiently large k, we must have an index getting out from the active set. This happens only if $\boldsymbol{d}_k = P_{J\setminus s}\boldsymbol{g}$ in the first step. It follows that for sufficiently large k, $\boldsymbol{a}_j^T \boldsymbol{d}_k = 0$ for $j \in J_{\varepsilon_k}(\boldsymbol{x}_k) \setminus J(\boldsymbol{x}_k)$. Thus, for sufficiently large k, we have $\bar{\lambda}_k \geq \varepsilon_k/\alpha$ where α is an upper bound for $|\boldsymbol{a}_j^T \boldsymbol{d}_k|$. This contradicts $\lim_{k\to\infty} \lambda_k = 0$. □

The global convergence of the RD method requires a weaker condition on the line search procedure. In fact, with some line search procedures that Rosen's method fail to have the global convergence property, the RD method still has global convergence. The counterexample in section 5.2 is such an example. This is an interesting phenomenon.

7.2 Rotating Tangent Plane

The technique for dealing with inexact active constraints also enables us to design an interesting algorithm for nonlinearly constrained optimization problems.

Consider the following problem:

$$\begin{array}{ll} \text{maximize} & f(\boldsymbol{x}) \\ \text{subject to} & h_j(\boldsymbol{x}) \geq 0, \quad j \in M. \end{array} \tag{7.6}$$

The feasible region is still denoted by Ω. We also assume the regularity condition, i.e. every feasible point is regular.

Denote $J_\varepsilon(\boldsymbol{x}) = \{j \mid h_j(\boldsymbol{x}) \leq \varepsilon\}$ and $H_J(\boldsymbol{x}) = (\nabla h_j(\boldsymbol{x}), j \in J)$. The following result corresponds to Theorem 7.1.4.

Theorem 7.2.1 *Suppose that Ω satisfies the regularity condition. Then for any bounded subset Q of Ω, there exists a positive number ε such that for every $\boldsymbol{x} \in Q$, $\nabla h_j(\boldsymbol{x})$, $j \in J_\varepsilon(\boldsymbol{x})$ are linearly independent.*

Proof. For contradiction, suppose that there is a sequence $\{\boldsymbol{x}_k\}$ in Q such that $\nabla h_j(\boldsymbol{x})$, $j \in J_{1/k}(\boldsymbol{x}_k)$ are linearly dependent. Since there are only finitely many index sets and Q is bounded, there exists an infinite set K of natural numbers and an index set J such that $J_{1/k}(\boldsymbol{x}_k) = J$ for all $k \in K$ and $\boldsymbol{x}_k \to \boldsymbol{x}^*$ as $k \to \infty$, $k \in K$. Thus for any $k \in K$,

$$\det(H_J(\boldsymbol{x}_k)^T H_J(\boldsymbol{x}_k)) = 0$$

and for any $j \in J$,

$$0 \leq h_j(\boldsymbol{x}_k) \leq 1/k.$$

Letting $k \to \infty$, we obtain

$$\det(H_J(\boldsymbol{x}^*)^T H_J(\boldsymbol{x}^*)) = 0 \tag{7.7}$$

and for any $j \in J$, $h_j(\boldsymbol{x}^*) = 0$. Thus, $J \subseteq J(\boldsymbol{x}^*)$. By the regularity condition, $\nabla h_j(\boldsymbol{x}^*)$, $j \in J$ are linearly independent, contradicting (7.7). □

Since Theorem 7.2.1 is a little weaker than Theorem 7.1.6, the corresponding ϵ-procedure has to be a little different from the ϵ-procedure L. The main difference is that ε_k is computed independently from ε_{k-1}.

ϵ-Procedure NL: *Let ε_0 be a positive number given at the initial stage. Then compute ε_k as follows:*

begin $\varepsilon := \varepsilon_0$; $J := J_\varepsilon(\boldsymbol{x}_k)$;
while $\det[H_J(\boldsymbol{x}_k)^T H_J(\boldsymbol{x}_k)] < \varepsilon$ **do** $\varepsilon := \varepsilon/2$ *and* $J := J_\varepsilon(\boldsymbol{x}_k)$;
$\varepsilon_k := \varepsilon$;
end.

Corollary 7.2.2 *Suppose that Ω satisfies the regularity condition. If $\{\varepsilon_k\}$ is generated by the ϵ-procedure NL for a sequence of feasible points $\{\boldsymbol{x}_k\}$, then for any bounded subsequence $\{\boldsymbol{x}_k\}_{k\in K}$, there exists a positive number ε such that for all $k \in K$, $\varepsilon_k > \varepsilon$ and $\det[H_{\hat{J}_k}(\boldsymbol{x}_k)^T H_{\hat{J}_k}(\boldsymbol{x}_k)] \geq \varepsilon$.*

Proof. It follows immediately from Theorem 7.2.1. □

When the feasible region is constrained nonlinearly, the gradient projection on tangent plane could lose the feasibility. Thus, it was suggested to choose the search direction in the following way: First, project the gradient to the tangent plane, and then pull the projection back to the feasible region. How is it pulled? There are many ways. Here, we introduce one of them: Rotate the tangent plane before projecting the gradient on it.

Consider a point $\boldsymbol{x} \in \Omega$ and an index set $J \supseteq J(\boldsymbol{x})$. Denote

$$\begin{aligned}
\boldsymbol{n}_j &= \nabla h_j(\boldsymbol{x})/\|\nabla h_j(\boldsymbol{x})\|, \\
N_J(\boldsymbol{x}) &= (\boldsymbol{n}_j(\boldsymbol{x}), j \in J), \\
\rho_J(\boldsymbol{x}) &= \det[N_J(\boldsymbol{x})^T N_J(\boldsymbol{x})]/(e|J|\,\|\boldsymbol{g}(\boldsymbol{x})\| + 1), \\
\boldsymbol{a}_j(\boldsymbol{x}) &= \boldsymbol{n}_j(\boldsymbol{x}) - \rho_J(\boldsymbol{x})\boldsymbol{g}(\boldsymbol{x}), \\
A_J(\boldsymbol{x}) &= (\boldsymbol{a}_j(\boldsymbol{x}), j \in J).
\end{aligned}$$

We will show that the plane $\{\boldsymbol{y} \mid A_J(\boldsymbol{x})^T\boldsymbol{y} = \boldsymbol{0}\}$ has properties similar to those of the tangent plane, and in addition, the gradient projection on this plane is feasible.

Lemma 7.2.3 *Suppose that $\boldsymbol{n}_j(\boldsymbol{x})$, $j \in J$ are linearly independent. Denote*

$$(\alpha_j(\boldsymbol{x}), j \in J)^T = [N_J(\boldsymbol{x})^T N_J(\boldsymbol{x})]^{-1} N_J(\boldsymbol{x})^T \boldsymbol{g}(\boldsymbol{x}).$$

Then

$$|\sum_{j\in J} \alpha_j(\boldsymbol{x})| < \frac{1}{\rho_J(\boldsymbol{x})}.$$

Moreover, if $\boldsymbol{g}(\boldsymbol{x}) = \sum_{j\in J} u_j \boldsymbol{n}_j(\boldsymbol{x})$, then $u_j = \alpha_j(\boldsymbol{x})$.

Proof. Note that

$$\begin{aligned} |\sum_{j\in J} \alpha_j(\boldsymbol{x})| &\leq \sqrt{|J|}\sqrt{\sum_{j\in J} \alpha_j(\boldsymbol{x})^2} \\ &= \sqrt{|J|}\, \|[N_J(\boldsymbol{x})^T N_J(\boldsymbol{x})]^{-1} N_J(\boldsymbol{x})^T \boldsymbol{g}(\boldsymbol{x})\| \\ &\leq \frac{\sqrt{|J|}}{\min_{i=1,\cdots,|J|} \lambda_i} \|N_J(\boldsymbol{x})^T \boldsymbol{g}(\boldsymbol{x})\| \end{aligned}$$

where λ_i, $i = 1, \cdots, |J|$ are the $|J|$ eigenvalues of $N_J(\boldsymbol{x})^T N_J(\boldsymbol{x})$. Clearly, $\lambda_i > 0$ for all i and

$$\prod_{i=1}^{|J|} \lambda_i = \det[N_J(\boldsymbol{x})^T N_J(\boldsymbol{x})],$$

$$\sum_{i=1}^{|J|} \lambda_i = \mathrm{tr}[N_J(\boldsymbol{x})^T N_J(\boldsymbol{x})] = |J|.$$

Without loss of generality, we assume $\lambda_1 \leq \lambda_2 \leq \cdots \leq \lambda_{|J|}$. Then

$$\prod_{i=2}^{|J|} \lambda_i \leq \left(\frac{\sum_{i=1}^{|J|} \lambda_i}{|J| - 1}\right)^{|J|-1} \leq \left(\frac{|J|}{|J| - 1}\right)^{|J|-1} < e.$$

Therefore

$$\lambda_1 \geq \det[N_J(\boldsymbol{x})^T N_J(\boldsymbol{x})]/e.$$

Moreover,

$$\begin{aligned}\|N_J(\boldsymbol{x})^T\boldsymbol{g}(\boldsymbol{x})\| &= \sqrt{\sum_{j\in J}[\boldsymbol{n}_j(\boldsymbol{x})^T\boldsymbol{g}(\boldsymbol{x})]^2} \\ &\leq \sqrt{\sum_{j\in J}\|\boldsymbol{n}_j(\boldsymbol{x})\|^2\|\boldsymbol{g}(\boldsymbol{x})\|^2} = \sqrt{|J|}\,\|\boldsymbol{g}(\boldsymbol{x})\|.\end{aligned}$$

Hence,

$$|\sum_{j\in J}\alpha_j(\boldsymbol{x})| \leq \frac{e|J|\,\|\boldsymbol{g}(\boldsymbol{x})\|}{\det[N_J(\boldsymbol{x})^TN_J(\boldsymbol{x})]} < \frac{1}{\rho_J(\boldsymbol{x})}.$$

For the second part, we represent $\boldsymbol{g}(\boldsymbol{x})$ as $\boldsymbol{g}(\boldsymbol{x}) = N_J(\boldsymbol{x})(u_j, j\in J)^T$. Then it is clear that $(\alpha_j(\boldsymbol{x}), j\in J)^T = (u_j, j\in J)^T$. □

Theorem 7.2.4 *$\boldsymbol{n}_j(\boldsymbol{x})$, $j\in J$ are linearly independent if and only if $\boldsymbol{a}_j(\boldsymbol{x})$, $j\in J$ are linearly independent.*

Proof. If $\boldsymbol{n}_j(\boldsymbol{x})$, $j\in J$ are not linearly independent, then $\rho_J(\boldsymbol{x}) = 0$. So $\boldsymbol{a}_j(\boldsymbol{x}) = \boldsymbol{n}_j(\boldsymbol{x})$. Hence, $\boldsymbol{a}_j(\boldsymbol{x})$, $j\in J$ are not linearly independent. Next, suppose that $\boldsymbol{n}_j(\boldsymbol{x})$, $j\in J$ are linearly independent. We show that so are $\boldsymbol{a}_j(\boldsymbol{x})$, $j\in J$. For contradiction, suppose that there exists $|J|$ numbers β_j with at least one nonzero such that

$$\sum_{j\in J}\beta_j\boldsymbol{a}_j(\boldsymbol{x}) = \boldsymbol{0}.$$

Substituting $\boldsymbol{a}_j(\boldsymbol{x}) = \boldsymbol{n}_j(\boldsymbol{x}) - \rho_j(\boldsymbol{x})\boldsymbol{g}(\boldsymbol{x})$ into it, we obtain

$$\sum_{j\in J}\beta_j\rho_J(\boldsymbol{x})\boldsymbol{g}(\boldsymbol{x}) = \sum_{j\in J}\beta_j\boldsymbol{n}_j(\boldsymbol{x}).$$

Since $\boldsymbol{n}_j(\boldsymbol{x})$, $j\in J$ are linearly independent, we must have $\sum_{j\in J}\beta_j\rho_J(\boldsymbol{x}) \neq 0$. Hence,

$$\boldsymbol{g}(\boldsymbol{x}) = \sum_{j\in J}\frac{\beta_j}{\rho_J(\boldsymbol{x})\sum_{i\in J}\beta_i}\boldsymbol{n}_j(\boldsymbol{x}).$$

It follows that

$$\alpha_j(\boldsymbol{x}) = \frac{\beta_j}{\rho_J(\boldsymbol{x})\sum_{i\in J}\beta_i}.$$

Therefore,

$$\sum_{j\in J}\alpha_j(\boldsymbol{x}) = 1/\rho_J(\boldsymbol{x}),$$

contradicting Lemma 7.2.3. □

Theorem 7.2.5 *Suppose that $J \supseteq J(\boldsymbol{x})$. Then $\boldsymbol{x}$ is a Kuhn-Tucker point if and only if there exist $u_j \le 0$ such that*

$$\boldsymbol{g}(\boldsymbol{x}) = \sum_{j \in J(\boldsymbol{x})} u_j \boldsymbol{a}_j(\boldsymbol{x}). \tag{7.8}$$

Proof. If $\boldsymbol{n}_j(\boldsymbol{x})$, $j \in J$ are linearly dependent, then $\boldsymbol{a}_j(\boldsymbol{x}) = \boldsymbol{n}_j(\boldsymbol{x})$, so the theorem is trivial. Next, we assume that $\boldsymbol{n}_j(\boldsymbol{x})$, $j \in J$ are linearly independent.

For sufficiency, suppose that (7.8) holds. Substituting $\boldsymbol{a}_j(\boldsymbol{x}) = \boldsymbol{n}_j(\boldsymbol{x}) - \rho_J(\boldsymbol{x})\boldsymbol{g}(\boldsymbol{x})$ into it, we obtain

$$\boldsymbol{g}(\boldsymbol{x}) = \sum_{j \in J(\boldsymbol{x})} \frac{u_j}{1 + \rho_J(\boldsymbol{x}) \sum_{i \in J(\boldsymbol{x})} u_i} \boldsymbol{n}_j(\boldsymbol{x}). \tag{7.9}$$

By Lemma 7.2.3, we have

$$\frac{1}{\rho_J(\boldsymbol{x})} > |\sum_{j \in J} \alpha_j(\boldsymbol{x})| = \left| \frac{\sum_{j \in J(\boldsymbol{x})} u_j}{1 + \rho_J(\boldsymbol{x}) \sum_{j \in J(\boldsymbol{x})} u_j} \right|.$$

Note that $u_j \le 0$ for all $j \in J(\boldsymbol{x})$. If $1 + \rho_J(\boldsymbol{x}) \sum_{j \in J(\boldsymbol{x})} u_j < 0$, then

$$\left| \frac{\sum_{j \in J(\boldsymbol{x})} u_j}{1 + \rho_J(\boldsymbol{x}) \sum_{j \in J(\boldsymbol{x})} u_j} \right| = \frac{|\sum_{j \in J(\boldsymbol{x})} u_j|}{\rho_J(\boldsymbol{x}) |\sum_{j \in J(\boldsymbol{x})} u_j| - 1} > \frac{1}{\rho_J(\boldsymbol{x})},$$

a contradiction. Thus, $1 + \rho_J(\boldsymbol{x}) \sum_{j \in J(\boldsymbol{x})} u_j > 0$. Hence, by (7.9), $\boldsymbol{x}$ is a Kuhn-Tucker point.

For necessity, suppose that there exist $\alpha_j \le 0$ such that

$$\boldsymbol{g}(\boldsymbol{x}) = \sum_{j \in J(\boldsymbol{x})} \alpha_j \boldsymbol{n}_j(\boldsymbol{x}).$$

Substituting $\boldsymbol{n}_j(\boldsymbol{x}) = \boldsymbol{a}_j(\boldsymbol{x}) + \rho_J(\boldsymbol{x})\boldsymbol{g}(\boldsymbol{x})$ into it, we can obtain

$$\left(1 - \rho_J(\boldsymbol{x}) \sum_{j \in J(\boldsymbol{x})} \alpha_j \right) \boldsymbol{g}(\boldsymbol{x}) = \sum_{j \in J(\boldsymbol{x})} \alpha_j \boldsymbol{a}_j(\boldsymbol{x}).$$

Since $1 - \rho_j(\boldsymbol{x}) \sum_{j \in J(\boldsymbol{x})} \alpha_j > 0$, we can obtain (7.8) by letting $u_j = \alpha_j / (1 - \rho_J(\boldsymbol{x}) \sum_{j \in J(\boldsymbol{x})} \alpha_j)$. □

Denote by $\hat{P}_J(\boldsymbol{x})$ the operator of the orthogonal projection into the subspace $\{\boldsymbol{y} \mid A_J(\boldsymbol{x})^T \boldsymbol{y} = \boldsymbol{0}\}$.

Theorem 7.2.6 *Suppose that $\boldsymbol{x}$ is regular. Let $J = J(\boldsymbol{x})$ and $\boldsymbol{g} = \boldsymbol{g}(\boldsymbol{x})$. Define*

$$\boldsymbol{d} = \begin{cases} \hat{P}_J(\boldsymbol{x})\boldsymbol{g} & \text{if } \|\hat{P}_J(\boldsymbol{x})\boldsymbol{g}\| > u_h(\boldsymbol{x}), \\ \hat{P}_{J\setminus h}(\boldsymbol{x})\boldsymbol{g} & \text{otherwise,} \end{cases}$$

where

$$(u_j(\boldsymbol{x}), j \in J)^T = (A_J(\boldsymbol{x})^T A_J(\boldsymbol{x}))^{-1} A_J(\boldsymbol{x})^T \boldsymbol{g},$$
$$u_h(\boldsymbol{x}) = \max\{u_j(\boldsymbol{x}) \mid j \in J\}.$$

Then $\boldsymbol{d} = \boldsymbol{0}$ if and only if $\boldsymbol{x}$ is a Kuhn-Tucker point. Moreover, if $\boldsymbol{d} \neq \boldsymbol{0}$, then $\boldsymbol{d}$ is an ascending feasible direction.

Proof. The first part follows immediately from Theorem 7.2.5. For the second part, we assume $\boldsymbol{d} \neq \boldsymbol{0}$. Since $\boldsymbol{g}^T\boldsymbol{d} = \|\boldsymbol{d}\|^2 > 0$, $\boldsymbol{d}$ is an ascending direction. It remains to show the feasibility of $\boldsymbol{d}$. If $\boldsymbol{d} = \hat{P}_J(\boldsymbol{x})\boldsymbol{g}$, then $\boldsymbol{a}_j(\boldsymbol{x})^T\boldsymbol{d} = 0$ for all $j \in J$. Thus, $\boldsymbol{n}_j(\boldsymbol{x})^T\boldsymbol{d} = \rho_J(\boldsymbol{x})\boldsymbol{g}^T\boldsymbol{d} > 0$, i.e. $\nabla h_j(\boldsymbol{x})^T\boldsymbol{d} > 0$. It follows that for sufficiently small λ,

$$h_j(\boldsymbol{x} + \lambda\boldsymbol{d}) = h_j(\boldsymbol{x}) + \lambda\nabla h_j(\boldsymbol{x})^T + o(\lambda) > 0$$

for all $j \in J$. Hence, $\boldsymbol{d}$ is a feasible direction. If $\boldsymbol{d} = \hat{P}_{J\setminus h}(\boldsymbol{x})\boldsymbol{g}$, then $\boldsymbol{a}_j(\boldsymbol{x})^T\boldsymbol{d} = 0$ for $j \in J \setminus h$ and $\boldsymbol{a}_h(\boldsymbol{x})^T\boldsymbol{d} = u_h(\boldsymbol{x})\|\hat{P}_{J\setminus h}\boldsymbol{a}_h(\boldsymbol{x})\|^2 > 0$. Thus, we also have $\boldsymbol{n}_j(\boldsymbol{x})^T\boldsymbol{d} > 0$ for all $j \in J$. By the same argument as above, we can prove that $\boldsymbol{d}$ is a feasible direction. □

Theorem 7.2.6 suggests an extension of Rosen's method to solve the nonlinearly constrained problem (7.6). For such an extension, we can prove that if a generated sequence is not convergent, then every cluster point of the sequence is a Kuhn-Tucker point. However, it is an open problem what happens if a generated sequence is convergent. In fact, it is not known how to extend the second slope lemma to this case. While the active set strategy has not successed completely, the ϵ-active set strategy works quite well. By extending the RD method, a globally convergent algorithm for (7.6) can been obtained. Next, we present an abstract algorithm which includes this case.

Consider a family of point-to-set mappings $\{F_J \mid J \subseteq \{1, 2, \cdots, m\}\}$. For example, define $F_J(\boldsymbol{x})$ by

$$\hat{P}_J(\boldsymbol{x})\boldsymbol{g} \in F_J(\boldsymbol{x}) \quad \text{iff} \quad \|\hat{P}_J(\boldsymbol{x})\boldsymbol{g}\| \geq -u_t(\boldsymbol{x})h_t(\boldsymbol{x}) \geq u_s(\boldsymbol{x}),$$
$$\hat{P}_{J\setminus s}(\boldsymbol{x})\boldsymbol{g} \in F_J(\boldsymbol{x}) \quad \text{iff} \quad u_t(\boldsymbol{x})h_t(\boldsymbol{x}) \leq u_s(\boldsymbol{x}),$$
$$\hat{P}_{J\setminus t}(\boldsymbol{x})\boldsymbol{g} \in F_J(\boldsymbol{x}) \quad \text{iff} \quad \|\hat{P}_J(\boldsymbol{x})\boldsymbol{g}\| \leq -u_t(\boldsymbol{x})h_t(\boldsymbol{x}) \text{ and } -u_t(\boldsymbol{x})h_t(\boldsymbol{x}) \geq u_s(\boldsymbol{x})$$

where

$$\begin{aligned}(u_j(\boldsymbol{x}), j \in J)^T &= (A_J(\boldsymbol{x})^T A_J(\boldsymbol{x}))^{-1} A_J(\boldsymbol{x})^T \boldsymbol{g}\\ u_t(\boldsymbol{x})h_t(\boldsymbol{x}) &= \min_{j\in J} u_j(\boldsymbol{x})h_j(\boldsymbol{x})\\ u_s(\boldsymbol{x}) &= \max_{j\in J} u_j(\boldsymbol{x}).\end{aligned}$$

Algorithm B: *Choose an initial point $\boldsymbol{x}_1$. At iteration k, perform the following steps:*

1. *Compute a positive number ε_k by the ϵ-procedure NL (or L if the constraints are all linear). Set $J := J_{\varepsilon_k}(\boldsymbol{x}_k)$. Choose a search direction $\boldsymbol{d}_k$ from $F_J(\boldsymbol{x}_k)$. If $\boldsymbol{d}_k$ is not an ascending feasible direction, then stop. Otherwise, go to the next step.*

2. *Compute a number $\bar{\lambda}_k$ such that $\boldsymbol{x}_k + \lambda \boldsymbol{d}_k$ for $0 \le \lambda \le \bar{\lambda}_k$ lies in the feasible region. Find a new point $\boldsymbol{x}_{k+1} = \boldsymbol{x}_k + \lambda_k \boldsymbol{d}_k$ $(0 < \lambda_k \le \bar{\lambda}_k)$ by a normal line search procedure.*

Theorem 7.2.7 *Let Ω be the feasible region of problem (7.6) satisfying the regularity condition that at each feasible points $\boldsymbol{x}$, $\nabla h_j(\boldsymbol{x})$ for $j \in J(\boldsymbol{x})$ are linearly independent. Let $\Gamma \subset \Omega$. Let $\{F_J\}_{J\subseteq M}$ be a family of point-to-set mappings from $\Omega \setminus \Gamma$ to $\mathcal{P}(D)$. Suppose that for every $J \supseteq J(\boldsymbol{x})$ and $\boldsymbol{x} \notin \Gamma$, if $\nabla h_j(\boldsymbol{x})$, $j \in J$ are linearly independent, then $F_J(\boldsymbol{x}) \neq \emptyset$, F_J is closed at $\boldsymbol{x}$ and*

$$\left.\begin{array}{l} J \supseteq J(\boldsymbol{x})\\ \boldsymbol{d} \in F_J(\boldsymbol{x})\\ \boldsymbol{x} \notin \Gamma \end{array}\right\} \Longrightarrow \begin{array}{l}\boldsymbol{d} \text{ is an ascending feasible direction at } \boldsymbol{x} \text{ such}\\ \text{that if } h_i(\boldsymbol{x}) \ge 0 \text{ is an active constraint}\\ \text{at } \boldsymbol{x}, \text{ then } \nabla h_i(\boldsymbol{x})^T \boldsymbol{d} > 0.\end{array} \tag{7.10}$$

Let $\bar{\lambda}_k$ be chosen such that for some constant and for every k, either $\bar{\lambda}_k$ is not less than the constant, or $\boldsymbol{x}_k + 2\bar{\lambda}_k \boldsymbol{d}_k$ is not feasible. Then Algorithm B either stops at a point in Γ, or generates an infinite sequence whose cluster points are all in Γ.

Proof. We first prove that for any subsequence $\{\boldsymbol{x}_k\}_{k\in K}$, we have (5.13). To do so, suppose that $\boldsymbol{x}_k \to \boldsymbol{x}^*$, $\boldsymbol{d}_k \to \boldsymbol{d}^*$ and $(\boldsymbol{g}^*)^T \boldsymbol{d}^* = 0$. Since the number of index sets is finite, there exists an infinite subset K' of K such that for $k \in K'$, $J_{\varepsilon_k}(\boldsymbol{x}_k) = J$. By Corollary 7.2.2, $J \supseteq J(\boldsymbol{x}^*)$ and $\det[H_J(\boldsymbol{x}^*)^T H_J(\boldsymbol{x}^*)] > 0$. If $\boldsymbol{x}^*$ is not in Γ, then F_J is closed on $\boldsymbol{x}^*$. Thus, $\boldsymbol{d}^* \in F_J(\boldsymbol{x}^*)$. By (5.13),

$\boldsymbol{d}^*$ is an ascending direction, contradicting $(\boldsymbol{g}^*)^T\boldsymbol{d}^* = 0$. Since $F_J(\boldsymbol{x}) \neq \emptyset$ for $J = J_{\varepsilon_k}(\boldsymbol{x}_k)$, it is easy to see from (5.13) that if the algorithm stops, then it stops at a point in Γ.

Next, suppose the algorithm generates an infinite sequence. If the sequence is not convergent, then by Theorem 5.3.2, every cluster point of the sequence is in Γ. So, we need only to consider a convergent generated sequence $\{\boldsymbol{x}_k\}$ with the limit $\boldsymbol{x}^*$. In this case, $\{\boldsymbol{d}_k\}$ is bounded and $\lim_{k\to\infty}\lambda_k = 0$. If there exist infinitely many k such that $\lambda_k < \bar{\lambda}_k$, then by Theorem 5.2.3, $\{\boldsymbol{d}_k\}$ has a cluster point $\boldsymbol{d}^*$ such that $(\boldsymbol{g}^*)^T\boldsymbol{d}^* = 0$. Thus, by (5.13), $\boldsymbol{x}^*$ is in Γ. Therefore, the last case is that $\lambda_k = \bar{\lambda}_k$ for sufficiently large k. Since $\boldsymbol{x}_k + 2\bar{\lambda}_k\boldsymbol{d}_k$ is not feasible, there exists an index i_k such that $\boldsymbol{x}_k + 2\bar{\lambda}_k\boldsymbol{d}_k$ does not satisfy the i_kth constraint, say $h_{i_k}(\boldsymbol{x}) \geq 0$. Choose an infinite set K such that $\lim_{k\to\infty,k\in K}\boldsymbol{d}_k = \boldsymbol{d}^*$ and $J = J_{\varepsilon_k}(\boldsymbol{x}_k)$ and $i_k = i$ for $k \in K$. If $\boldsymbol{x}^*$ is not in Γ, then F_J is closed at $\boldsymbol{x}^*$. So, $\boldsymbol{d}^* \in F_J(\boldsymbol{x}^*)$. By Corollary 7.2.2, we also have $J \supseteq J_\varepsilon(\boldsymbol{x}^*)$ for some positive number ε. Since $h_i(\boldsymbol{x}_k + 2\bar{\lambda}_k\boldsymbol{d}_k) < 0$. Letting k go to ∞ through K, we obtain $h_i(\boldsymbol{x}^*) \leq 0$. Thus, $h_i(\boldsymbol{x}) \geq 0$ is an active constraint at $\boldsymbol{x}^*$. In addition, since $0 \leq h_i(\boldsymbol{x}_k) - h_i(\boldsymbol{x}_k + 2\bar{\lambda}_k\boldsymbol{d}_k) = -2\bar{\lambda}_k\nabla h_i(\boldsymbol{x}_k + \mu_k\boldsymbol{d}_k)^T\boldsymbol{d}_k$ for some μ_k in $[0, 2\bar{\lambda}_k]$ and $k \in K$, we have $\nabla h_i(\boldsymbol{x}_k + \mu_k\boldsymbol{d}_k)^T\boldsymbol{d}_k < 0$ for $k \in K$. Letting $k \to \infty$, we obtain $\nabla h_i(\boldsymbol{x}^*)^T\boldsymbol{d}^* \leq 0$, contradicting (7.10). □

7.3 Reduced Gradient

The reduced gradient method is very efficient in practice. It's global convergence can be obtained successfully with the techniques described in the ε-active set strategy.

Consider the following problem:

$$\begin{array}{ll} \text{maximize} & f(\boldsymbol{x}) \\ \text{subject to} & A\boldsymbol{x} = \boldsymbol{b}, \\ & \boldsymbol{x} \geq \boldsymbol{0}, \end{array} \tag{7.11}$$

where A is an $m \times n$ matrix of rank m, $\boldsymbol{b} \in R^m$ and $\boldsymbol{x} \in R^n$. Given a feasible point $\boldsymbol{x}$, we partition its components into two groups

$$\boldsymbol{x} = \begin{pmatrix} \boldsymbol{x}_I \\ \boldsymbol{x}_{\bar{I}} \end{pmatrix}$$

where $\boldsymbol{x}_I \in R^m$ and $\boldsymbol{x}_{\bar{I}} \in R^{n-m}$. This partition is required to have the following properties:

(a) All components in $\boldsymbol{x}_I$ are positive.

(b) $|I| = m$ and the square submatrix A_I has rank m.

Corresponding to such a partition, $\boldsymbol{x}_I$ can be expressed as a function of $\boldsymbol{x}_{\bar{I}}$, i.e. $\boldsymbol{x}_I = A_I^{-1}\boldsymbol{b} - A_I^{-1}A_{\bar{I}}\boldsymbol{x}_{\bar{I}}$. So, the problem (7.11) can be represented as follows:

$$\begin{array}{ll} \text{maximize} & \bar{f}(\boldsymbol{x}_{\bar{I}}) = f(A_I^{-1}\boldsymbol{b} - A_I^{-1}A_{\bar{I}}\boldsymbol{x}_{\bar{I}}, \boldsymbol{x}_{\bar{I}}) \\ \text{subject to} & A_I^{-1}\boldsymbol{b} - A_I^{-1}A_{\bar{I}}\boldsymbol{x}_{\bar{I}} \geq \boldsymbol{0}, \\ & \boldsymbol{x}_{\bar{I}} \geq \boldsymbol{0}. \end{array} \tag{7.12}$$

Denote $\boldsymbol{r} = \nabla\bar{f}(\boldsymbol{x}_{\bar{I}})$. Note that at the considered point, the active constraints of (7.12) are all in $\boldsymbol{x}_{\bar{I}}$. Thus, the projection of the gradient $\boldsymbol{r}$ on the tangent plane is $\sum_{x_i>0} r_i\boldsymbol{e}_i$ where $\boldsymbol{e}_i$ is a m-dimensional vector with all components 0 except the ith component being 1. Since

$$\boldsymbol{r} = \sum_{x_i>0} r_i\boldsymbol{e}_i + \sum_{x_i=0} r_i\boldsymbol{e}_i,$$

r_i's for $x_i = 0$ are Lagrange multipliers. Therefore, $\boldsymbol{x}_{\bar{I}}$ is a Kuhn-Tucker point of (7.12) (i.e. $\boldsymbol{x}$ is a Kuhn-Tucker point of problem (7.11)) if and only if $r_i = 0$ for all $x_i > 0$ and $r_i \leq 0$ for all $x_i = 0$.

Suppose that $\boldsymbol{x}_{\bar{I}}$ is not a Kuhn-Tucker point. Define that for $i \in \bar{I}$,

$$d_i = \begin{cases} 0 & \text{if } r_i \leq 0 \text{ and } x_i = 0, \\ r_i & \text{otherwise.} \end{cases} \tag{7.13}$$

Then $\boldsymbol{d}_{\bar{I}} \neq \boldsymbol{0}$, $\boldsymbol{x}_{\bar{I}} + \lambda\boldsymbol{d}_{\bar{I}} \geq \boldsymbol{0}$ for $\lambda \geq 0$, and

$$\boldsymbol{r}^T\boldsymbol{d}_{\bar{I}} = \sum_{x_i>0} r_i^2 + \sum_{x_i=0, r_i>0} r_i^2.$$

This means that $\boldsymbol{d}_{\bar{I}}$ is a feasible ascending direction at $\boldsymbol{x}_{\bar{I}}$. Moreover, $\boldsymbol{g}(\boldsymbol{x})^T\boldsymbol{d} = \boldsymbol{r}^T\boldsymbol{d}_{\bar{I}}$. It turns out that

$$\boldsymbol{d} = \begin{pmatrix} -A_I^{-1}A_{\bar{I}}\boldsymbol{d}_{\bar{I}} \\ \boldsymbol{d}_{\bar{I}} \end{pmatrix} \tag{7.14}$$

is a feasible ascending direction at $\boldsymbol{x}$.

The index set I satisfying (b) is called a *basis* and it is called a *reducible basis* at the point $\boldsymbol{x}$ if it also satisfies (a). The gradient $\boldsymbol{r}$ is called a *reduced gradient*. The above analysis suggests the following method:

Reduced Gradient Method: *Choose an initial point* $\boldsymbol{x}_1$. *At iteration* $k = 1, 2, \cdots$, *the following steps are carried out.*

1. *Set $\boldsymbol{x} := \boldsymbol{x}_k$. Find a reducible basis I at $\boldsymbol{x}$. Compute $\boldsymbol{r} = \nabla \bar{f}(\boldsymbol{x}_{\bar{I}})$ and compute $\mathbf{d}$ by (7.14) and*

$$d_i = \begin{cases} r_i & \text{if } -r_i \le x_i, \\ -x_i & \text{otherwise.} \end{cases} \tag{7.15}$$

If $\mathbf{d} = \mathbf{0}$, then stop ($\boldsymbol{x}_k$ is a Kuhn-Tucker point). Otherwise go to the next step.

2. *Define*

$$\bar{\lambda}_k = \begin{cases} 1 & \text{if } d_i \ge 0 \text{ for all } i, \\ \min\left\{ \frac{x_i}{-d_i} \,\middle|\, d_i < 0 \right\} & \text{otherwise.} \end{cases}$$

Find a new point $\boldsymbol{x}_{k+1} = \boldsymbol{x}_k + \lambda_k \mathbf{d}$ ($0 < \lambda_k \le \bar{\lambda}_k$) by a normal line search procedure.

Finding a reducible basis is a key step in the reduced gradient method. We still have two questions to consider:

(1) Does a reducible basis exist for any feasible point?

(2) How can we find a reducible basis?

We will answer them below:

Consider a basis I. If $A_I^{-1}\boldsymbol{b} \ge \mathbf{0}$, then I is called a *feasible basis*. For a feasible basis I, setting $\boldsymbol{x}_I = A_I^{-1}\boldsymbol{b}$ and $\boldsymbol{x}_{\bar{I}} = \mathbf{0}$ gives a feasible point $\boldsymbol{x}$. This point is called a *basic feasible solution*. A basic feasible solution is said to be *degenerate* if $A_I^{-1}\boldsymbol{b}$ has a zero component. Clearly, for a degenerate basic feasible solution, there does not exist a reducible basis. Therefore, with the reduced gradient method, we have to make the following assumption.

Nondegeneracy Assumption. Every basic feasible solution is nondegenerate.

The next theorem shows that the above assumption is also sufficient for the existence of a reducible basis at any point. In fact, it gives even a stronger result.

Theorem 7.3.1 *Suppose that the nondegeneracy assumption holds for $\Omega = \{\boldsymbol{x} \mid A\boldsymbol{x} = \boldsymbol{b}, \boldsymbol{x} \ge \mathbf{0}\}$. Then there exists a positive number ε such that for every $\boldsymbol{x} \in \Omega$, rank$(A_{I_\varepsilon(\boldsymbol{x})}) = m$ where $I_\varepsilon(\boldsymbol{x}) = \{i \mid x_i > \varepsilon\}$.*

To prove the theorem, let us first prove two lemmas. A polyhedron is called a *pointed* polyhedron if no line lies entirely in it. A point $\boldsymbol{x}$ in a polyhedron Q is called a *extreme point* if there do not exist $\boldsymbol{y}$ and $\boldsymbol{z}$ in Q such that $\boldsymbol{x}$ lies in the interior of segment $[\boldsymbol{y}, \boldsymbol{z}]$.

Lemma 7.3.2 *Let $\Omega = \{\boldsymbol{x} \mid A\boldsymbol{x} = \boldsymbol{b}, \boldsymbol{x} \geq \boldsymbol{0}\}$. Then Ω is pointed. Moreover, if Ω satisfies the nondegeneracy assumption, then $\boldsymbol{x}$ in Ω is an extreme point if and only if $\boldsymbol{x}$ is a basic feasible solution.*

Proof. Since $\boldsymbol{x} \geq \boldsymbol{0}$ for all $\boldsymbol{x} \in \Omega$, Ω is pointed. To prove the second part, we note that a feasible point $\boldsymbol{x}$ is not an extreme point if and only if there exists a nonzero vector $\boldsymbol{d}$ such that both $\boldsymbol{d}$ and $-\boldsymbol{d}$ are feasible directions at $\boldsymbol{x}$. The latter one is equivalent to the existence of nonzero solution for the equation $A_{I(\boldsymbol{x})}\boldsymbol{d}_{I(\boldsymbol{x})} = \boldsymbol{0}$ where $I(\boldsymbol{x}) = \{i \mid x_i > 0\}$. That is, $rank(A_{I(\boldsymbol{x})}) < |I(\boldsymbol{x})|$. With the nondegeneracy assumption, if $\boldsymbol{x}$ is a basic feasible direction, then clearly, $rank(A_{I(\boldsymbol{x})}) = |I(\boldsymbol{x})| = m$, so $\boldsymbol{x}$ is an extreme point. If $\boldsymbol{x}$ is an extreme point, then $rank(A_{I(\boldsymbol{x})}) = |I(\boldsymbol{x})|$. So, all column vectors of $A_{I(\boldsymbol{x})}$ are linearly independent. It follows that there is a basis I such that $I(\boldsymbol{x}) \subseteq I$. Clearly, $x_I = A_I^{-1}\boldsymbol{b}$ and $\boldsymbol{x}_{\bar{I}} = \boldsymbol{0}$. So $\boldsymbol{x}$ is a basic feasible solution. □

Lemma 7.3.3 *Let Q be a pointed polyhedron. Let $\boldsymbol{x}_1, \cdots, \boldsymbol{x}_u$ be all of its extreme points. Then there exist $\boldsymbol{d}_1, \cdots, \boldsymbol{d}_v$ in R^n such that $\boldsymbol{x} \in Q$ if and only if*

$$\boldsymbol{x} = \sum_{i=1}^{u} \lambda_i \boldsymbol{x}_i + \sum_{j=1}^{v} \mu_j \boldsymbol{d}_j \tag{7.16}$$

for some $\lambda_i \geq 0, \mu_j \geq 0$ and $\sum_{i=1}^{u} \lambda_i = 1$.

Proof. By Theorem 7.1.3, there exist $\boldsymbol{x}_i$'s and $\boldsymbol{d}_j$'s such that (7.16) holds. Our aim is to show that if $\boldsymbol{x}_u$ is not an extreme point of Q, then $\boldsymbol{x}_u$ can be deleted from the statement. Suppose that $\boldsymbol{x}_u$ is not an extreme point. Note that Q does not contain any line. Thus, we can find $\boldsymbol{y}$ and $\boldsymbol{z}$ in Q such that $\boldsymbol{x}_u$ is in the interior of $[\boldsymbol{y}, \boldsymbol{z}]$ and $\boldsymbol{y} + \lambda(\boldsymbol{y} - \boldsymbol{x}_u)$ is not in Q for any positive λ. Denote

$$Q' = \left\{ \sum_{i=1}^{u-1} \lambda_i \boldsymbol{x}_i + \sum_{j=1}^{v} \mu_j \boldsymbol{d}_j \;\middle|\; \lambda_i \geq 0, \mu_j \geq 0, \sum_{j=1}^{u-1} \lambda_i = 1 \right\}.$$

We claim $\boldsymbol{y} \in Q'$. For contradiction, suppose $\boldsymbol{y} \notin Q'$. Then

$$\boldsymbol{y} = \sum_{i=1}^{u} \lambda_i \boldsymbol{x}_i + \sum_{j=1}^{v} \mu_j \boldsymbol{d}_j$$

with $\lambda_u > 0$, $\lambda_i \geq 0$, $\sum_{i=1}^{u} \lambda_i = 1$ and $\mu_j \geq 0$. Thus,

$$\boldsymbol{y} + \frac{\lambda_u}{2}(\boldsymbol{y} - \boldsymbol{x}_u) = \frac{1}{1 - \frac{\lambda_u}{2}} \left(\sum_{i=1}^{u-1} \lambda_i \boldsymbol{x}_i + \frac{\lambda_u}{2} \boldsymbol{x}_u + \sum_{j=1}^{v} \mu_j \boldsymbol{d}_j \right)$$

belongs to Q, contradicting the assumption on $\boldsymbol{y}$. Hence $\boldsymbol{y} \in Q'$. Next, we show $\boldsymbol{z} \in Q'$. Since $\boldsymbol{z} \in Q$, we have

$$\boldsymbol{z} = \sum_{i=1}^{u} \lambda_i \boldsymbol{x}_i + \sum_{j=1}^{v} \mu_j \boldsymbol{d}_j \tag{7.17}$$

for some $\lambda_i \geq 0$, $\mu_j \geq 0$ and $\sum_{i=1}^{u} \lambda_i = 1$. Note that $\boldsymbol{x} = (1-\alpha)\boldsymbol{y} + \alpha \boldsymbol{z}$ for some $0 < \alpha < 1$. So $\boldsymbol{z} = [\boldsymbol{x} - (1-\alpha)\boldsymbol{y}]/\alpha$. Substituting it into (7.17), we obtain

$$\boldsymbol{z} = \frac{1 - \lambda_u}{1 - \lambda_u \alpha} \boldsymbol{x}' + \frac{\lambda_u (1 - \alpha)}{1 - \lambda_u \alpha} \boldsymbol{y}$$

where

$$\boldsymbol{x}' = \frac{1}{1 - \lambda_u} \left(\sum_{i=1}^{u-1} \lambda_i \boldsymbol{x}_i + \sum_{j=1}^{v} \mu_j \boldsymbol{d}_j \right).$$

Clearly, $\boldsymbol{x}' \in Q'$. Since Q' is convex, $\boldsymbol{z} \in Q'$ and $\boldsymbol{x}_u \in Q'$. It follows that $Q = Q'$. □

Proof of Theorem 7.3.1. Using Lemma 7.3.3 instead of Theorem 7.1.3 in the proof of Theorem 7.1.4, we can obtain the existence of $\varepsilon > 0$ such that for any $\boldsymbol{x} \in \Omega$, there exists an extreme point $\boldsymbol{x}_i$ such that $I_\varepsilon(\boldsymbol{x}) \supseteq I(\boldsymbol{x}_i)$ where $I(\boldsymbol{x}_u) = I_0(\boldsymbol{x}_u)$. By the nondegeneracy assumption, $rank(I(\boldsymbol{x}_u)) = m$. Hence $rank(I_\varepsilon(\boldsymbol{x})) = m$. □

There are several ways for finding a reducible basis. Here, we introduce one of them.

Yue-Han's Pivot: *Let I_{k-1} be a basis obtained from the previous iteration. Denote $T(I) = (A_I)^{-1}A$ and denote by $T(I)_{ij}$ the element of $T(I)$ at the ith row and the jth column. Then a reducible basis I_k for $\boldsymbol{x}_k$ is computed as follows:*

begin
$\boldsymbol{x} := \boldsymbol{x}_k$; $I := I_{k-1}$;
repeat until $x_t \geq \frac{1}{2}x_s$ *(where* $x_t = \min\{x_i \mid i \in I\}$ *and* $x_s = \max\{x_i \mid i \in \bar{I}, T(I)_{ti} \neq 0\}$*)*
do $I := (I \cup s) \setminus t$;
$I_k := I$;
end.

The Yue-Han's pivot method has several nice properties. Let us show some of them in the following.

Proposition 7.3.4 *The loop in Yue-Han's pivot method operates in finitely many times.*

Proof. To keep the loop operating, we must have $x_s > 2x_t$. This means that $\sum_{i \in I} x_i$ is strictly increasing as the loop keeps going. However, there are only finitely many such sums. Thus, the loop must halt in finitely many times. □

Proposition 7.3.5 *Let ε be the positive number in Theorem 7.3.1. Then for every $i \in I_k$, $x_i > \varepsilon/2$.*

Proof. Consider $I = I_k$. Suppose $x_t = \min\{x_i \mid i \in I\} \leq \varepsilon/2$. Then we must have $T(I)_{ti} = 0$ for $i \in \bar{I} \cap I_\varepsilon(\boldsymbol{x})$. Moreover, $T(I)_{ti} = 0$ for every $i \in I \setminus t$. Hence $T(I)_{ti} = 0$ for every $i \in I_\varepsilon(\boldsymbol{x})$, that is, the t-th row of $A_I^{-1}A_{I_\varepsilon(\boldsymbol{x})}$ is a zero vector. Therefore,

$$rank(A_{I_\varepsilon(\boldsymbol{x})}) = rank(A_I^{-1}A_{I_\varepsilon(\boldsymbol{x})}) < m$$

contradicting Theorem 7.3.1. □

Proposition 7.3.6 *Let $\{\boldsymbol{x}_k\}$ be a sequence converging to $\boldsymbol{x}^*$. Let $\{I_k\}$ be the sequence of reducible bases obtained by Yue-Han's pivot. Then there exists k' such that for $k \geq k'$, $I_k = I_{k+1}$.*

Proof. For contradiction, suppose that such a number k' does not exist. Since the number of indices is finite, there exists an index i_0 such that $i_0 \in I_k \setminus I_{k+1}$ for infinitely many k. Let K be the set of such k's. For each $k \in K$, let s_k denote the index which gets in the basis I_k while i_0 gets out from I_k. There must exists an infinite subset K' of K such that $s_k = i_1$ for all $k \in K'$ and a certain index i_1. From the definition of s_k, we have

$$x_{ki_0} < \frac{1}{2} x_{ki_1} \text{ for } k \in K' \tag{7.18}$$

where x_{ki} is the ith component of $\boldsymbol{x}_k$. Letting k go to ∞ through K', we obtain $x^*_{i_0} \leq \frac{1}{2} x^*_{i_1}$. Since i_1 gets in the reducible basis for infinitely many times, it must also get out for infinitely many times. Thus, by the above argument, we can find an index i_2 such that $x^*_{i_1} \leq \frac{1}{2} x^*_{i_2}$. In this way we can find an infinite sequence of indices $i_0, i_1, \cdots$ such that

$$x^*_{i_0} \leq \frac{1}{2} x^*_{i_1} \leq \cdots \leq (\frac{1}{2})^j x^*_{i_j} \leq \cdots.$$

Thus, for any j,

$$0 \leq x^*_{i_0} \leq (\frac{1}{2})^j \max\{x^*_i \mid i = 1, \cdots, n\}.$$

Letting $j \to \infty$, we obtain $x^*_{i_0} = 0$. However, since $i_0 \in I_k$ for $k \in K$, by Proposition 7.3.5, we have $x_{ki_0} > \varepsilon/2$ for $k \in K$. Letting $k \to \infty$, we obtain $x_{ki_0} \geq \varepsilon/2 > 0$, a contradiction. □

Next, we study the convergence of the reduced gradient method with Yue-Han's pivot. Clearly, if the computation stops at $\boldsymbol{x}_k$, then $\boldsymbol{x}_k$ is a Kuhn-Tucker point. Next, we consider the case where an infinite sequence is generated.

Lemma 7.3.7 *The reduced gradient method satisfies the following:*

$$\left.\begin{array}{l} \boldsymbol{x}_k \to \boldsymbol{x}^* \\ \boldsymbol{d}_k \to \boldsymbol{d}^* \\ (\boldsymbol{g}^*)^T \boldsymbol{d}^* = 0 \end{array}\right\} \Longrightarrow \boldsymbol{x}^* \textit{ is a Kuhn-Tucker point.} \tag{7.19}$$

Proof. Suppose that $\{\boldsymbol{x}_k\}_{k\in K}$ is a subsequence such that $\lim_{k\to\infty,k\in K} \boldsymbol{x}_k = \boldsymbol{x}^*$ and $\lim_{k\to\infty,k\in K} \boldsymbol{d}_k = \boldsymbol{d}^*$. Then we can find an infinite subset K' of K such that for $k \in K'$, $I_k = I$ and $J(\boldsymbol{x}_k) = J$ for certain I and J. Denote $\boldsymbol{r}_k = \nabla \bar{f}((\boldsymbol{x}_k)_{\bar{I}})$ and $\boldsymbol{r}^* = \nabla \bar{f}(\boldsymbol{x}^*_{\bar{I}})$. Then for $r^*_i > 0$, $r_{ki} > 0$ for sufficiently large $k \in K$. Thus, for $r^*_i > 0$, $r^*_i = \lim_{k\to\infty,k\in K'} d_{ki} = d^*_i$. Similarly, for $r^*_i < 0$ and $j \notin J$, we also have $r^*_i = d^*_i$. If $\|\boldsymbol{d}^*\|^2 = (\boldsymbol{g}^*)^T \boldsymbol{d}^* = 0$, then $r^*_i \leq 0$ for any $i \in \bar{I}$ and $r^*_i = 0$ for any $i \notin J$. Note that $\boldsymbol{r}^* = \boldsymbol{g}^*_{\bar{I}} - (A_I^{-1} A_{\bar{I}})^T \boldsymbol{g}^*_I$ and $0 = \boldsymbol{g}^*_I - (A_I^{-1} A_I) \boldsymbol{g}^*_I$. Thus, $\boldsymbol{g}^* = \sum_{i\in J} r^* \boldsymbol{e}_i + A^T (A_I^{-1})^T \boldsymbol{g}^*_I$ and $r^*_i \leq 0$. So, $\boldsymbol{x}^*$ is a Kuhn-Tucker point. □

Theorem 7.3.8 *The reduced gradient method with Yue-Han's pivot either stops at a Kuhn-Tucker point or generates an infinite sequence whose cluster points are all Kuhn-Tucker points.*

Proof. Obviously, if the computation stops at $\boldsymbol{x}_k$, then $\boldsymbol{x}_k$ is a Kuhn-Tucker point. If the sequence is not convergent, then its cluster points are all Kuhn-Tucker points. Thus, we have to discuss the case where the sequence converges to a point $\boldsymbol{x}^*$. In this case, the lengths of search directions are bounded by a number B and the search step length λ_k converges to 0. Moreover, by Proposition 7.3.5, we have

$$\bar{\lambda}_k \geq \min(1, \frac{\varepsilon}{2B})$$

where ε is a positive number. Hence, for sufficiently large k, $\lambda_k < \bar{\lambda}_k$. By Theorem 5.2.3, there exists a subsequence satisfying the conditions on the left hand side of (7.19). Thus, $\boldsymbol{x}^*$ is a Kuhn-Tucker point. □

There is a special interesting property of the reduced gradient method.

Proposition 7.3.9 *Suppose that the objective function f is concave. Then the sequence generated by the reduced gradient method with the Yue-Han's pivot and Armijo's rule, is convergent if and only if for sufficiently large k, $I_k = I_{k+1}$.*

Proof. By Proposition 7.3.6, we need only to prove the backward direction. When I_k is fixed, we may assume without loss of generality that the problem is in the form of (7.11). Therefore, (7.15) is equivalent to

$$d_i = \begin{cases} g_i & \text{if } -g_i \leq x_i, \\ -x_i & \text{otherwise}, \end{cases}$$

where $\boldsymbol{g} = \boldsymbol{g}(\boldsymbol{x})$. We first claim that for any feasible point $\boldsymbol{y}$,

$$\|\boldsymbol{d} - \boldsymbol{g}\| \leq \|\boldsymbol{y} - \boldsymbol{x} - \boldsymbol{g}\|. \tag{7.20}$$

In fact,

$$\|\boldsymbol{d} - \boldsymbol{g}\|^2 = \sum_{x_i + g_i < 0} (x_i + g_i)^2 \leq \sum_{x_i + g_i < 0} (y_i - x_i - g_i)^2 \leq \|\boldsymbol{y} - \boldsymbol{x} - \boldsymbol{g}\|^2.$$

Denote $\boldsymbol{z} = \boldsymbol{x} + \boldsymbol{d}$. Then $\boldsymbol{z}$ is a feasible point. Thus, for any feasible point $\boldsymbol{y}$ and $0 \leq \lambda \leq 1$, $\boldsymbol{z} + \lambda(\boldsymbol{y} - \boldsymbol{z})$ is a feasible point. By (7.20),

$$\|\boldsymbol{d} - \boldsymbol{g}\| \leq \|\lambda(\boldsymbol{y} - \boldsymbol{z}) + \boldsymbol{d} - \boldsymbol{g}\|$$

for $0 \leq \lambda \leq 1$. It follows that

$$(\boldsymbol{d} - \boldsymbol{g})^T(\boldsymbol{z} - \boldsymbol{y}) \leq \frac{\lambda}{2}\|\boldsymbol{y} - \boldsymbol{z}\|^2.$$

Letting $\lambda \to 0$, we obtain $(\boldsymbol{d}-\boldsymbol{g})^T(\boldsymbol{z}-\boldsymbol{y}) \leq 0$, that is, $\boldsymbol{d}^T(\boldsymbol{z}-\boldsymbol{y}) \leq \boldsymbol{g}^T(\boldsymbol{z}-\boldsymbol{y})$. Now, we consider the ith iteration. Let $\boldsymbol{z}_k = \boldsymbol{x}_k + \boldsymbol{d}_k$. Then we have

$$\begin{aligned}
\|\boldsymbol{x}_{k+1} - \boldsymbol{y}\|^2 &= \|\boldsymbol{x}_k - \boldsymbol{y}\|^2 + 2\lambda_k \boldsymbol{d}_k^T(\boldsymbol{x}_k - \boldsymbol{y}) + \lambda_k^2\|\boldsymbol{d}_k\|^2 \\
&\leq \|\boldsymbol{x}_k - \boldsymbol{y}\|^2 + 2\lambda_k \boldsymbol{d}_k^T(\boldsymbol{z}_k - \boldsymbol{y}) \\
&\leq \|\boldsymbol{x}_k - \boldsymbol{y}\|^2 + 2\lambda_k \boldsymbol{g}_k^T(\boldsymbol{z}_k - \boldsymbol{y}) \\
&= \|\boldsymbol{x}_k - \boldsymbol{y}\|^2 + 2\lambda_k \boldsymbol{g}_k^T \boldsymbol{d}_k + 2\lambda_k \boldsymbol{g}_k^T(\boldsymbol{x}_k - \boldsymbol{y}) \\
&\leq \|\boldsymbol{x}_k - \boldsymbol{y}\|^2 + 4[f(\boldsymbol{x}_{k+1}) - f(\boldsymbol{x}_k)] + 2\lambda_k[f(\boldsymbol{x}_k) - f(\boldsymbol{y})].
\end{aligned}$$

The last inequality follows from the concavity of f. Let $\boldsymbol{x}^*$ be a cluster point of $\{\boldsymbol{x}_k\}$. Then $\lim_{k\to\infty} f(\boldsymbol{x}_k) = f(\boldsymbol{x}^*)$ and $f(\boldsymbol{x}_k) \leq f(\boldsymbol{x}^*)$ for all k. Thus

$$\|\boldsymbol{x}_{k+1} - \boldsymbol{x}^*\|^2 \leq \|\boldsymbol{x}_k - \boldsymbol{x}^*\|^2 + 4[f(\boldsymbol{x}_{k+1}) - f(\boldsymbol{x}_k)],$$

that is,

$$\|\boldsymbol{x}_{k+1} - \boldsymbol{x}^*\|^2 - 4f(\boldsymbol{x}_{k+1}) \leq \|\boldsymbol{x}_k - \boldsymbol{x}^*\|^2 - 4f(\boldsymbol{x}_k).$$

This means that $\{\|\boldsymbol{x}_k - \boldsymbol{x}^*\|^2 - 4f(\boldsymbol{x}_k)\}$ is a monotonic sequence. Moreover, this sequence has a subsequence converging to $-4f(\boldsymbol{x}^*)$. Therefore, the sequence itself should converge to $-4f(\boldsymbol{x}^*)$. If follows that $\|\boldsymbol{x}_k - \boldsymbol{x}^*\|$ converges to 0. □

Exercises

1. Prove that for any linearly constrained set Ω, there exists a positive number ε such that for any $\boldsymbol{x} \in \Omega$ and any $J \subseteq J_\varepsilon(\boldsymbol{x})$, the equation $A_J^T \boldsymbol{y} = \boldsymbol{b}_J$ is consistent.

2. Consider the following line search procedure: At iteration k, $\boldsymbol{x}_{k+1}$ is chosen to satisfy $f(\boldsymbol{x}_{k+1}) > f(\boldsymbol{x}_k)$ and $f(\boldsymbol{x}_{k+1}) \geq \max_{0\leq\lambda\leq\bar{\lambda}_k} f(\boldsymbol{x}_k + \lambda \boldsymbol{d}_k) - 1/k$.

(a) Prove the global convergence of the RD method with the above line search procedure.

(b) Disprove the global convergence of Rosen's method with the above line search procedure.

3. Extend the RD method to design an algorithm for (7.6). Prove the global convergence of the extension.

4. Consider the problem (4.1) and the following ε-procedure:

begin
$J := \hat{J}_{k-1}$;
repeat until $\boldsymbol{a}_t^T\boldsymbol{x}_k - b_t \geq \frac{1}{2}(\boldsymbol{a}_s^T\boldsymbol{x}_k - b_s)$ **do** $J := (J \cup t) \setminus s$;
(where

$$\begin{aligned}&\boldsymbol{a}_t^T\boldsymbol{x}_k - b_t = \min\{\boldsymbol{a}_j^T\boldsymbol{x}_k - b_j \mid j \notin J\},\\&(\mu_j, j \in J)^T = (A_J^TA_J)^{-1}A_J^T\boldsymbol{a}_t,\\&\boldsymbol{a}_s^T\boldsymbol{x}_k - b_s = \max\{\boldsymbol{a}_j^T\boldsymbol{x} - b_j \mid \mu_j \neq 0, j \in J\};)\end{aligned}$$

$\hat{J}_k := J$;
end.

Prove that for a convergent sequence $\{\boldsymbol{x}_k\}$, $\hat{J}_{k-1} = \hat{J}_k$ for k sufficiently large.

5. There are several variations of reduced gradient methods in the literature. The earliest variation is given by formula (7.13) instead of (7.15). For this variation, show that $\boldsymbol{g}^T\boldsymbol{d} = \|\boldsymbol{d}\|^2$ and that if the algorithm generates an infinit sequence not convergent, then every cluster point of the sequence is a Kuhn-Tucker point.

6. Consider a simple case (without equality constraints) in the following form:

$$\begin{array}{ll}\text{maximize} & f(\boldsymbol{x})\\ \text{subject to} & \boldsymbol{x} \geq \boldsymbol{0}.\end{array} \tag{7.21}$$

When the reduced gradient method with formula (7.13) is applied to this case, the search direction $\boldsymbol{d}_k$ at $\boldsymbol{x}_k$ is determined as follows:

$$d_{ki} = \begin{cases} 0 & \text{if } x_{ki} = 0 \text{ and } g_{ki} < 0,\\ g_{ki} & \text{otherwise.}\end{cases}$$

Prove that if a generated infinite sequence $\{\boldsymbol{x}_k\}$ convergent to $\boldsymbol{x}^*$ with $g_i^* \neq 0$ for all $x^* = 0$, then $\boldsymbol{x}^*$ is a Kuhn-Tucker point.

7. Prove that if the reduced gradient method with formula (7.13) generates a convergent infinite sequence, then it converges to either a Kuhn-Tucker point or a feasible point at which for some reducible basis the reduced gradient has a zero component with index in its active set.

8. Show by a counterexample that the reduced gradient method with formula (7.13) may generate a sequence convergent to a point which is not a Kuhn-Tucker point.

9. Prove that if formula (7.15) is replaced by

$$d_i = \begin{cases} r_i & \text{if } r_i \geq 0, \\ r_i x_i & \text{otherwise} \end{cases} \tag{7.22}$$

then the resulting variation is still globally convergent.

10. The reduced gradient method allows us to release more active constraints than Rosen's method does at each iteration. This may give an advantage in practice. By this observation, a combination of gradient projection and reduced gradient methods have been proposed as follows:

DS Method: *Given a positive number ε and an initial feasible point $\boldsymbol{x}_1$, at each iteration $k = 1, 2, \cdots$, the following steps are carried out:*

Step 1 *Compute a positive number ε_k by the ϵ-procedure L. Let $J := J_{\varepsilon_k}(\boldsymbol{x}_k)$, $\boldsymbol{x} := \boldsymbol{x}_k$ and $\boldsymbol{g} := \boldsymbol{g}_k$. Then compute the search direction $\boldsymbol{d}_k$ as follows:*

$$B_J = (A_J^T A_J)^{-1} A_J^T,$$
$$(u_j, j \in J)^T = B_J \boldsymbol{g},$$
$$\boldsymbol{d}_k = P_J \boldsymbol{g} + B_J^T \boldsymbol{v}$$

where $\boldsymbol{v} = (v_j, j \in J)^T$ is defined by

$$v_j = \begin{cases} u_j & \text{if } u_j \geq 0 \text{ and } j \in J \setminus M', \\ b_j - \boldsymbol{a}_j^T \boldsymbol{x} & \text{otherwise.} \end{cases}$$

If $\boldsymbol{d}_k = \mathbf{0}$, *then stops; otherwise, go to the next step.*

Step 2 *Defined* $\bar{\lambda}_k$ *as shown in Rosen's method. Find a new point* $\boldsymbol{x}_{k+1} = \boldsymbol{x}_k + \lambda_k \boldsymbol{d}_k$ $(0 < \lambda_k \leq \bar{\lambda}_k)$ *by a normal line search procedure.*

Prove the following:

(a) Under the regularity condition, the DS method either stops at a Kuhn-Tucker point or generates an infinite sequence whose cluster points are all Kuhn-Tucker points.

(b) This method can also be extended by the rotating tangent plane method to solve the nonlinearly constrained problem (7.6).

Chapter 8

Efficiency

The really efficient laborer
will be found not to crowed his day with work, ...
THOREAU

Global convergence indicates whether an algorithm is able to find an optimal or locally optimal solution for a nonlinear optimization problem. The rate of convergence is another mathematical measure which evaluates the computational efficiency of an iterative algorithm.

8.1 Rate of Convergence

Consider a sequence $\{\boldsymbol{x}_k\}$ with $\boldsymbol{x}_k \in R^n$ that converges to a point $\boldsymbol{x}^*$. Suppose $\boldsymbol{x}_k \neq \boldsymbol{x}^*$ for sufficiently large k. If there exists a number r such that for sufficiently large k

$$\alpha \leq \frac{\|\boldsymbol{x}_{k+1} - \boldsymbol{x}^*\|}{\|\boldsymbol{x}_k - \boldsymbol{x}^*\|^r} \leq \beta$$

for some positive numbers α and β, then r is called the *order* of convergence of sequence $\{\boldsymbol{x}_k\}$. When $r = 1$, we say that the rate of convergence is *linear*. When $r = 2$, the rate of convergence is *quadratic*. The rate of convergence is said to be *superlinear* if

$$\lim_{k \to \infty} \frac{\|\boldsymbol{x}_{k+1} - \boldsymbol{x}^*\|}{\|\boldsymbol{x}_k - \boldsymbol{x}^*\|} = 0.$$

An algorithm is said to be *linearly* convergent if every sequence generated by the algorithm has at least linear rate of convergence. Similarly, an algorithm is said to be *quadratically* convergent if every sequence generated by the algorithm has at least quadratic rate of convergence, and is said to be *superlinearly* convergent if every sequence generated by the algorithm has at least superlinear rate of convergence.

Let us look at an example. *Newton's method* is a classical technique for solving nonlinear equations. Suppose we want to find a root of $g(x) = 0$ with $x \in R^1$. Newton's method can be described as follows:
Initially, we guess a solution at x_1. For each iteration with x_k, we linearize $f(x)$ by the tangent line at x_k

$$g(x_k) + g'(x_k)(x - x_k) = 0.$$

Then x_{k+1} is given by the root of this equation

$$x_{k+1} = x_k - \frac{g(x_k)}{g'(x_k)}.$$

Suppose x_k converges to x^*. Then x^* is a fixed point of the function

$$\phi(x) = x - \frac{g(x)}{g'(x)}.$$

By the mean value theorem,

$$x_{k+1} - x^* = \phi(x_k) - \phi(x^*) = \phi(\tau_k)(x_k - x^*)$$

for some τ_k lying between x_k and x^*. Note that

$$\phi(\tau_k) = \frac{g''(\tau_k)g(\tau_k)}{g'(\tau_k)^2}$$

and

$$|g(\tau_k)| = |g(\tau_k) - g(x^*)| = |g'(\zeta_k)| \cdot |\tau_k - x^*| \leq |g'(\zeta_k)| \cdot |x_k - x^*|$$

for some ζ_k lying between τ_k and x^*. Therefore,

$$|x_{k+1} - x^*| \leq \frac{|g''(\tau_k)g'(\zeta_k)|}{g'(\tau_k)^2}|x_k - x^*|^2.$$

If g satisfies the condition that

$$\frac{|g''(\tau)g'(\zeta)|}{g'(\tau)^2}$$

is bounded for τ and ζ in a certain neighborhood of x^*, then

$$\eta = sup_{1 \le k < \infty} \frac{|g''(\tau_k)g'(\zeta_k)|}{g'(\tau_k)^2}$$

is bounded and

$$|x_{k+1} - x^*| \le \eta |x_k - x^*|^2.$$

This means that Newton's method is quadratically convergent under certain conditions.

Newton's method can be applied to equation $f'(x) = 0$ for solving maximization problem $\max f(x)$.

8.2 Linear Rate

All algorithms introduced in previous chapters are linearly convergent. Here, we consider Rosen's method as an example to show this fact. Other methods can be shown in similar ways.

A regular Kuhn-Tucker point $\boldsymbol{x}^*$ satisfies the *strict complementary slackness condition* if $\boldsymbol{g}^* = \sum_{j \in J^*} u_j \boldsymbol{a}_j$ and $u_j < 0$ for $j \in J^* \setminus M'$. Clearly, if $\boldsymbol{x}^*$ satisfies the strict complementary slackness condition, then for any proper subset J of J^*, $P_J \boldsymbol{g}^* \neq \mathbf{0}$.

Lemma 8.2.1 *Suppose that for all k, $\alpha \le c_k \le \beta$ for two positive numbers α and β. Let $\{\boldsymbol{x}_k\}$ be a sequence generated by Rosen's method with a normal line search procedure. Suppose that the sequence converges to a point $\boldsymbol{x}^*$ satisfying the strict complementary slackness condition. Then there exists k' such that for $k > k'$, $J_k = J^*$. Moreover, $P_{J^*} \boldsymbol{g}^* = \mathbf{0}$.*

Proof. We first show that there exists k' such that $k > k'$, $J_k = J$ for a certain J. For contradiction, suppose that such a k' does not exist. Then there are infinitely many k such that $J_k \neq J_{k+1}$. It follows that there are infinitely many k such that $J_k \setminus J_{k+1} \neq \emptyset$. Let K be the set of all such k's. We consider the following cases:

Case 1. There are infinitely many $k \in K$ such that $\lambda_k \neq \bar{\lambda}_k$. Since the line search procedure is normal, we can find $K' \subseteq K$ such that $\{\boldsymbol{d}_k\}_{k \in K'}$

converges to a vector $\boldsymbol{d}^*$ and $(\boldsymbol{g}^*)^T\boldsymbol{d}^* = 0$. However, $k \in K$ implies $\boldsymbol{d}_k = P_{J_k \setminus h_k}$. Without loss of generality, we may assume that $J_k = J$ and $h_k = h$ for $k \in K'$. Then $\boldsymbol{d}^* = P_{J\setminus h}\boldsymbol{g}$. Note that $J \subseteq J^*$. So, $J \setminus h$ is a proper subset of J^*. Hence $0 = (\boldsymbol{g}^*)^T\boldsymbol{d}^* = \|P_{J\setminus h}\boldsymbol{g}^*\|^2 > 0$, a contradiction.

Case 2. For sufficiently large $k \in K$, $\lambda_k = \bar{\lambda}_k$. Since $\lim_{k\to\infty} \boldsymbol{x}_k = \boldsymbol{x}^*$, $\lim_{k\to\infty} \lambda_k = 0$. Thus for sufficiently large $k \in K$, $J_{k+1} \setminus J_k \neq \emptyset$. Note that for $k \notin K$, $|J_k| \leq |J_{k+1}|$. It follows that for sufficiently large k, $|J_k| \leq |J_{k+1}|$. Thus there exists k' such that for $k > k'$, $|J_k| = |J_{k+1}| = s$ for a certain number s. Let J' and J'' be two index sets such that there are infinitely many $k \in K$ such that $J_k = J'$ and $J_{k+1} = J''$. Then $J' \subseteq J^*$, $J'' \subseteq J^*$ and $J' \neq J''$. Hence $|J^*| \geq |J' \cup J''| \geq s+1$. Next, consider any cluster point $\boldsymbol{d}^*$ of $\{\boldsymbol{d}_k\}$. Clearly, there exists a subsequence $\{\boldsymbol{d}_k\}_{k\in K'}$ converging to $\boldsymbol{d}^*$ such that for $k \in K'$, $J_k = J$ and $h_k = h$. Thus $J \subseteq J^*$, $|J| = s$ and $\boldsymbol{d}^* = P_J\boldsymbol{g}^*$ or $P_{J\setminus h}\boldsymbol{g}^*$. That is, J is a proper subset of J^* and hence $\boldsymbol{d}^* \neq \boldsymbol{0}$, contradicting Corollary 6.2.3.

Now, we have proved that there exists k' such that $k > k'$, $J_k = J$ for a certain J. Clearly, $J \subseteq J^*$. If $J \subset J^*$, then we can derive a contradiction from Corollary 6.2.3. Therefore $J = J^*$. By Corollary 6.2.3, we also see $P_{J^*}\boldsymbol{g}^* = \boldsymbol{0}$. □

Theorem 8.2.2 *Suppose that f is continuously second order differentiable and strictly concave function. Let $\{\boldsymbol{x}_k\}$ be an infinite sequence generated by Rosen's method with Armijo's rule as its line search procedure. If $\{\boldsymbol{x} \mid f(\boldsymbol{x}) \geq f(\boldsymbol{x}_1)\}$ is a bounded set, then the sequence converges to a Kuhn-Tucker point $\boldsymbol{x}^*$. Furthermore, if $\boldsymbol{x}^*$ satisfies the strict complementary slackness condition, then*

$$\limsup_{k\to\infty} \frac{f(\boldsymbol{x}^*) - f(\boldsymbol{x}_{k+1})}{f(\boldsymbol{x}^*) - f(\boldsymbol{x}_k)} \leq 1 - \frac{ab}{(a+b)^2}$$

where a and b are, respectively, the smallest and the largest eigenvalues of the Hession matrix $-H(\boldsymbol{x}^*)$ at $\boldsymbol{x}^*$ restricted to the subspace D_{J^*}.

Proof. By Theorem 6.2.1, every cluster point of $\{\boldsymbol{x}_k\}$ is a Kuhn-Tucker point. Since f is strictly concave, there exists at most one Kuhn-Tucker point. Moreover the sequence is bounded. Thus it converges to a Kuhn-Tucker point $\boldsymbol{x}^*$. Let k' be the number in Lemma 8.2.1. Since $\boldsymbol{x}_k \to \boldsymbol{x}^*$ implies $\lambda_k \to 0$, we can assume without loss of generality that for $k > k'$, $\lambda_k < \bar{\lambda}_k$. Since $\boldsymbol{g}_k - \boldsymbol{g}^* = \int_{\boldsymbol{x}^*}^{\boldsymbol{x}_k} H(\boldsymbol{x})d\boldsymbol{x}$ and $P_{J^*}\boldsymbol{g}^* = \boldsymbol{0}$, we have that for

$k > k'$,

$$\begin{aligned} \boldsymbol{d}_k &= P_{J^*}\boldsymbol{g}_k = \int_{\boldsymbol{x}^*}^{\boldsymbol{x}_k} P_{J^*}H(\boldsymbol{x})d\boldsymbol{x} = \int_{\boldsymbol{x}^*}^{\boldsymbol{x}_k} P_{J^*}H(\boldsymbol{x}^*)P_{J^*}d\boldsymbol{x} \\ &= P_{J^*}H(\boldsymbol{x}_k)P_{J^*}(\boldsymbol{x}_k - \boldsymbol{x}^*)(1 + o(1)). \end{aligned}$$

Since f is strictly concave, $H(\boldsymbol{x}_k)$ is invertible. The matrix $P_{J^*}H(\boldsymbol{x}_k)P_{J^*}$ may not be invertible. However, for any $\boldsymbol{y}$ the equation $P_{J^*}H(\boldsymbol{x}_k)P_{J^*}\boldsymbol{z} = \boldsymbol{y}$ has a unique solution in the subspace D_{J^*}. This means that $P_{J^*}H(\boldsymbol{x}_k)P_{J^*}$ is invertible in the subspace D_{J^*}. Denote $L(\boldsymbol{x}_k)$ to be the linear operator $-P_{J^*}H(\boldsymbol{x}_k)P_{J^*}$ restricted to D_{J^*} and $L^{-1}(\boldsymbol{x}_k)$ its inverse. Then we have

$$\boldsymbol{x}^* - \boldsymbol{x}_k = L^{-1}(\boldsymbol{x}_k)\boldsymbol{d}_k/(1 + o(1)).$$

Moreover, since $\boldsymbol{x}^* - \boldsymbol{x}_k \in D_{J^*}$ and $\boldsymbol{d}_k = P_{J^*}\boldsymbol{g}_k$, we have

$$\boldsymbol{g}_k^T(\boldsymbol{x}^* - \boldsymbol{x}_k) = \boldsymbol{d}_k^T L^{-1}(\boldsymbol{x}_k)\boldsymbol{d}_k/(1 + o(1)).$$

Hence

$$\begin{aligned} \frac{f(\boldsymbol{x}_{k+1}) - f(\boldsymbol{x}_k)}{f(\boldsymbol{x}^*) - f(\boldsymbol{x}_k)} &\geq \frac{\frac{1}{2}\lambda_k \boldsymbol{g}_k^T \boldsymbol{d}_k}{\boldsymbol{g}_k^T(\boldsymbol{x}^* - \boldsymbol{x}_k)(1 + O(\|\boldsymbol{x}^* - \boldsymbol{x}_k\|))} \\ &= \frac{\lambda_k}{2} \frac{\|\boldsymbol{d}_k\|^2(1 + o(1))}{\boldsymbol{d}_k^T L^{-1}(\boldsymbol{x}_k)\boldsymbol{d}_k(1 + O(\|\boldsymbol{x}^* - \boldsymbol{x}_k\|))}. \end{aligned}$$

Since $\lambda_k < \bar{\lambda}_k$, we have

$$\begin{aligned} \lambda_k \boldsymbol{g}_k^T \boldsymbol{d}_k &\geq f(\boldsymbol{x}_k + 2\lambda_k \boldsymbol{d}_k) - f(\boldsymbol{x}_k) \\ &= 2\lambda_k \boldsymbol{g}_k^T \boldsymbol{d}_k + 2\lambda_k^2 \boldsymbol{d}_k^T H(\boldsymbol{x}_k)\boldsymbol{d}_k(1 + o(1)). \end{aligned}$$

Thus

$$\lambda_k \geq \frac{\boldsymbol{g}_k^T \boldsymbol{d}_k}{-2\boldsymbol{d}_k^T H(\boldsymbol{x}_k)\boldsymbol{d}_k(1 + o(1))} = \frac{\|\boldsymbol{d}_k\|^2}{2\boldsymbol{d}_k^T L(\boldsymbol{x}_k)\boldsymbol{d}_k(1 + o(1))},$$

so

$$\begin{aligned} \frac{\lambda_k}{2} \frac{\|\boldsymbol{d}_k\|^2}{\boldsymbol{d}_k^T L^{-1}(\boldsymbol{x}_k)\boldsymbol{d}_k} &= \frac{\|\boldsymbol{d}_k\|^4}{4(\boldsymbol{d}_k^T L(\boldsymbol{x}_k)\boldsymbol{d}_k)(\boldsymbol{d}_k^T L^{-1}(\boldsymbol{x}_k)\boldsymbol{d}_k)(1 + o(1))} \\ &\geq \frac{a_k b_k}{(a_k + b_k)^2(1 + o(1))} \end{aligned}$$

where a_k and b_k are the smallest and the largest eigenvalue of $L(\boldsymbol{x}_k)$. The last inequality is the well-known Kontorovich inequality, which will be proved in the next section. Hence

$$\begin{aligned}\frac{f(\boldsymbol{x}^*) - f(\boldsymbol{x}_{k+1})}{f(\boldsymbol{x}^*) - f(\boldsymbol{x}_k)} &= 1 - \frac{f(\boldsymbol{x}_{k+1}) - f(\boldsymbol{x}_k)}{f(\boldsymbol{x}^*) - f(\boldsymbol{x}_k)} \\ &\leq 1 - \frac{a_k b_k}{(a_k + b_k)^2}(1 + o(1)).\end{aligned}$$

Note that $P_{J^*}H(\boldsymbol{x}_k)P_{J^*}$ restricted to D_{J^*} is the same as $H(\boldsymbol{x}_k)$ restricted to D_{J^*}. Letting $k \to \infty$, we obtain

$$\limsup_{k\to\infty} \frac{f(\boldsymbol{x}^*) - f(\boldsymbol{x}_{k+1})}{f(\boldsymbol{x}^*) - f(\boldsymbol{x}_k)} \leq 1 - \frac{ab}{(a+b)^2}.$$

□

8.3 Kantorovich Inequality

In this section we show Kontorovich inequality. Suppose $\phi(x)$ is a convex function on an interval $[a, b]$. Let $\lambda_1, \cdots, \lambda_n$ be n numbers in $[a, b]$. It is well-known that $\phi(\frac{1}{n}\sum_{i=1}^n \lambda_i) \leq \frac{1}{n}\sum_{i=1}^n \phi(\lambda_i)$. This means that 1 is a lower bound for the ratio $\frac{1}{n}\sum_{i=1}^n \phi(\lambda_i)/\phi(\sum_{i=1}^n \lambda_i)$. Now we are interested in establishing an upper bound.

Theorem 8.3.1 *Consider a positive convex function ϕ on an interval $[a, b]$ such that $\phi(a) \neq \phi(b)$. Let $\lambda_1, \cdots, \lambda_n$ be n numbers in $[a, b]$. Denote by ξ the number in $[a, b]$ such that the tangent line at point C $(\xi, \phi(\xi))$ intersects the x-axis at point D which is the intersection point of the x-axis and a line through A $(a, \phi(a))$ and B $(b, \phi(b))$ (see Figure 8.1). Then*

$$\frac{\sum_{i=1}^n \alpha_i \phi(\lambda_i)}{\phi(\sum_{i=1}^n \alpha_i \lambda_i)} \leq \frac{\phi(b) - \phi(a)}{\phi'(\xi)(b-a)}$$

where $\alpha_i \geq 0$ and $\sum_{i=1}^n \alpha_i = 1$.

Proof. Let S be the region bounded by the curve $y = \phi(x)$ and the chord between A and B. Since $\sum_{i=1}^n \alpha_i\lambda_i$ is a number in $[a, b]$, E $(\sum_{i=1}^n \alpha_i\lambda_i, \phi(\sum_{i=1}^n \alpha_i\lambda_i))$ is a point on the curve. Note, F $(\sum_{i=1}^n \alpha_i\lambda_i, \sum_{i=1}^n \alpha_i\phi(\lambda_i))$

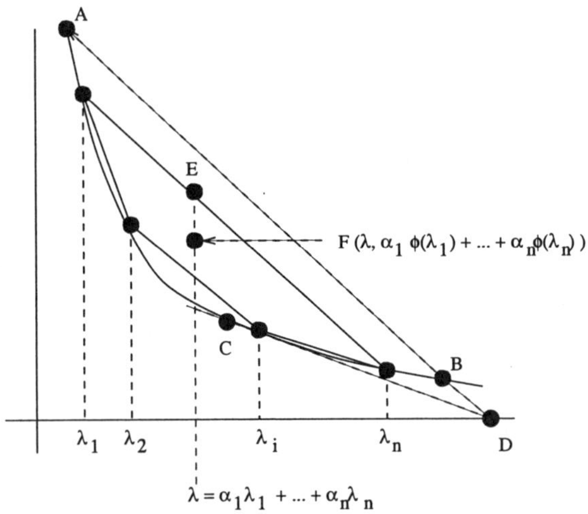

Figure 8.1: Kontorovich inequality.

is a convex linear combination of points on the curve and must lie in the region. Thus

$$\frac{\sum_{i=1}^n \alpha_i \phi(\lambda_i)}{\phi(\sum_{i=1}^n \alpha_i \lambda_i)} \leq \frac{|\text{ slope of line } FD|}{|\text{ slope of line } ED|}.$$

However, for any point X in the region S, we have

$$|\text{ slope of line } CD| \leq |\text{ slope of line } XD| \leq |\text{ slope of line } AD|.$$

Moreover, the slopes of the lines CD and AD have the same sign. Therefore

$$\frac{\sum_{i=1}^n \alpha_i \phi(\lambda_i)}{\phi(\sum_{i=1}^n \alpha_i \lambda_i)} \leq \frac{|\text{ slope of line } AD|}{|\text{ slope of line } CD|} = \frac{\phi(b) - \phi(a)}{\phi'(\xi)(b-a)}.$$

□

It is easy to verify that the number ξ in Theorem 8.3.1 should satisfy the following equation:

$$\frac{\phi(\xi)}{\phi'(\xi)} = \xi - \frac{\phi(b)a - \phi(a)b}{\phi(b) - \phi(a)}.$$

When $\phi(x) = x^k$ ($k \geq -1$ and $k \neq 0$), the equation is easily solved. For $k = -1$, we obtain $\xi = (a+b)/2$.

Corollary 8.3.2 *Let Q be a positive definite symmetric $n \times n$ matrix. Then for any vector $\boldsymbol{x}$, there holds*

$$(\boldsymbol{x}^T Q\boldsymbol{x})(\boldsymbol{x}^T Q^{-1}\boldsymbol{x}) \leq \frac{(a+b)^2}{4ab}(\boldsymbol{x}^T\boldsymbol{x})^2$$

where a and b are the smallest and the largest eigenvalues of Q, respectively.

Proof. Let $\lambda_1, \cdots, \lambda_n$ be eigenvalues of Q with $a = \lambda_1 \leq \cdots \leq \lambda_n = b$. Clearly, there exists an orthogonal matrix P such that $Q = P^T diag(\lambda_1, \cdots, \lambda_n)P$. Denote $\boldsymbol{y} = P\boldsymbol{x}$. We obtain that

$$\begin{aligned} \frac{(\boldsymbol{x}^T Q\boldsymbol{x})(\boldsymbol{x}^T Q^{-1}\boldsymbol{x})}{(\boldsymbol{x}^T\boldsymbol{x})^2} &= \frac{(\sum_{i=1}^n \lambda_i \boldsymbol{y}_{i=1}^2)(\sum_{i=1}^n \boldsymbol{y}_i^2/\lambda_i)}{(\sum_{i=1}^n \boldsymbol{y}_i^2)^2} \\ &\leq -(\frac{a+b}{2})^2 \frac{1/b - 1/a}{b-a} = \frac{(a+b)^2}{4ab}. \end{aligned}$$

Exercises

1. Prove the same result as Theorem 8.2.2 for other line search procedures.

2. Determine convergence rate of the following sequences:

   (a) $x_k = (1/2)^k$.

   (b) $x_k = (1/k)^k$.

   (c) $x_k = (1/2)^{2^k}$.

   (d) $x_k = (1/2)^{2^{2^k}}$.

3. Let $f(x)$ be a continuously second differentiable function on $[x_1, x_2]$. Suppose there exists x^* in $[x_1, x_2]$ such that $f(x^*) = 0$ and $f'(x^*) \neq 0$. Consider the sequence $\{x_k\}$ generated by the following iteration

$$x_{k+1} = x_k - f(x_k) \cdot \frac{x_k - x_{k-1}}{f(x_k) - f(x_{k-1})}.$$

   Show that if $|x_2 - x_1|$ is sufficiently small, then x_k converges to x^* as k goes to infinity. Moreover, the convergence rate of the sequence is $\frac{1+\sqrt{5}}{2}$.

Chapter 9

Variable Metric Methods

One half of the world cannot understand
the pleasure of the other.
JANE AUSTEN

We learned in the last chapter that Newton's method has higher rate of convergence. Could we extend Newton's method to higher dimensions? We study this question in this chapter.

9.1 Newton's Method

Consider a twice differentiable real function $f(\boldsymbol{x})$ on R^n and a point $\boldsymbol{x}_k$ in R^n. By Taylor's theorem,

$$f(\boldsymbol{x}) = f(\boldsymbol{x}_k) + \boldsymbol{g}_k^T \boldsymbol{\sigma} + \frac{1}{2}\boldsymbol{\sigma}^T G_k \boldsymbol{\sigma} + o(\|\boldsymbol{\sigma}\|^2)$$

where $\boldsymbol{\sigma} = \boldsymbol{x} - \boldsymbol{x}_k$, $\boldsymbol{g}_k = \nabla f(\boldsymbol{x}_k)$, and $G_k = \nabla^2 f(\boldsymbol{x}_k)$. This means that $f(x)$ can be approximated by

$$q(\boldsymbol{\sigma}) = f(\boldsymbol{x}_k) + \boldsymbol{g}_k^T \boldsymbol{\sigma} + \frac{1}{2}\boldsymbol{\sigma}^T G_k \boldsymbol{\sigma}$$

when $\boldsymbol{\sigma}$ is small. Any stationary point of $q(\boldsymbol{\sigma})$ would satisfy

$$\nabla q(\boldsymbol{\sigma}) = \boldsymbol{g}_k + G_k \boldsymbol{\sigma} = \boldsymbol{0}.$$

Therefore, $\boldsymbol{\sigma} = -G_k^{-1}\boldsymbol{g}_k$. Set

$$\boldsymbol{x}_{k+1} = \boldsymbol{x}_k - G_k^{-1}\boldsymbol{g}_k.$$

This formula suggests an algorithm to find a minimum of f as follows.

Newton's Method: *Choose an initial point* $\boldsymbol{x}_1$ *and set* $k = 1$.

1. *Compute* $\boldsymbol{g}_k$. *If* $\boldsymbol{g}_k = \mathbf{0}$, *then stop; else, go to* **2**.
2. *Compute* G_k *and* $\boldsymbol{\sigma} = -G_k^{-1}\boldsymbol{g}_k$. *Set* $\boldsymbol{x}_{k+1} = \boldsymbol{x}_k + \boldsymbol{\sigma}$ *and* $k := k + 1$. *Go to* **1**.

When the initial point is chosen nearby a maximal point, Newton's method would produces a sequence of points convergent quickly to the maximal point.

Theorem 9.1.1 *Let* $f(\boldsymbol{x})$ *be a twice continuously differentiable real function on* R^n *whose Hessian satisfes the Lipschitz condition, i.e. there exists a constant* α *such that for every* $\boldsymbol{x}$ *and* $\boldsymbol{y}$ *in* R^n,

$$\|G(\boldsymbol{x}) - G(\boldsymbol{y})\| \le \alpha\|\boldsymbol{x} - \boldsymbol{y}\|$$

where $G(\boldsymbol{x}) = \nabla^2 f(\boldsymbol{x})$. *Suppose* $\boldsymbol{x}^*$ *is a maximal point of* $f(\boldsymbol{x})$ *and* $G(\boldsymbol{x}^*)$ *is negative definite. Then when* $\boldsymbol{x}_1$ *is sufficiently close to* $\boldsymbol{x}^*$, *Newton's method generates a sequence of points* $\{\boldsymbol{x}_k\}$ *quadratically convergent to* $\boldsymbol{x}^*$, *i.e.*

$$\frac{\|\boldsymbol{x}_{k+1} - \boldsymbol{x}^*\|}{\|\boldsymbol{x}_k - \boldsymbol{x}^*\|^2} = O(1).$$

Proof. First, we claim that

$$\boldsymbol{g}(\boldsymbol{x}_k + \boldsymbol{\sigma}) = \boldsymbol{g}_k + G_k\boldsymbol{\sigma} + O(\|\boldsymbol{\sigma}\|^2).$$

In fact, by the mean value theorem, for each vector $\boldsymbol{a}$ we have

$$\boldsymbol{a}^T\boldsymbol{g}(\boldsymbol{x}_k + \boldsymbol{\sigma}) = \boldsymbol{a}^T\boldsymbol{g}_k + \boldsymbol{a}^T G(\boldsymbol{x}_k + \theta\boldsymbol{\sigma})\boldsymbol{\sigma}$$

for some $\theta \in (0, 1)$. Thus,

$$\begin{aligned} & \boldsymbol{a}^T(\boldsymbol{g}(\boldsymbol{x}_k + \boldsymbol{\sigma}) - \boldsymbol{g}_k - G_k\boldsymbol{\sigma}) \\ = \; & \boldsymbol{a}^T(G(\boldsymbol{x}_k + \theta\boldsymbol{\sigma}) - G_k)\boldsymbol{\sigma} \\ \le \; & \|\boldsymbol{a}\| \cdot \|G(\boldsymbol{x}_k + \theta\boldsymbol{\sigma}) - G_k\| \cdot \|\boldsymbol{\sigma}\| \\ \le \; & \alpha\|\boldsymbol{a}\| \cdot \|\boldsymbol{\sigma}\|^2. \end{aligned}$$

Now, let $e_i^T = (\underbrace{0, \cdots, 0, 1}_{i}, 0, \cdots, 0)$. Then

$$\begin{aligned} & \|g(x_k + \sigma) - g_k - G_k\sigma\| \\ \leq \ & \sum_{i=1}^{n} e_i^T (g(x_k + \sigma) - g_k - G_k\sigma) \\ \leq \ & n\alpha\|\sigma\|^2. \end{aligned}$$

Since f is twice continuously differentiable and $G(x^*)$ is negative definite, we can choose a positive number β such that for x with $\|x - x^*\| \leq \beta$,

$$G(x) \text{ is negative definite,} \tag{9.1}$$

$$\|G(x)^{-1}\| \leq 2\|G(x^*)^{-1}\|, \tag{9.2}$$

$$2n\beta\alpha\|G(x^*)^{-1}\| < 1. \tag{9.3}$$

Now, choose x_1 in $B(\beta) = \{x \mid \|x - x^*\| \leq \beta\}$. We will prove by induction on k that $x_k \in B(\beta)$ and

$$\frac{\|x_{k+1} - x^*\|}{\|x_k - x^*\|} \leq 2n\alpha\|G(x^*)^{-1}\|. \tag{9.4}$$

Let us first show that if $x_k \in B(\beta)$, then (9.4) holds. Note that

$$0 = g^* = g_k + G_k(x^* - x_k) + r$$

where $\|r\| \leq n\alpha\|x^* - x_k\|^2$. By (9.2) and (9.3),

$$0 = G_k^{-1}g_k + (x^* - x_k) + G_k^{-1}r$$

where $\|G_k^{-1}r\| \leq 2n\alpha\|G(x^*)^{-1}\| \cdot \|x^* - x_k\|^2$. Note that $x_{k+1} = x_k - G_k^{-1}g_k$. Therefore,

$$\frac{\|x_{k+1} - x^*\|}{\|x_k - x^*\|^2} \leq 2n\alpha\|G(x^*)^{-1}\|.$$

Now, we show that $x_k \in B(\beta)$ implies $x_{k+1} \in B(\beta)$. In fact, by (9.4) and (9.3),

$$\|x_{k+1} - x^*\| \leq 2n\alpha\|G(x^*)^{-1}\| \cdot \|x_k - x^*\|^2 < \|x_k - x^*\| \leq \beta.$$

□

Newton's method is locally convergent, but not globally convergent. In fact, when the initial point is chosen far from a local minimum, Newton's method may generate a sequence of points going to nowhere. The sequence $\{\boldsymbol{x}_k\}$ generated by Newton's method may not be convergent and the sequence $\{f(\boldsymbol{x}_k)\}$ may not be increasing. A simple way to improve Newton's method is to use a limited-step formula

$$\boldsymbol{x}_{k+1} = \boldsymbol{x}_k - \theta G_k^{-1}\boldsymbol{g}_k$$

where θ is chosen such that $f(\boldsymbol{x}_{k+1}) > f(\boldsymbol{x}_k)$ or is chosen by a line search procedure.

Limited-Step Newton's Method: *Choose an initial point* $\boldsymbol{x}_1$ *and set* $k = 1$.

1. *Compute* $\boldsymbol{g}_k$. *If* $\boldsymbol{g}_k = \mathbf{0}$, *then stop; else, go to* **2**.
2. *Compute* G_k *and* $\mathbf{d}_k = -G_k^{-1}\boldsymbol{g}_k$. *Set* $\boldsymbol{x}_{k+1} = \boldsymbol{x}_k + \theta\mathbf{d}_k$ *and* $k := k+1$, *where* θ *is chosen by a line search. Go to* **1**.

Theorem 9.1.2 *Suppose* $f(\boldsymbol{x})$ *is a twice continuously differentiable and strictly concave real function on* R^n. *If* f *has a minimum, then the limited-step Newton method with a normal line search either stops at the maximum point or generates a sequence of points convergent to the maximum point.*

Proof. It is clear that the algorithm must stop at the maximum point. Next, suppose $\boldsymbol{x}^*$ is a cluster point of the infinite sequence $\{\boldsymbol{x}_k\}$ generated by the algorithm. Then there exists a subsequence $\{\boldsymbol{x}_k\}_{k\in K}$ convergent to $\boldsymbol{x}^*$. Since $f(\boldsymbol{x})$ is twice continuously differentiable, $\{\boldsymbol{d}_k\}_{k\in K}$ is convergent to $\boldsymbol{d}^* = -(G^*)^{-1}\boldsymbol{g}^*$ where $G^* = G(\boldsymbol{x}^*)$ and $\boldsymbol{g}^* = \nabla f(\boldsymbol{x}^*)$. By the normal property of the line search, we must have $-(\boldsymbol{g}^*)^T G^*\boldsymbol{g}^* = (\boldsymbol{g}^*)^T\boldsymbol{d}^* \leq 0$. Since $f(\boldsymbol{x})$ is strictly concave, G^* is negative definite. Therefore, $\boldsymbol{g}^* = 0$ and hence $\boldsymbol{x}^*$ is the maximum point.

Note that a strictly concave real function $f(\boldsymbol{x})$ has at most one maximum point and if its maximum exists, it has exactly one maximum point. Thus, if the generated sequence $\{\boldsymbol{x}_k\}$ is bounded, then it must converge to the maximum point. Next, we show that the sequence $\{\boldsymbol{x}_k\}$ is indeed bounded. Actually, we can prove that $L(\boldsymbol{x}_1) = \{\boldsymbol{x} \mid f(\boldsymbol{x}) \geq f(\boldsymbol{x}_1)\}$ is bounded.

In fact, since f is strictly concave, $L(\boldsymbol{x}_1)$ is a convex closed set. By contradiction, suppose $L(\boldsymbol{x}_1)$ is not bounded. Then we can find a sequence of points, $\{\boldsymbol{y}_k\}$, in $L(\boldsymbol{x}_1)$ such that $\|\boldsymbol{y}_k - \boldsymbol{x}^*\| > k$. Now, choose $\boldsymbol{z}_k$ on the segment $[\boldsymbol{x}^*, \boldsymbol{y}_k]$ such that $\|\boldsymbol{x}^* - \boldsymbol{z}_k\| = 1$. Then $\{\boldsymbol{z}_k\}$ is bounded and hence has a cluster point z^*. Clearly, $\|\boldsymbol{x}^* - \boldsymbol{z}^*\| = 1$. We claim that for any $\alpha \geq 0$, $\boldsymbol{x}^* + \alpha(\boldsymbol{z}^* - \boldsymbol{x}^*) \in L(\boldsymbol{x}_1)$. In fact, for sufficiently large k, $\boldsymbol{x}^* + \alpha(\boldsymbol{z}_k - \boldsymbol{x}^*) \in L(\boldsymbol{x}_1)$. Note that $\boldsymbol{x}^* + \alpha(\boldsymbol{z}^* - \boldsymbol{x}^*)$ must be a cluster point of $\boldsymbol{x}^* + \alpha(\boldsymbol{z}_k - \boldsymbol{x}^*)$. Thus, $\boldsymbol{x}^* + \alpha(\boldsymbol{z}^* - \boldsymbol{x}^*) \in L(\boldsymbol{x}_1)$. Now, since f is strictly concave, we have that for $\alpha > 1$,

$$f(\boldsymbol{x}_1) \leq f(\boldsymbol{x}^* + \alpha(\boldsymbol{z}^* - \boldsymbol{x}^*) < f(\boldsymbol{x}^*) + \alpha(f(\boldsymbol{z}^*) - f(\boldsymbol{x}^*).$$

Since $f(\boldsymbol{x}^*) > f(\boldsymbol{z}^*)$, the right hand of the above inequality goes to $-\infty$ as α goes to $+\infty$, a contradiction. □

In Theorem 9.1.2, the strict concavity of the objective function f is assumed. Without this assumption, the search direction $G_k^{-1}\boldsymbol{g}_k$ may not be an ascend direction and the algorithm is cycles. Of course, in such a case it is meaningless to study the global convergence of the limited-step Newton method. For example, consider a function

$$f(x_1, x_2) = x_1^4 + x_1 x_2 + (1 + x_2)^2.$$

Choose initial points $\boldsymbol{x}_1 = (0, 0)^T$. Then

$$\boldsymbol{g}_1 = \begin{pmatrix} 0 \\ 2 \end{pmatrix}, \quad G_1 = \begin{pmatrix} 0 & 1 \\ 1 & 2 \end{pmatrix}, \quad G_1^{-1} = \begin{pmatrix} -2 & 1 \\ 1 & 0 \end{pmatrix}.$$

Thus,

$$\boldsymbol{d}_1 = -G_1^{-1}\boldsymbol{g}_1 = \begin{pmatrix} -2 \\ 0 \end{pmatrix}$$

and

$$f(\boldsymbol{x}_1 + \theta \boldsymbol{d}_1) = 16\theta^4 + 1.$$

No $\theta > 0$ exists to satisfy

$$f(\boldsymbol{x}_1 + \theta \boldsymbol{d}_1) > f(\boldsymbol{x}_1).$$

9.2 Quasi-Newton Methods

When the objective function is not concave, the Hessian matrix may not be negative definite and then the search direction in the limited-step Newton method may not be ascent. What should we do in such a situation? In addition, the inverse of the Hessian matrix is very expensive to compute when the number of variables is really large. How should we get rid of this inverse? To overcome these troubles, we may use a positive definite matrix to approximate the negation of the inverse of the Hessian matrix and use this approximation to compute a search direction. It would also be the best if this approximation can be computed recursively.

Let H_k be an approximation of $-G_k^{-1}$ where $G_k = \nabla^2 f(\boldsymbol{x}_k)$. We next study how to determine H_{k+1} from H_k. First, note that

$$\boldsymbol{g}_k \approx \boldsymbol{g}_{k+1} - G_{k+1}\boldsymbol{\sigma}_k$$

where $\boldsymbol{\sigma}_k = \boldsymbol{x}_{k+1} - \boldsymbol{x}_k$, $\boldsymbol{g}_k = \nabla f(\boldsymbol{x}_k)$. Thus,

$$\boldsymbol{\sigma}_k \approx G_{k+1}^{-1}\boldsymbol{y}_k$$

where $\boldsymbol{y}_k = \boldsymbol{g}_{k+1} - \boldsymbol{g}_k$. To let H_{k+1} simulate $-G_{k+1}^{-1}$ closely, we ask H_{k+1} to satisfy

$$\boldsymbol{\sigma}_k = -H_{k+1}\boldsymbol{y}_k.$$

This equation is called *quasi-Newton equation.* The following is the general form of quasi-Newton method.

Quasi-Newton Method: *Choose an initial point $\boldsymbol{x}_1$ and an initial positive definite symmetric matrix H_1. Set $k = 1$.*

1. *Compute $\boldsymbol{g}_k$. If $\boldsymbol{g}_k = \boldsymbol{0}$, then stop; else, go to step 2.*

2. *If $k > 1$, then compute a positive definite symmetric matrix H_k with an updating formula such that the quasi-Newton equation holds. Compute a search direction $\boldsymbol{d}_k = H_k\boldsymbol{g}_k$. Set $\boldsymbol{x}_{k+1} = \boldsymbol{x}_k + \theta\boldsymbol{d}_k$ and $k := k+1$, where θ is chosen by a line search. Go to step 1.*

Clearly, we would like to have H_{k+1} computed easily. This requires $H_{k+1} - H_k$ to be a simple matrix. Suppose that $H_{k+1} - H_k$ has rank one, that is,

$$H_{k+1} - H_k = \boldsymbol{a}\boldsymbol{b}^T$$

for some $\boldsymbol{a}, \boldsymbol{b} \in R^n$. By quasi-Newton equation, we have

$$\boldsymbol{\sigma}_k = -H_k\boldsymbol{y}_k - \boldsymbol{a}(\boldsymbol{b}^T\boldsymbol{y}_k)$$

that is,

$$\boldsymbol{a} = -\frac{\boldsymbol{\sigma}_k + H_k \boldsymbol{y}_k}{\boldsymbol{b}^T \boldsymbol{y}_k}.$$

If $\boldsymbol{a} = \boldsymbol{0}$, then $H_{k+1} = H_k$. This is not what we want. Thus, we may assume $\boldsymbol{\sigma}_k + H_k \boldsymbol{y}_k \neq 0$, i.e. H_k does not satisfy the quasi-Newton equation. Now, we have

$$H_{k+1} = H_k - \frac{(\boldsymbol{\sigma}_k + H_k \boldsymbol{y}_k)\boldsymbol{b}^T}{\boldsymbol{b}^T \boldsymbol{y}_k}.$$

To keep H_{k+1} symmetric, we need to set $\boldsymbol{b}$ parallel to $\boldsymbol{\sigma}_k + H_k \boldsymbol{y}_k$. Thus,

$$H_{k+1} = H_k - \frac{(\boldsymbol{\sigma}_k + H_k \boldsymbol{y}_k)(\boldsymbol{\sigma}_k + H_k \boldsymbol{y}_k)^T}{(\boldsymbol{\sigma}_k + H_k \boldsymbol{y}_k)^T \boldsymbol{y}_k}.$$

This is called the *symmetric rank one updating formula.*

This formula cannot guarantee that H_{k+1} is positive definite. For example, consider the following quadratic function

$$f(\boldsymbol{x}) = -2x_1^2 - \frac{1}{4}x_2^2.$$

Choose $H_1 = I$ where I is the identity matrix. If $\boldsymbol{\sigma}_1^T = (1, 2)$, then $\boldsymbol{y}_1^T = G\boldsymbol{\sigma}_1 = (-2, -1)$. Thus,

$$H_2 = \begin{pmatrix} 0 & 1 \\ 1 & 0 \end{pmatrix}$$

is not positive definite. When H_{k+1} is not positive definite, we cannot guarantee that $H_{k+1}\boldsymbol{g}_{k+1}$ is an ascent direction. This may cause the algorithm to halt before it reaches a stationary point.

Since the symmetric rank one updating formula has only one choice, we have to consider possible rank two updating formula. Suppose

$$H_{k+1} = H_k + (\boldsymbol{a}, \boldsymbol{b})(\boldsymbol{u}, \boldsymbol{v})^T.$$

By the quasi-Newton equation, we know

$$-\boldsymbol{\sigma}_k = H_{k+1}\boldsymbol{y}_k = H_k \boldsymbol{y}_k + (\boldsymbol{a}, \boldsymbol{b})(\boldsymbol{u}^T \boldsymbol{y}_k, \boldsymbol{v}^T \boldsymbol{y}_k)^T.$$

Thus,

$$(\boldsymbol{u}^T \boldsymbol{y}_k)\boldsymbol{a} + (\boldsymbol{v}^T \boldsymbol{y})\boldsymbol{b} = -\boldsymbol{\sigma}_k - H_k \boldsymbol{y}_k.$$

A simple way is to choose $\boldsymbol{a} = \boldsymbol{\sigma}_k$ and $\boldsymbol{b} = H_k\boldsymbol{y}_k$. Then, by symmetry, we may write the rank two formula as

$$H_{k+1} = H_k + (\boldsymbol{\sigma}_k, H_k\boldsymbol{y}_k)\begin{pmatrix} a_{11} & a_{12} \\ a_{12} & a_{22} \end{pmatrix}(\boldsymbol{\sigma}_k, H_k\boldsymbol{y}_k)^T. \tag{9.5}$$

Again, by the quasi-Newton equation, we obtain

$$-\boldsymbol{\sigma}_k - H_k\boldsymbol{y}_k = (a_{11}\boldsymbol{\sigma}_k^T\boldsymbol{y}_k + a_{12}\boldsymbol{y}_k^T H_k\boldsymbol{y}_k)\boldsymbol{\sigma}_k + (a_{12}\boldsymbol{\sigma}_k^T\boldsymbol{y}_k + a_{22}\boldsymbol{y}_k^T H_k\boldsymbol{y}_k)H_k\boldsymbol{y}_k.$$

Assume $\boldsymbol{\sigma}_k$ and $H_k\boldsymbol{y}_k$ are not parallel. Then

$$\begin{aligned} a_{11}\boldsymbol{\sigma}_k^T\boldsymbol{y}_k + a_{12}\boldsymbol{y}_k^T H_k\boldsymbol{y}_k &= -1 \\ a_{12}\boldsymbol{\sigma}_k^T\boldsymbol{y}_k + a_{22}\boldsymbol{y}_k^T H_k\boldsymbol{y}_k &= -1. \end{aligned}$$

Thus,

$$\begin{aligned} a_{11} &= -\frac{1 + a_{12}\boldsymbol{y}_k^T H_k\boldsymbol{y}_k}{\boldsymbol{\sigma}_k^T\boldsymbol{y}_k} \\ a_{12} &= -\frac{1 + a_{12}\boldsymbol{\sigma}_k^T\boldsymbol{y}_k}{\boldsymbol{y}_k^T H_k\boldsymbol{y}_k}. \end{aligned}$$

Substituting them back to (9.5), we obtain

$$H_{k+1} = H_k - \frac{\boldsymbol{\sigma}_k\boldsymbol{\sigma}_k^T}{\boldsymbol{\sigma}_k^T\boldsymbol{y}_k} - \frac{H_k\boldsymbol{y}_k(H_k\boldsymbol{y}_k)^T}{\boldsymbol{y}_k^T H_k\boldsymbol{y}_k} + W$$

where

$$\begin{aligned} W &= a_{12}(\boldsymbol{\sigma}_k, H_k\boldsymbol{y}_k)\begin{pmatrix} -\frac{\boldsymbol{y}_k^T H_k\boldsymbol{y}_k}{\boldsymbol{\sigma}_k^T\boldsymbol{y}_k} & 1 \\ 1 & -\frac{\boldsymbol{\sigma}_k^T\boldsymbol{y}_k}{\boldsymbol{y}_k^T H_k\boldsymbol{y}_k} \end{pmatrix}\begin{pmatrix} \boldsymbol{\sigma}_k \\ H_k\boldsymbol{y}_k \end{pmatrix} \\ &= -a_{12}(\boldsymbol{\sigma}_k^T\boldsymbol{y}_k)(\boldsymbol{y}_k^T H_k\boldsymbol{y}_k) \\ &\quad (\boldsymbol{\sigma}_k, H_k\boldsymbol{y}_k)\begin{pmatrix} \frac{1}{(\boldsymbol{\sigma}_k^T\boldsymbol{y}_k)^2} & \frac{-1}{(\boldsymbol{y}_k^T H_k\boldsymbol{y}_k)(\boldsymbol{\sigma}_k^T\boldsymbol{y}_k)} \\ \frac{-1}{(\boldsymbol{y}_k^T H_k\boldsymbol{y}_k)(\boldsymbol{\sigma}_k^T\boldsymbol{y}_k)} & \frac{1}{(\boldsymbol{y}_k^T H_k\boldsymbol{y}_k)^2} \end{pmatrix}\begin{pmatrix} \boldsymbol{\sigma}_k \\ H_k\boldsymbol{y}_k \end{pmatrix} \\ &= -a_{12}(\boldsymbol{\sigma}_k^T\boldsymbol{y}_k)(\boldsymbol{y}_k^T H_k\boldsymbol{y}_k) \\ &\quad (\boldsymbol{\sigma}_k, H_k\boldsymbol{y}_k)\begin{pmatrix} \frac{1}{\boldsymbol{\sigma}_k^T\boldsymbol{y}_k} \\ \frac{-1}{\boldsymbol{y}_k^T H_k\boldsymbol{y}_k} \end{pmatrix}\left(\frac{1}{\boldsymbol{\sigma}_k^T\boldsymbol{y}_k}, \frac{-1}{\boldsymbol{y}_k^T H_k\boldsymbol{y}_k}\right)\begin{pmatrix} \boldsymbol{\sigma}_k \\ H_k\boldsymbol{y}_k \end{pmatrix} \\ &= -a_{12}(\boldsymbol{\sigma}_k^T\boldsymbol{y}_k)(\boldsymbol{y}_k^T H_k\boldsymbol{y}_k)\left(\frac{\boldsymbol{\sigma}_k}{\boldsymbol{\sigma}_k^T\boldsymbol{y}_k} - \frac{H_k\boldsymbol{y}_k}{\boldsymbol{y}_k^T H_k\boldsymbol{y}_k}\right)\left(\frac{\boldsymbol{\sigma}_k}{\boldsymbol{\sigma}_k^T\boldsymbol{y}_k} - \frac{H_k\boldsymbol{y}_k}{\boldsymbol{y}_k^T H_k\boldsymbol{y}_k}\right)^T. \end{aligned}$$

Therefore

$$H_{k+1} = H_k - \frac{\sigma_k \sigma_k^T}{\sigma_k^T y_k} - \frac{H_k y_k (H_k y_k)^T}{y_k^T H_k y_k} + w v v^T \tag{9.6}$$

where w is a parameter and

$$v = \frac{\sigma_k}{\sigma_k^T y_k} - \frac{H_k y_k}{y_k^T H_k y_k}.$$

Theorem 9.2.1 *Suppose H_k is positive definite and $\sigma_k^T y_k < 0$. If $w \geq 0$, then H_{k+1} determined by (9.6) is also positive definite.*

Proof. Since H_k is positive definite, H_k can be decomposed as $H_k = LL^T$ where L is a nonsingular square matrix of order n. For and $z \in R^n$,

$$z^T H_{k+1} z = z^T H_k z - \frac{(z^T \sigma_k)^2}{\sigma_k^T y_k} - \frac{(z^T H_k y_k)^2}{y_k^T H_k y_k} + w(v^T z)^2.$$

Denote $a = L^T z$ and $b = L^T y_k$. Then

$$z^T H_{k+1} z = \|a\|^2 - \frac{(a^T b)^2}{\|b\|^2} - \frac{(z^T \sigma_k)^2}{\sigma_k^T y_k} - w(v^T z)^2.$$

Note that

$$\|a\|^2 \cdot \|b\|^2 \geq (a^T b)^2$$

and the equality sign holds if and only if a and b are parallel. Since $\sigma_k^T y_k < 0$, we have $z^T H_{k+1} z \geq 0$ and hence H_{k+1} is positive semi-definite. To show H_{k+1} to be positive definite, let $z^T H_{k+1} z = 0$. Then $\sigma_k^T z = 0$ and

$$\|a\|^2 \cdot \|b\|^2 = (a^T b)^2.$$

Hence, a and b are parallel, i.e. $a = \zeta b$ for some scaler ζ. Since L is nonsingular, $z = \zeta y_k$. Therefore, $0 = \sigma_k^T z = \zeta \sigma_k^T y_k$. Since $\sigma_k^T y_k < 0$, we have $\zeta = 0$. Therefore, $z = \mathbf{0}$. This means that H_{k+1} is positive definite. □

When does the assumption $\sigma_k^T y_k$ hold? This depends on what line search the algorithm uses and the objective function $f(x)$. Let us consider unbounded line search procedures, i.e. in the line search procedures stated in section 4.2, we set $\bar{\lambda} = +\infty$.

Corollary 9.2.2 *Suppose H_k is positive definite and x_{k+1} is obtained from x_k along direction $H_k g_k$ with exact line search or the Wolfe test. If $w \geq 0$, then H_{k+1} obtained from (9.6) is also positive definite.*

Proof. Since the algorithm does not stop at $\boldsymbol{x}_k$, we have $\boldsymbol{g}_k \neq \boldsymbol{0}$. Since H_k is positive definite, $H_k\boldsymbol{g}_k$ is an ascent direction at $\boldsymbol{x}_k$. Thus, along direction $H_k\boldsymbol{g}_k$, we will obtain a point $\boldsymbol{x}_{k+1} \neq \boldsymbol{x}_k$ with exact line search or the Wolfe test.

If the line search is exact, then $\boldsymbol{g}_{k+1}^T H_k \boldsymbol{g}_k = 0$. Therefore,

$$\boldsymbol{\sigma}_k^T \boldsymbol{y}_k = \lambda_k \boldsymbol{g}_k^T H_k \boldsymbol{y}_k = -\lambda_k \boldsymbol{g}_k^T H_k \boldsymbol{g}_k < 0.$$

If the line search is the Wolfe test, then

$$(1-\alpha)\boldsymbol{g}_k^T H_k \boldsymbol{g}_k \geq \boldsymbol{g}_{k+1}^T H_k \boldsymbol{g}_k$$

for $\alpha \in (0, 1/2)$. Therefore,

$$\boldsymbol{\sigma}_k^T \boldsymbol{y}_k = \lambda_k \boldsymbol{y}_k^T H_k \boldsymbol{g}_k \leq -\lambda_k \alpha \boldsymbol{g}_k^T H_k \boldsymbol{g}_k < 0.$$

□

Corollary 9.2.3 *Suppose H_k is positive definite and the objective function is strictly concave. If $w \geq 0$ and $\boldsymbol{\sigma}_k \neq \boldsymbol{0}$, then H_{k+1} obtained from (9.6) is also positive definite.*

Proof. Since the objective function $f(\boldsymbol{x})$ is strictly concave, we have

$$\boldsymbol{g}_k^T \boldsymbol{\sigma}_k > f(\boldsymbol{x}_k + 1) - f(\boldsymbol{x}_k) > \boldsymbol{g}_{k+1}^T \boldsymbol{\sigma}_k.$$

Therefore, $\boldsymbol{y}_k^T \boldsymbol{\sigma}_k < 0$. □

The above results tell us that there are many choices for the rank two updating formula. Setting $w = 0$ in (9.6), we obtain

$$H_{k+1} = H_k - \frac{\boldsymbol{\sigma}_k \boldsymbol{\sigma}_k^T}{\boldsymbol{\sigma}_k^T \boldsymbol{y}_k} - \frac{H_k \boldsymbol{y}_k (H_k \boldsymbol{y}_k)^T}{\boldsymbol{y}_k^T H_k \boldsymbol{y}_k}.$$

This is called the DFP (Davidon-Fletcher-Powell) formula. The quasi-Newton method with the DFP formula is called the DFP method. Setting $w = (\boldsymbol{y}_k^T H_k \boldsymbol{y}_k)$ in (9.6), we obtain BFGS (Broyden-Fletcher-Goldfarb-Shanno) formula

$$H_{k+1} = H_k - (1 - \frac{\boldsymbol{y}_k^T H_k \boldsymbol{y}_k}{\boldsymbol{\sigma}_k^T \boldsymbol{y}_k})\frac{\boldsymbol{\sigma}_k \boldsymbol{\sigma}_k^T}{\boldsymbol{\sigma}_k^T \boldsymbol{y}_k} - \frac{H_k \boldsymbol{y}_k \boldsymbol{\sigma}_k^T + \boldsymbol{\sigma}_k \boldsymbol{y}_k^T H_k}{\boldsymbol{\sigma}_k^T \boldsymbol{y}_k}.$$

When $w = (1-\alpha)(\boldsymbol{y}_k^T H_k \boldsymbol{y}_k)$ for $\alpha \in [0, 1]$, (9.6) gives a family of rank two formulas, this family is called Broyden's family. In Broyden's family, every member can be represented by $H_{k+1} = \alpha H_{k+1}^{DFP} + (1-\alpha) H_{k+1}^{BFGS}$ for some $\alpha \in [0, 1]$.

9.3 Huang's Family

If we give up the symmetry of H_k's and relax the quasi-Newton equation to

$$H_{k+1}\boldsymbol{y}_k = -\omega\boldsymbol{\sigma}_k,$$

then in a similar way to that in the last section, we can obtain a family of formula as follows

$$H_{k+1} = H_k + \boldsymbol{\sigma}_k\boldsymbol{u}^T + H_k\boldsymbol{y}_k\boldsymbol{v}^T$$

where

$$\begin{aligned}\boldsymbol{u} &= a_{11}\boldsymbol{\sigma}_k + a_{12}H_k^T\boldsymbol{y}_k,\\ \boldsymbol{v} &= a_{21}\boldsymbol{\sigma}_k + a_{22}H_k^T\boldsymbol{y}_k\end{aligned}$$

satisfying

$$\begin{aligned}\boldsymbol{u}^T\boldsymbol{y}_k &= -\omega,\\ \boldsymbol{v}^T\boldsymbol{y}_k &= -1.\end{aligned}$$

This family is called Huang's family. Using the updating formula in Huang's family, the resulting algorithms are called variable metric methods as follows.

Variable Metric Method: *Choose an initial point $\boldsymbol{x}_1$ and an intial matrix H_1. Set $k = 1$.*

1. *Compute $\boldsymbol{g}_k$. If $\boldsymbol{g}_k = \mathbf{0}$, then stop; else, go to* **2**.

2. *If $k > 1$, the compute a positive definite symmetric matrix H_k with a updating formula in Huang's family. Compute a search direction $\boldsymbol{d}_k = H_k^T\boldsymbol{g}_k$. Set $\boldsymbol{x}_{k+1} = \boldsymbol{x}_k + \theta\boldsymbol{d}_k$ and $k := k+1$, where θ is chosen by a line search. Go to* **1**.

Every quasi-Newton method is a variable metric method. However, not very variabe metric method is a quasi-Newton method. Huang's family has several interesting properties. The first one is called *quadratic termination* as stated in the following theorem.

Theorem 9.3.1 *Consider a concave quadratic function on R^n*

$$f(\boldsymbol{x}) = a + \boldsymbol{b}^T\boldsymbol{x} + \frac{1}{2}\boldsymbol{x}^T G\boldsymbol{x}$$

where G is negative definite. Then every variable metric method in Huang's family with exact line search stops within $n+1$ iterations. Moreover, if m iterations are performed, then for all $1 \le j < i \le m$,

$$\begin{aligned} H_i\boldsymbol{y}_j &= -\omega\boldsymbol{\sigma}_j \\ \boldsymbol{\sigma}_i^T G\boldsymbol{\sigma}_j &= 0 \\ \boldsymbol{g}_i^T\boldsymbol{\sigma}_j &= 0. \end{aligned}$$

Proof. We prove the three sets of equations by induction on i. For $i = 2$, since $\boldsymbol{\sigma}_k = -\omega H_{k+1}\boldsymbol{y}_k$, we have

$$H_2\boldsymbol{y}_1 = -\omega\boldsymbol{\sigma}_1$$

and

$$\begin{aligned} \boldsymbol{\sigma}_2^T G\boldsymbol{\sigma}_1 = (\lambda_2 H_2^T\boldsymbol{g}_2)^T\boldsymbol{y}_1 &= \lambda_2\boldsymbol{g}_2^T(H_2\boldsymbol{y}_1) \\ &= -\lambda_2\omega\boldsymbol{g}_2^T\boldsymbol{\sigma}_1 = 0. \end{aligned}$$

Here, we note that

$$\boldsymbol{g}_2^T\boldsymbol{\sigma}_1 = 0$$

since the line search is exact.

Assume that for $i \le k$ the theorem holds. We consider $i = k+1$. Note that

$$\boldsymbol{x}_{k+1} = \boldsymbol{x}_1 + \boldsymbol{\sigma}_1 + \cdots + \boldsymbol{\sigma}_k.$$

We first show that

$$\boldsymbol{g}_{k+1}^T\boldsymbol{\sigma}_j = 0$$

for $j = 1, 2, \cdots, k$. Since the line search is exact, we have

$$\boldsymbol{g}_{k+1}^T\boldsymbol{\sigma}_k = 0.$$

For $j \le k-1$, by the induction hypothesis,

$$\boldsymbol{g}_k^T\boldsymbol{\sigma}_j = 0$$

and

$$(\boldsymbol{g}_{k+1} - \boldsymbol{g}_k)^T\boldsymbol{\sigma}_k = \boldsymbol{\sigma}_k^T\boldsymbol{\sigma}_j = 0.$$

Hence

$$g_{k+1}^T \sigma_k = 0.$$

Now, for $j = 1, 2, \cdots, k$, we have

$$\begin{aligned}
\sigma_{k+1}^T G \sigma_j &= (\lambda_{k+1} H_{k+1}^T g_{k+1})^T y_j \\
&= \lambda_{k+1} g_{k+1}^T (H_{k+1} y_j) \\
&= -\omega \lambda_{k+1} g_{k+1}^T \sigma_j \\
&= 0.
\end{aligned}$$

and

$$\begin{aligned}
y_{k+1}^T H_{k+1} y_j &= -\omega y_{k+1}^T \sigma_j \\
&= -\omega (G \sigma_{k+1})^T \sigma_j \\
&= -\omega \sigma_{k+1}^T G \sigma_j \\
&= 0.
\end{aligned}$$

Thus, for $j = 1, 2, \cdots, k$,

$$\begin{aligned}
(H_{k+2} - H_{k+1}) y_j &= \sigma_{k+1} u^T y_j + H_{k+1} y_{k+1} v^T y_j \\
&= a_{11} \sigma_{k+1} (\sigma_{k+1}^T y_j) + a_{21} H_{k+1} y_{k+1} (\sigma_{k+1}^T y_j) \\
&= \mathbf{0}
\end{aligned}$$

since $\sigma_{k+1}^T y_j = \sigma_{k+1}^T G \sigma_j = 0$. It follows that for $j = 1, 2, \cdots, k$,

$$H_{k+2} y_j = H_{k+1} y_j = -\omega \sigma_j$$

In addition,

$$H_{k+2} y_{k+1} = -\omega \sigma_{k+1}.$$

This completes our induction argument.

Next, we show that $\sigma_1, \cdots, \sigma_m$ are linearly independent. Suppose $\sum_{j=1}^m \alpha_j \sigma_j = \mathbf{0}$. Then

$$(\sum_{j=1}^m \alpha_j \sigma_j)^T G (\sum_{j=1}^m \alpha_j \sigma_j) = 0.$$

It follows that

$$\sum_{j=1}^m \alpha_j^2 \|\sigma_j\|^2 = 0.$$

Therefore, $\alpha_j = 0$ for all $j = 1, \cdots, m$.

Finally, we note that when $m = n$,

$$(H_{n+1}G + \omega I)\boldsymbol{\sigma}_j = (H_{n+1}\boldsymbol{y}_j + \omega\boldsymbol{\sigma}_j = \mathbf{0}$$

for $j = 1, 2, \cdots, n$. Since $\boldsymbol{\sigma}_1, \cdots, \boldsymbol{\sigma}_n$ are are linearly independent, $H_{n+1}G = -\omega I$. Thus, $H_{n+1} = -\omega G^{-1}$. This means that one more exact line search would find the minimum. □

The nonzero vectors $\boldsymbol{\sigma}_1, \cdots, \boldsymbol{\sigma}_m$ are called *conjugate* directions if they satisfy

$$\boldsymbol{\sigma}_i^T G \boldsymbol{\sigma}_j = 0$$

for $1 \leq j < i \leq m$. From the proof of the above theorem, we know that the conjugate directions are linearly independent.

The second interesting property of Huang's family is that for nonquadratic objective functions, with exact line search all members with a fixed ω in the family generates the same sequence if they start with the same initial point and the same initial matrix. To show such a property, we first prove two lemmas.

Lemma 9.3.2 *Let $f(\boldsymbol{x})$ be a differentiable real function on R^n. If $\boldsymbol{x}_k$ and $\boldsymbol{x}_{k+1}$ satisfy*

$$\boldsymbol{g}_k^T \boldsymbol{\sigma}_k \neq 0, \quad \boldsymbol{g}_{k+1}^T \boldsymbol{\sigma}_k = 0,$$

and H_{k+1} is obtained from H_k through an updating formula in Huang's family, then

$$H_{k+1}^T \boldsymbol{g}_{k+1} = (1 + a_{22}\boldsymbol{y}_k^T H_k^T \boldsymbol{g}_{k+1}) \left[1 - \frac{\boldsymbol{\sigma}_k \boldsymbol{y}_k^T}{\boldsymbol{\sigma}_k^T \boldsymbol{y}_k} \right] H_k^T \boldsymbol{g}_{k+1}.$$

Proof. Note that in Huang's family, we have $\boldsymbol{y}_k^T \boldsymbol{v} = \boldsymbol{v}^T \boldsymbol{y}_k = -1$, that is,

$$a_{21}\boldsymbol{y}_k^T \boldsymbol{\sigma}_k + a_{22}\boldsymbol{y}_k^T H_k^T \boldsymbol{y}_k = -1.$$

Moreover,

$$H_k^T \boldsymbol{y}_k = H_k^T(\boldsymbol{g}_{k+1} - \boldsymbol{g}_k) = H_k^T \boldsymbol{g}_{k+1} - \frac{1}{\lambda_k}\boldsymbol{\sigma}_k.$$

Thus,

$$\begin{aligned} a_{21} &= -(1 + a_{22}\boldsymbol{y}_k^T H_k^T \boldsymbol{y}_k) \\ &= -\frac{(1 + a_{22}\boldsymbol{y}_k^T H_k^T \boldsymbol{g}_{k+1})}{\boldsymbol{y}_k^T \boldsymbol{\sigma}_k} + \frac{a_{22}}{\lambda_k}. \end{aligned}$$

Hence,

$$\begin{aligned}
& H_{k+1}^T \boldsymbol{g}_{k+1} \\
= \; & H_k^T \boldsymbol{g}_{k+1} + \boldsymbol{u}\boldsymbol{\sigma}_k^T \boldsymbol{g}_{k+1} + \boldsymbol{v}(H_k \boldsymbol{y}_k)^T \boldsymbol{g}_{k+1} \\
= \; & H_k^T \boldsymbol{g}_{k+1} + \boldsymbol{v}(H_k \boldsymbol{y}_k)^T \boldsymbol{g}_{k+1} \\
= \; & H_k^T \boldsymbol{g}_{k+1} + (a_{21}\boldsymbol{\sigma}_k + a_{22} H_k^T \boldsymbol{y}_k)(H_k \boldsymbol{y}_k)^T \boldsymbol{g}_{k+1} \\
= \; & (1 + a_{22}\boldsymbol{y}_k^T H_k^T \boldsymbol{g}_{k+1}) H_k^T \boldsymbol{g}_{k+1} - a_{22}(\boldsymbol{y}_k^T H_k^T \boldsymbol{g}_{k+1}) \frac{\boldsymbol{\sigma}_k}{\lambda_k} \\
& + a_{21}(\boldsymbol{y}_k^T H_k^T \boldsymbol{g}_{k+1}) \boldsymbol{\sigma}_k \\
= \; & (1 + a_{22}\boldsymbol{y}_k^T H_k^T \boldsymbol{g}_{k+1}) H_k^T \boldsymbol{g}_{k+1} + (a_{21} - \frac{a_{22}}{\lambda_k})(\boldsymbol{y}_k^T H_k^T \boldsymbol{g}_{k+1}) \boldsymbol{\sigma}_k \\
= \; & (1 + a_{22}\boldsymbol{y}_k^T H_k^T \boldsymbol{g}_{k+1}) \left[I - \frac{\boldsymbol{\sigma}_k \boldsymbol{y}_k^T}{\boldsymbol{\sigma}_k^T \boldsymbol{y}_k} \right] H_k^T \boldsymbol{g}_{k+1}.
\end{aligned}$$

□

Note that the scale coefficient does not affect the search direction. Thus, from the lemma, we see that the search direction in the $(k+1)$th iteration can also be represented as

$$R_k H_k \boldsymbol{g}_{k+1}$$

where

$$R_k = I - \frac{\boldsymbol{\sigma}_k \boldsymbol{y}_k^T}{\boldsymbol{\sigma}_k^T \boldsymbol{y}_k}.$$

The next lemma shows an expression of $R_k H_k^T$.

Lemma 9.3.3 *Suppose $\{\boldsymbol{x}_k\}$ is a sequence generated by variable metric methods in Huang's family such that for all k,*

$$\boldsymbol{\sigma}_k^T \boldsymbol{g}_k \neq 0, \quad \boldsymbol{\sigma}_k^T \boldsymbol{g}_{k+1} = 0, \quad a_{21}^{(k)} - \frac{a_{22}^{(k)}}{\lambda_k} \neq 0$$

where $a_{ij}^{(k)}$ is the value of parameter a_{ij} in Huang's family used in the kth iteration and $\boldsymbol{x}_{k+1} = \boldsymbol{x}_k + \lambda_k H_k^T \boldsymbol{g}_k$. Then

$$R_k H_k^T = \left(\prod_{i=1}^{k} R_i \right) H_1^T - \omega \left[\sum_{i=1}^{k-1} \left(\prod_{j=i+1}^{k} R_j \right) \frac{\boldsymbol{\sigma}_i \boldsymbol{\sigma}_i^T}{\boldsymbol{\sigma}_i^T \boldsymbol{y}_i} \right].$$

Proof. Note that $R_k\boldsymbol{\sigma}_k = \mathbf{0}$. By Lemma 9.3.2

$$R_{k+1}\boldsymbol{\sigma}_{k+1} = \lambda_{k+1}H_{k+1}^T\boldsymbol{g}_{k+1} = (1 + a_{22}^{(k)}\boldsymbol{y}_k^T H_k^T\boldsymbol{g}_{k+1})R_{k+1}R_kH_k^T\boldsymbol{g}_{k+1} = \mathbf{0}.$$

However,

$$1 + a_{22}^{(k)}\boldsymbol{y}_k^T H_k^T\boldsymbol{g}_{k+1} = -(a_{21}^{(k)} - \frac{a_{22}^{(k)}}{\lambda_k})\boldsymbol{y}_k^T\boldsymbol{\sigma}_k \neq 0.$$

Thus,

$$R_{k+1}R_kH_k^T\boldsymbol{g}_{k+1} = \mathbf{0}.$$

Moreover, from the updating formula, we know that

$$\begin{aligned} R_k\boldsymbol{u}^{(k)} &= \boldsymbol{u}^{(k)} - \frac{\boldsymbol{\sigma}_k\boldsymbol{y}_k^T\boldsymbol{u}^{(k)}}{\boldsymbol{\sigma}_k^T\boldsymbol{y}_k} \\ &= \boldsymbol{u}^{(k)} + \frac{\omega\boldsymbol{\sigma}_k}{\boldsymbol{\sigma}_k^T\boldsymbol{y}_k} \end{aligned}$$

that is,

$$\boldsymbol{u}^{(k)} = R_k\boldsymbol{u}^{(k)} - \frac{\omega\boldsymbol{\sigma}_k}{\boldsymbol{\sigma}_k^T\boldsymbol{y}_k}. \tag{9.7}$$

Note that $R_k\boldsymbol{\sigma}_k = \mathbf{0}$ implies

$$R_kH_k^T\boldsymbol{g}_k = R_k\boldsymbol{\sigma}_k/\lambda_k = \mathbf{0}.$$

It follows that

$$R_{k+1}R_k\boldsymbol{u}^{(k)} = a_{12}^{(k)}R_{k+1}R_kH_k^T\boldsymbol{y}_k = a_{12}^{(k)}R_{k+1}R_kH_k^T\boldsymbol{g}_{k+1} = \mathbf{0}.$$

By (9.7),

$$R_{k+1}\boldsymbol{u}^{(k)} = -\omega\frac{R_{k+1}\boldsymbol{\sigma}_k}{\boldsymbol{\sigma}_k^T\boldsymbol{y}_k}. \tag{9.8}$$

Similarly, we can prove that

$$R_{k+1}\boldsymbol{v}^{(k)} = -\frac{R_{k+1}\boldsymbol{\sigma}_k}{\boldsymbol{\sigma}_k^T\boldsymbol{y}_k}. \tag{9.9}$$

Thus,

$$\begin{aligned} R_{k+1}H_{k+1}^T &= R_{k+1}H_k^T + (R_{k+1}\boldsymbol{u}^{(k)})\boldsymbol{\sigma}_k^T + (R_{k+1}\boldsymbol{v}^{(k)})(H_k\boldsymbol{y}_k)^T \\ &= R_{k+1}H_k^T - \omega\frac{R_{k+1}\boldsymbol{\sigma}_k}{\boldsymbol{\sigma}_k^T\boldsymbol{y}_k}\boldsymbol{\sigma}_k^T - \frac{R_{k+1}\boldsymbol{\sigma}_k}{\boldsymbol{\sigma}_k^T\boldsymbol{y}_k}(H_k\boldsymbol{y}_k)^T \\ &= R_{k+1}\left[I - \frac{\boldsymbol{\sigma}_k\boldsymbol{y}_k^T}{\boldsymbol{\sigma}_k^T\boldsymbol{y}_k}\right]H_k^T - \omega\frac{R_{k+1}\boldsymbol{\sigma}_k}{\boldsymbol{\sigma}_k^T\boldsymbol{y}_k}\boldsymbol{\sigma}_k^T \\ &= R_{k+1}R_kH_k^T - \omega\frac{R_{k+1}\boldsymbol{\sigma}_k}{\boldsymbol{\sigma}_k^T\boldsymbol{y}_k}\boldsymbol{\sigma}_k^T. \end{aligned}$$

The proof is completed by applying this recursive equation repeatedly. □

The following theorem is an immediate consequence of Lemma 9.3.3.

Theorem 9.3.4 *Let f be a differentiable real function on R^n. Suppose that the initial point $\boldsymbol{x}_1$ and initial matrix H_1 are given. Then the sequence $\{\boldsymbol{x}_k\}$ generated by a variable metric method that uses updating formulas in Huang's family depends only on parameter ω, provided that for all k,*

$$\boldsymbol{\sigma}_k^T \boldsymbol{g}_k \neq 0, \quad \boldsymbol{\sigma}_k^T \boldsymbol{g}_{k+1} = 0, \quad a_{21}^{(k)} - \frac{a_{22}^{(k)}}{\lambda_k} \neq 0$$

where $a_{ij}^{(k)}$ is the value of parameter a_{ij} in Huang's family used in the kth iteration and λ_k is unique positive number such that $\boldsymbol{x}_{k+1} = \boldsymbol{x}_k + \lambda_k H_k^T \boldsymbol{g}_k$ satisfies the above conditions. Then, as long as the initial point $\boldsymbol{x}_1$ and initial matrix H_1 are fixed, the sequence $\{\boldsymbol{x}_k\}$ is determined by ω.

Proof. By Lemma 9.3.3, the search direction in each iteration is determined by the initial point, initial matrix, and parameter ω. □

Corollary 9.3.5 *Let f be a differentiable real function on R^n. Then for any given initial point and initial matrix, Broyden's family with exact line search generates identical sequence of points provided each line search finds a unique point.*

Proof. By Corollary 9.2.2, H_k is positive definite for all k. Thus, $\boldsymbol{\sigma}_k^T \boldsymbol{g}_k = \lambda_k \boldsymbol{g}_k^T H_k \boldsymbol{g}_k > 0$ and $\boldsymbol{\sigma}_k^T \boldsymbol{g}_{k+1} = 0$. In addition, we claim that $a_{21}^{(k)} - \frac{a_{22}^{(k)}}{\lambda_k} \neq 0$. In fact,

$$\begin{aligned} & a_{21}^{(k)} - \frac{a_{22}^{(k)}}{\lambda_k} \\ = & -\frac{1 + a_{22}^{(k)} \boldsymbol{y}_k^T H_k^T \boldsymbol{g}_{k+1}}{\boldsymbol{y}_k^T \boldsymbol{\sigma}_k} \\ = & -\frac{1 + a_{22}^{(k)} \boldsymbol{g}_{k+1}^T H_k^T \boldsymbol{g}_{k+1}}{\boldsymbol{y}_k^T \boldsymbol{\sigma}_k}. \end{aligned}$$

If $a_{21}^{(k)} - \frac{a_{22}^{(k)}}{\lambda_k} = 0$, then we must have $a_{22} = -1$ and hence $a_{21} < 0$. However, in Broyden's family, $a_{21} = a_{12} \geq 0$, a contradiction.

Finally, this corollary is proved by Theorem 9.3.4. □

Exercises

1. Let f be a quadratic function on R^n with a negative definite Hessian matrix. Then the sequence $\{\boldsymbol{x}_k\}$ generated by a variable metric method that uses updating formulas in Huang's family is determined by the initial point $\boldsymbol{x}_1$ and the initial matrix H_1.

2. Show that BFGS formula can also written as
$$H_{k+1} = (I + \frac{\boldsymbol{\sigma}_k \boldsymbol{y}_k^T}{\boldsymbol{\sigma}_k^T \boldsymbol{y}_k}) H_k (I + \frac{\boldsymbol{\sigma}_k \boldsymbol{y}_k^T}{\boldsymbol{\sigma}_k^T \boldsymbol{y}_k})^T - \frac{\boldsymbol{\sigma}_k \boldsymbol{\sigma}_k^T}{\boldsymbol{\sigma}_k^T \boldsymbol{y}_k}.$$

3. Consider a formula $H_{k+1} = h(H_k, \boldsymbol{\sigma}_k, \boldsymbol{y}_k)$ in Hung's family. For any sequence $\{\boldsymbol{\sigma}_k\}$, a positive definite symmetric matrix A, and an initial positive definite symmetric matrix H_0, sequences $\{\boldsymbol{y}_k\}$ and $\{H_k\}$ can be computed through $\boldsymbol{y}_k = A\boldsymbol{\sigma}_k$ and considered formula. This considered formula called *F-monotone* if for any sequence $\{H_k\}$ obtained in such a way, $\{A^{1/2}H_kA^{1/2}\}$ has the following property: For any $a, b > 0$, if $[1-a, 1+b]$ contains k eigenvalues of $A^{1/2}H_kA^{1/2}$, then $[1-a, 1-b]$ contains at least k eigenvalues of $A^{1/2}H_{k+1}A^{1/2}$. Show that for $\alpha \in [0, 1]$, every formula in Broyden's family is F-monotone.

Chapter 10

Powell's Conjecture

I try not to guess.
JOSELF CAMPBELL

There are several interesting long-standing open problems around Powell's conjecture regarding the global convergence of quasi-Newton methods. In this chapter, we review these open problems and existing related results.

10.1 Inverse and Eigenvalues

There is a major difference in establishing global convergence of quasi-Newton methods and Rosen's gradient projection method. When the generated sequence $\{\boldsymbol{x}_k\}$ is convergent and the line search is exact and unbounded ($\bar{\lambda}_k = +\infty$), it will not be too hard to prove that $\boldsymbol{g}_k^T \boldsymbol{d}_k \to 0$ as $k \to \infty$. That is,

$$\lim_{k\to\infty} \boldsymbol{g}_k^T H_k \boldsymbol{g}_k = 0.$$

However, this cannot imply $\lim_{k\to\infty} \|\boldsymbol{g}_k\| = 0$ since the smallest eigenvalue of H_k may go to zero as $k \to \infty$. This means that information about eigenvalues of H_k is very important. Note that the trace $\text{Tr}(H)$ of a matrix H is the sum of all eigenvalues of H and $\det(H)$ is the product of all eigenvalues of H. Moreover, the eigenvalues of H^{-1} are the inverse of eigenvalues of H. Therefore, the study of $\text{Tr}(H_{k+1})$, $\det(H_{k+1})$, and $\text{Tr}(H_{k+1}^{-1})$ play important role in the proof of global convergence of quasi-Newton methods. In this section, we look at several useful tools.

First, we study the inverse of H_k.

Recall that Broyden's family has the following updating formula:

$$H_{k+1} = H_k - \frac{\boldsymbol{\sigma}_k\boldsymbol{\sigma}_k^T}{\boldsymbol{\sigma}_k^T\boldsymbol{y}_k} - \frac{H_k\boldsymbol{y}_k(H_k\boldsymbol{y}_k)^T}{\boldsymbol{y}_k^T H_k\boldsymbol{y}_k} + (1-\alpha)(\boldsymbol{y}_k^T H_k\boldsymbol{y}_k)\boldsymbol{v}\boldsymbol{v}^T$$

where $\alpha \in [0,1]$ and

$$\boldsymbol{v} = \frac{\boldsymbol{\sigma}_k}{\boldsymbol{\sigma}_k^T\boldsymbol{y}_k} - \frac{H_k\boldsymbol{y}_k}{\boldsymbol{y}_k^T H_k\boldsymbol{y}_k}.$$

To obtain the inverse of H_{k+1}, we first show a formula.

Lemma 10.1.1 (Sherman-Morrison formula) *If the matrix A is nonsingular and $1+\boldsymbol{b}^T A^{-1}\boldsymbol{a} \neq 0$, then $A+\boldsymbol{a}\boldsymbol{b}^T$ is nonsingular and its inverse can be represented as*

$$(A+\boldsymbol{a}\boldsymbol{b}^T)^{-1} = A^{-1} - \frac{A^{-1}\boldsymbol{a}\boldsymbol{b}^T A^{-1}}{1+\boldsymbol{b}^T A^{-1}\boldsymbol{a}}.$$

Proof.

$$\begin{aligned} & (A+\boldsymbol{a}\boldsymbol{b}^T)(A^{-1} - \frac{A^{-1}\boldsymbol{a}\boldsymbol{b}^T A^{-1}}{1+\boldsymbol{b}^T A^{-1}\boldsymbol{a}}) \\ = \; & I - \frac{A^{-1}\boldsymbol{a}\boldsymbol{b}^T}{1+\boldsymbol{b}^T A^{-1}\boldsymbol{a}} \\ & + A^{-1}\boldsymbol{a}\boldsymbol{b}^T - (\boldsymbol{b}^T A^{-1}\boldsymbol{a})\frac{A^{-1}\boldsymbol{a}\boldsymbol{b}^T}{1+\boldsymbol{b}^T A^{-1}\boldsymbol{a}} \\ = \; & I. \end{aligned}$$

□

Denote by $B_k = H_k^{-1}$. By Sherman-Morrison formula,

$$(H_k - \frac{\boldsymbol{\sigma}_k\boldsymbol{\sigma}_k^T}{\boldsymbol{\sigma}_k^T\boldsymbol{y}_k})^{-1} = B_k + \frac{B_k\boldsymbol{\sigma}_k\boldsymbol{\sigma}_k^T B_k}{\boldsymbol{\sigma}_k^T\boldsymbol{y}_k - \boldsymbol{\sigma}_k^T B_k\boldsymbol{\sigma}_k}$$

and

$$B_{k+1}^{DFP} = B_k - (1 - \frac{\boldsymbol{\sigma}_k^T B_k\boldsymbol{\sigma}_k}{\boldsymbol{y}_k^T\boldsymbol{\sigma}_k})\frac{\boldsymbol{y}_k\boldsymbol{y}_k^T}{\boldsymbol{y}_k^T\boldsymbol{\sigma}_k} - \frac{B_k\boldsymbol{\sigma}_k\boldsymbol{y}_k^T + \boldsymbol{y}_k\boldsymbol{\sigma}_k^T B_k}{\boldsymbol{y}_k^T\boldsymbol{\sigma}_k}.$$

Compared with BFGS updating formula, we may find a duality between DFP formula and BFGS formula as follows:

$$\begin{array}{ccc} H_k^{DFP} & \leftrightarrow & B_k^{BFGS} \\ H_k^{BFGS} & \leftrightarrow & B_k^{DFP} \\ \boldsymbol{\sigma}_k & \leftrightarrow & \boldsymbol{y}_k \\ \boldsymbol{y}_k & \leftrightarrow & \boldsymbol{\sigma}_k. \end{array}$$

From this duality, we see that

$$B_{k+1}^{BFGS} = B_k - \frac{\boldsymbol{y}_k \boldsymbol{y}_k^T}{\boldsymbol{y}_k^T \boldsymbol{\sigma}_k} - \frac{B_k \boldsymbol{\sigma}_k \boldsymbol{\sigma}_k^T B_k}{\boldsymbol{\sigma}_k^T B_k \boldsymbol{\sigma}_k}.$$

For $H_{k+1}^{\alpha} = \alpha H_{k+1}^{DFP} + (1-\alpha) H_{k+1}^{BFGS}$,

$$(H_{k+1}^{\alpha})^{-1} = \beta B_{k+1}^{BFGS} + (1-\beta) B_{k+1}^{DFP}$$

where

$$\beta = \frac{(1-\alpha)(\boldsymbol{y}_k^T H_k \boldsymbol{y}_k)(\boldsymbol{\sigma}_k^T B_k \boldsymbol{\sigma}_k)}{-\alpha(\boldsymbol{\sigma}_k^T \boldsymbol{y}_k)^2 + (1-\alpha)(\boldsymbol{\sigma}_k^T B_k \boldsymbol{\sigma}_k)(\boldsymbol{y}_k^T H_k \boldsymbol{y}_k)}.$$

Now, we study $\det(H_{k+1})$. The following is a useful tool.

Lemma 10.1.2 *Let $\boldsymbol{x}, \boldsymbol{y}, \boldsymbol{a}, \boldsymbol{b} \in R^n$. Then*

$$\det(I + \boldsymbol{x}\boldsymbol{y}^T + \boldsymbol{a}\boldsymbol{b}^T) = (1 + \boldsymbol{x}^T \boldsymbol{y})(1 + \boldsymbol{a}^T \boldsymbol{b}) - (\boldsymbol{b}^T \boldsymbol{x})(\boldsymbol{a}^T \boldsymbol{y}).$$

Proof. Note that for any nonsingular matrix T,

$$\det(T^{-1} A T) = \det(A).$$

Therefore,

$$\det(I + \boldsymbol{x}\boldsymbol{y}^T + \boldsymbol{a}\boldsymbol{b}^T) = \det(I + T^{-1}\boldsymbol{x} \cdot \boldsymbol{y}^T T + T^{-1}\boldsymbol{a} \cdot \boldsymbol{b}^T T).$$

Choose T such that

$$\boldsymbol{y}^T T = \boldsymbol{e}_1^T, \boldsymbol{b}^T T = \boldsymbol{e}_2^T$$

where $\boldsymbol{e}_i = (\underbrace{0, \cdots, 0, 1}_{i}, 0, \cdots, 0)^T$. Denote $T^{-1}\boldsymbol{x} = \boldsymbol{u}$ and $T^{-1}\boldsymbol{a} = \boldsymbol{w}$. Then $\det(I + \boldsymbol{x}\boldsymbol{y}^T + \boldsymbol{a}\boldsymbol{b}^T)$ has the form

$$|\boldsymbol{e}_1 + \boldsymbol{u}, \boldsymbol{e}_2 + \boldsymbol{w}, \boldsymbol{e}_3, \cdots, \boldsymbol{e}_n|.$$

Thus,

$$\det(I + \boldsymbol{x}\boldsymbol{y}^T + \boldsymbol{a}\boldsymbol{b}^T) = (1 + u_1)(1 + w_2) - u_2 w_1$$

where

$$\begin{aligned} a_1 &= \boldsymbol{e}_1^T \boldsymbol{u} = \boldsymbol{e}_1^T T^{-1} \boldsymbol{x} = \boldsymbol{y}^T \boldsymbol{x}, \\ a_2 &= \boldsymbol{e}_2^T \boldsymbol{u} = \boldsymbol{e}_2^T T^{-1} \boldsymbol{x} = \boldsymbol{b}^T \boldsymbol{x}, \\ b_1 &= \boldsymbol{e}_1^T \boldsymbol{w} = \boldsymbol{e}_1^T T^{-1} \boldsymbol{a} = \boldsymbol{y}^T \boldsymbol{a}, \\ b_2 &= \boldsymbol{e}_2^T \boldsymbol{w} = \boldsymbol{e}_2^T T^{-1} \boldsymbol{a} = \boldsymbol{b}^T \boldsymbol{a}. \end{aligned}$$

□

10.2 Powell's Conjecture

By Corollary 9.3.5, given an initial point and an initial matrix the quasi-Newton method with updating formula in Broyden's family and with exact line search produces identifical sequence of points. Thus, to study the global convergence of quasi-Newton method in Broyden's family with exact line search, it suffices to investigate one member. In this section, we will study only the DFP method with exact line search. the reader should note that all results also hold for all quasi-Newton methods in Broyden's family.

In 1971, Powell proved that if the objective function is uniformly concave, then DFP method with exact line search would generate a sequence convergent to a maximum point. However, he made the following conjecture: There exists a differentiable objective function f such that the sequence generated by DFP method with exact line search converges to a point which is not a stationary point.

In 1972, Powell showed that this conjecture is false in the two-dimensional space. He also improved his result from uniformly concave objective functions to concave functions. Based on new evidences, Powell believed that his conjecture in 1971 fails. Today, we would like to call Powell's conjecture as follows:

Weak Powell's Conjecture. If the DFP method with exact line search generates an infinite sequence $\{\boldsymbol{x}_k\}$, then

$$\liminf \|\boldsymbol{g}_k\| = 0.$$

Strong Powell's Conjecture. Every cluster point of the sequence generated by DFP method with exact line search is a stationary point.

There are two results about Powell's conjecture.

Theorem 10.2.1 *Let f be a twice continuously differentiable concave real function on R^n. Suppose that the level set $\{\boldsymbol{x} \mid f(\boldsymbol{x}) \geq f(\boldsymbol{x}_1)\}$ is bounded. Then the DFP method with exact line search and initial point $\boldsymbol{x}_1$ either stops at the maximum or generates an infinite sequence whose function value converges to the maximum value of $f(\boldsymbol{x})$.*

Theorem 10.2.2 *Suppose DFP method generates an infinite sequence $\{\boldsymbol{x}_k\}$ converging to $\boldsymbol{x}^*$. If the objective function f belongs to the class $C^{1,1}$, that is, there exists a constant L such that for every $\boldsymbol{x}$ and $\boldsymbol{y}$*

$$\|\boldsymbol{g}(\boldsymbol{x}) - \boldsymbol{g}(\boldsymbol{y})\| \leq L\|\boldsymbol{x} - \boldsymbol{y}\|,$$

then

$$\liminf_{k\to\infty} \|\boldsymbol{g}_k\| = 0.$$

These two theorems do not cover each other. They both support Powell's conjecture in 1972. It seems cumbersome to include both proofs here. Therefore, we introduce only the proof of Theorem 10.2.2 in the following.

To show Theorem 10.2.2 by contradiction, we suppose that for all k,

$$\|\boldsymbol{g}_k\| \geq c > 0 \tag{10.1}$$

and consider the following three sequences of real numbers:

$$\begin{aligned}
u_k &= \sum_{j=1}^{k} \|\boldsymbol{\sigma}_j\|^2/\lambda_j = \sum_{j=1}^{k} \lambda_j \|H_j \boldsymbol{g}_j\|^2, \\
v_k &= \|H_{k+1}\boldsymbol{g}_{k+1}\| + \sum_{j=1}^{k} \|H_{j+1}\boldsymbol{g}_{j+1} - H_j\boldsymbol{g}_j\| + \|H_1\boldsymbol{g}_1\|, \\
w_k &= \boldsymbol{g}_{k+1}^T H_{k+1}\boldsymbol{g}_{k+1}(\mathrm{Tr}(H_{k+1}) + \mathrm{Tr}(H_{k+1}^{-1})) + \boldsymbol{g}_1^T H_1 \boldsymbol{g}_1(\mathrm{Tr}(H_1) + \mathrm{Tr}(H_1^{-1})) \\
&\quad + \sum_{j=1}^{k} \boldsymbol{g}_j^T H_j \boldsymbol{g}_j (\boldsymbol{y}_j^T H_j \boldsymbol{y}_j)^{-1} (\|H_j \boldsymbol{y}_j\|^2 + \boldsymbol{g}_j^T H_j \boldsymbol{g}_j \mathrm{Tr}(H_{j+1}^{-1})).
\end{aligned}$$

The main idea of this proof is to study three ratios v_k/w_k, w_k/u_k and u_k/v_k. Two of them will be shown to be bounded and the third will be shown to be infinitely small. It will contradict the fact $(v_k/w_k)(w_k/u_k)(u_k/v_k) = 1$. We first show some useful equalities.

Multiplying by $\boldsymbol{g}_{k+1}$ to the DFP updating formula from right and noticing $\boldsymbol{\sigma}_k^T \boldsymbol{g}_{k+1} = 0$, we obtain

$$H_{k+1}\boldsymbol{g}_{k+1} = H_k\boldsymbol{g}_{k+1} - \frac{H_k\boldsymbol{y}_k\boldsymbol{y}_k^T H_k\boldsymbol{g}_{k+1}}{\boldsymbol{y}_k^T H_k\boldsymbol{y}_k}.$$

Thus

$$H_{k+1}\boldsymbol{g}_{k+1} - H_k\boldsymbol{g}_k = H_k\boldsymbol{y}_k - \frac{H_k\boldsymbol{y}_k\boldsymbol{y}_k^T H_k\boldsymbol{g}_{k+1}}{\boldsymbol{y}_k^T H_k\boldsymbol{y}_k}.$$

But

$$\boldsymbol{y}_k^T H_k\boldsymbol{y}_k - \boldsymbol{y}_k^T H_k\boldsymbol{g}_{k+1} = -\boldsymbol{y}_k^T H_k\boldsymbol{g}_k = \boldsymbol{g}_k^T H_k\boldsymbol{g}_k.$$

Hence

$$H_{k+1}\boldsymbol{g}_{k+1} - H_k\boldsymbol{g}_k = H_k\boldsymbol{y}_k\frac{\boldsymbol{g}_k^T H_k\boldsymbol{g}_k}{\boldsymbol{y}_k^T H_k\boldsymbol{y}_k}. \tag{10.2}$$

Multiplying $\boldsymbol{g}_{k+1}$ to (10.2) from left and noticing $\boldsymbol{g}_{k+1}^T H_k\boldsymbol{g}_k = 0$, we obtain

$$\boldsymbol{g}_{k+1}^T H_{k+1}\boldsymbol{g}_{k+1} - \boldsymbol{g}_k^T H_k\boldsymbol{g}_k = \boldsymbol{g}_k^T H_k\boldsymbol{g}_k\left(\frac{\boldsymbol{g}_{k+1}^T H_k\boldsymbol{y}_k}{\boldsymbol{y}_k^T H_k\boldsymbol{y}_k} - 1\right).$$

Since $\boldsymbol{g}_{k+1}^T H_k\boldsymbol{y}_k - \boldsymbol{y}_k^T H_k\boldsymbol{y}_k = -\boldsymbol{g}_k^T H_k\boldsymbol{g}_k$, we have

$$\boldsymbol{g}_{k+1}^T H_{k+1}\boldsymbol{g}_{k+1} - \boldsymbol{g}_k^T H_k\boldsymbol{g}_k = -\frac{(\boldsymbol{g}_k^T H_k\boldsymbol{g}_k)^2}{\boldsymbol{g}_k^T H_k\boldsymbol{g}_k}. \tag{10.3}$$

The next equality follows immediately from the DFP updating formula,

$$\mathrm{Tr}(H_{k+1}) = \mathrm{Tr}(H_k) - \frac{\|H_k\boldsymbol{y}_k\|^2}{\boldsymbol{y}_k^T H_k\boldsymbol{y}_k} - \frac{\|\boldsymbol{\sigma}_k\|^2}{\boldsymbol{\sigma}_k^T\boldsymbol{y}_k}. \tag{10.4}$$

From the DFP updating formula, we can also obtain

$$H_{k+1}^{-1} = \left(I - \frac{\boldsymbol{y}_k\boldsymbol{\sigma}_k^T}{\boldsymbol{\sigma}_k^T\boldsymbol{y}_k}\right)H_k^{-1}\left(I - \frac{\boldsymbol{\sigma}_k\boldsymbol{y}_k^T}{\boldsymbol{\sigma}_k^T\boldsymbol{y}_k}\right) - \frac{\boldsymbol{y}_k\boldsymbol{y}_k^T}{\boldsymbol{\sigma}_k^T\boldsymbol{y}_k},$$

from which we can derive

$$\mathrm{Tr}(H_{k+1}^{-1}) = \mathrm{Tr}(H_k^{-1}) + \frac{\|\boldsymbol{g}_{k+1}\|^2 - \|\boldsymbol{g}_k\|^2}{\boldsymbol{g}_k^T H_k\boldsymbol{g}_k} - \frac{\|\boldsymbol{y}_k\|^2}{\boldsymbol{\sigma}_k^T\boldsymbol{y}_k}. \tag{10.5}$$

Next, we study the three ratios.

Lemma 10.2.3 *If $f \in C^{1.1}$, then there exists a constant α such that for all k, $w_k/u_k \le \alpha$.*

Proof. By (10.4),

$$\boldsymbol{g}_j^T H_j \boldsymbol{g}_j(\mathrm{Tr}(H_{j+1}) - \mathrm{Tr}(H_j)) + \frac{\boldsymbol{g}_j^T H_j \boldsymbol{g}_j}{\boldsymbol{y}_j^T H_j \boldsymbol{y}_j}\|H_j \boldsymbol{g}_j\|^2 = \|\boldsymbol{\sigma}_j\|^2/\lambda_j. \qquad (10.6)$$

Taking summation of (10.6) for $j = 1, 2, \cdots, k$, using Abel formula for summation of product, and noticing (10.3), we have

$$\begin{aligned}
&\boldsymbol{g}_{k+1}^T H_{k+1} \boldsymbol{g}_{k+1} \mathrm{Tr}(H_{k+1}) + \sum_{j=1}^{k} (\boldsymbol{g}_j^T H_j \boldsymbol{g}_j)^2 (\boldsymbol{y}_j^T H_j \boldsymbol{y}_j)^{-1} \mathrm{Tr}(H_{j+1}) \\
&-\boldsymbol{g}_1^T H_1 \boldsymbol{g}_1 \mathrm{Tr}(H_1) + \sum_{j=1}^{k} (\boldsymbol{g}_j^T H_j \boldsymbol{g}_j)(\boldsymbol{y}_j^T H_j \boldsymbol{y}_j)^{-1} \|H_j \boldsymbol{y}_j\|^2 \\
&= \sum_{j=1}^{k} \|\boldsymbol{\sigma}_j\|^2/\lambda_j = u_k.
\end{aligned}$$

Thus

$$\begin{aligned}
&\boldsymbol{g}_{k+1}^T H_{k+1} \boldsymbol{g}_{k+1} \mathrm{Tr}(H_{k+1}) + \sum_{j=1}^{k} \frac{\boldsymbol{g}_j^T H_j \boldsymbol{g}_j}{\boldsymbol{y}_j^T H_j \boldsymbol{y}_j} \|H_j \boldsymbol{y}_j\|^2 \\
&\le u_k + \boldsymbol{g}_1^T H_1 \boldsymbol{g}_1 \mathrm{Tr}(H_1). \qquad (10.7)
\end{aligned}$$

Similarly, from (10.5), we see

$$\boldsymbol{g}_j^T H_j \boldsymbol{g}_j (\mathrm{Tr}(H_{j+1}^{-1}) - \mathrm{Tr}(H_j^{-1})) = \|\boldsymbol{g}_{j+1}\|^2 - \|\boldsymbol{g}_j\|^2 + \|\boldsymbol{y}_j\|^2/\lambda_j,$$

from which we can obtain

$$\begin{aligned}
&\boldsymbol{g}_{k+1}^T H_{k+1} \boldsymbol{g}_{k+1} \mathrm{Tr}(H_{k+1}^{-1}) - \boldsymbol{g}_1^T H_1 \boldsymbol{g}_1 \mathrm{Tr}(H_1^{-1}) + \sum_{j=1}^{k} \frac{(\boldsymbol{g}_j^T H_j \boldsymbol{g}_j)^2}{\boldsymbol{y}_j^T H_j \boldsymbol{y}_j} \mathrm{Tr}(H_{j+1}^{-1}) \\
\le\ & \sum_{j=1}^{k} \|\boldsymbol{y}_j\|^2/\lambda_j + \|\boldsymbol{g}_{k+1}\|^2 - \|\boldsymbol{g}_1\|^2 \\
\le\ & L^2 u_k + \|\boldsymbol{g}_{k+1}\|^2,
\end{aligned}$$

where L is the Lipschitz constant of $\boldsymbol{g}$. Adding (10.7) to it, we obtain

$$w_k \le (L^2+1)u_k + 2\boldsymbol{g}_1^T H_1 \boldsymbol{g}_1 (\mathrm{Tr}(H_1) + \mathrm{Tr}(H_1^{-1})) + \|\boldsymbol{g}_{k+1}\|^2.$$

Note that u_k is increasing and $\|\boldsymbol{g}_{k+1}\|$ is bounded. Therefore, there exists a constant α such that

$$w_k/u_k \leq (L^2+1) + (2\boldsymbol{g}_1^T H_1 \boldsymbol{g}_1(\mathrm{Tr}(H_1) + \mathrm{Tr}(H_1^{-1})) + \|\boldsymbol{g}_{k+1}\|^2)/u_k \leq \alpha.$$

□

Lemma 10.2.4 *If (10.1) holds, then for every k, $v_k/w_k \leq (2c)^{-1}$.*

Proof. By (10.1), for every k,

$$c^2 \leq \|\boldsymbol{g}_k\|^2 = (H_k^{1/2}\boldsymbol{g}_k)^T H_k^{-1}(H_k^{1/2}\boldsymbol{g}_k) \leq \boldsymbol{g}_k^T H_k \boldsymbol{g}_k \mathrm{Tr}(H_k^{-1}).$$

Moreover, for every k,

$$\|H_k\boldsymbol{g}_k\|^2 = (H_k^{1/2}\boldsymbol{g}_k)^T H_k (H_k^{1/2}\boldsymbol{g}_k) \leq \boldsymbol{g}_k^T H_k \boldsymbol{g}_k \mathrm{Tr}(H_k).$$

Thus

$$2c\|H_1\boldsymbol{g}_1\| \leq c^2 + \|H_1\boldsymbol{g}_1\|^2 \leq \boldsymbol{g}_1^T H_1 \boldsymbol{g}_1(\mathrm{Tr}(H_1) + \mathrm{Tr}(H_1^{-1})) \tag{10.8}$$

and

$$2c\|H_{k+1}\boldsymbol{g}_{k+1}\| \leq \boldsymbol{g}_{k+1}^T H_{k+1}\boldsymbol{g}_{k+1}(\mathrm{Tr}(H_{k+1}) + \mathrm{Tr}(H_{k+1}^{-1})). \tag{10.9}$$

Note that as a consequence of (10.3), $\{\boldsymbol{g}_k^T H_k \boldsymbol{g}_k\}$ is a decreasing sequence. By (10.2), we have

$$\begin{aligned} & 2c\|H_{j+1}\boldsymbol{g}_{j+1} - H_j\boldsymbol{g}_j\| = 2c\|H_j\boldsymbol{y}_j\| \frac{\boldsymbol{g}_j^T H_j \boldsymbol{g}_j}{\boldsymbol{y}_j^T H_j \boldsymbol{y}_j} \\ \leq\ & (c^2 + \|H_j\boldsymbol{g}_j\|^2) \frac{\boldsymbol{g}_j^T H_j \boldsymbol{g}_j}{\boldsymbol{y}_j^T H_j \boldsymbol{y}_j} \\ \leq\ & (\boldsymbol{g}_{j+1}^T H_{j+1}\boldsymbol{g}_{j+1}\mathrm{Tr}(H_{j+1}^{-1}) + \|H_j\boldsymbol{y}_j\|^2) \frac{\boldsymbol{g}_j^T H_j \boldsymbol{g}_j}{\boldsymbol{y}_j^T H_j \boldsymbol{y}_j} \\ \leq\ & ((\boldsymbol{g}_j^T H_j \boldsymbol{g}_j)\mathrm{Tr}(H_{j+1}^{-1}) + \|H_j\boldsymbol{y}\|^2) \frac{\boldsymbol{g}_j^T H_j \boldsymbol{g}_j}{\boldsymbol{y}_j^T H_j \boldsymbol{g}_j}. \end{aligned}$$

Taking summation for $j = 1, 2, \cdots, k$ and adding (10.8), (10.9) and the summation together, we obtain $2cv_k \leq w_k$. □

Lemma 10.2.5 *If (10.1) holds and $f \in C^{1.1}$, then $\lim_{k\to\infty} u_k = \infty$.*

Proof. Since $\boldsymbol{\sigma}_j^T \boldsymbol{y}_j = \lambda \boldsymbol{g}_j^T H_j \boldsymbol{g}_j$, we have

$$\|\boldsymbol{\sigma}_j\|^2(\lambda_j \boldsymbol{g}_j^T H_j \boldsymbol{g}_j)^{-1} = \|\boldsymbol{\sigma}_j\|^2/(\boldsymbol{\sigma}_j^T \boldsymbol{y}_j) \geq \|\boldsymbol{\sigma}_j\|/\|\boldsymbol{y}_j\| \geq L^{-1}.$$

Thus

$$\sum_{j=1}^{\infty} \|\boldsymbol{\sigma}_j\|^2(\lambda_j \boldsymbol{g}_j^T H_j \boldsymbol{g}_j)^{-1} = \infty.$$

Denote $p_j = \|\boldsymbol{\sigma}_j\|^2(\lambda_j \boldsymbol{g}_j^T H_j \boldsymbol{g}_j)^{-1}$ and $q_j = (\boldsymbol{g}_j^T H_j \boldsymbol{g}_j)^{-1}$. Then

$$\|\boldsymbol{g}_j\|^2 q_j \leq \mathrm{Tr}(H_j^{-1}).$$

From (10.5), we have

$$\mathrm{Tr}(H_{i+1}^{-1}) = \mathrm{Tr}(H_i^{-1}) + \frac{\|\boldsymbol{g}_{i+1}\|^2 - c^2}{\boldsymbol{g}_i^T H_i \boldsymbol{g} - i} - \frac{\|\boldsymbol{g}_i\|^2 - c^2}{\boldsymbol{g}_i^T H_i \boldsymbol{g}_i} - \frac{\|\boldsymbol{y}_i\|^2}{\boldsymbol{\sigma}_i^T \boldsymbol{y}_i}.$$

Taking summation and using Abel formula for summation of products, we obtain

$$\begin{aligned}
\mathrm{Tr}(H_j^{-1}) &= \mathrm{Tr}(H_1^{-1}) + \frac{\|\boldsymbol{g}_j\|^2 - c^2}{\boldsymbol{g}_j^T H_j \boldsymbol{g}_j} - \frac{\|\boldsymbol{g}_1\|^2 - c^2}{\boldsymbol{g}_1^T H_1 \boldsymbol{g}_1} - \sum_{i=1}^{j-1} \frac{\|\boldsymbol{y}_i\|^2}{\boldsymbol{\sigma}_i^T \boldsymbol{y}_i} \\
&\quad + \sum_{i=1}^{j-1} (\|\boldsymbol{g}_{i+1}\|^2 - c^2)\left(\frac{1}{\boldsymbol{g}_i^T H_i \boldsymbol{g}_i} - \frac{1}{\boldsymbol{g}_{i+1}^T H_{i+1} \boldsymbol{g}_{i+1}}\right) \\
&\leq \mathrm{Tr}(H_1^{-1}) + (\|\boldsymbol{g}_j\|^2 - c^2) q_j - L^2 \sum_{i=1}^{j-1} \frac{\|\boldsymbol{\sigma}_i\|^2}{\boldsymbol{\sigma}_i^T \boldsymbol{y}_i}.
\end{aligned}$$

Thus

$$\|\boldsymbol{g}_j\|^2 q_j \leq \mathrm{Tr}(H_1^{-1}) + (\|\boldsymbol{g}_j\|^2 - c^2) q_j + L^2 \sum_{i=1}^{j-1} p_i,$$

so,

$$q_j \leq \frac{1}{c^2}\left(\mathrm{Tr}(H_1^{-1}) + L^2 \sum_{i=1}^{j-1} p_i\right).$$

It follows that for any $k < k'$,

$$\sum_{j=k}^{k'} p_j/q_j \geq c^2 \frac{\sum_{j=k}^{k'} p_j}{\mathrm{Tr}(H_1^{-1}) + L^2 \sum_{j=k}^{k'} p_j}.$$

Letting $k' \to \infty$, we obtain that for any k,

$$\sum_{j=k}^{\infty} p_j/q_j \geq c^2/L^2 > 0.$$

Thus, $\sum_{j=1}^{\infty} p_j/q_j = \infty$, i.e. $\lim_{k\to\infty} u_k = \infty$. □

We complete the proof of Theorem 10.2.2 by proving the following lemma.

Lemma 10.2.6 *If (10.1) holds, $f \in C^{1.1}$ and $\{\boldsymbol{x}_k\}$ is convergent, then*

$$\lim_{k\to\infty} u_k/v_k = 0.$$

Proof. Suppose that $\boldsymbol{x}_k$ converges to $\boldsymbol{x}^*$. Applying Abel formula for summation of inner products, we obtain that for $k > k'$,

$$\begin{aligned}
u_k &= \sum_{j=1}^{k} \|\boldsymbol{\sigma}_j\|^2/\lambda_j = \sum_{j=1}^{k}((\boldsymbol{x}_{j+1} - \boldsymbol{x}^*) - (\boldsymbol{x}_j - \boldsymbol{x}^*))^T H_j \boldsymbol{g}_j \\
&= -(\boldsymbol{x}_{k+1} - \boldsymbol{x}^*)^T H_{k+1} \boldsymbol{g}_{k+1} + (\boldsymbol{x}_1 - \boldsymbol{x}^*)^T H_1 \boldsymbol{g}_1 \\
&\quad + \sum_{j=1}^{k} (\boldsymbol{x}_{j+1} - \boldsymbol{x}^*)^T (H_{j+1}\boldsymbol{g}_{j+1} - H_j \boldsymbol{g}_j) \\
&\leq \|\boldsymbol{x}_{k+1} - \boldsymbol{x}^*\| \|H_{k+1}\boldsymbol{g}_{k+1}\| + \|\boldsymbol{x}_1 - \boldsymbol{x}^*\| \|H_1 \boldsymbol{g}_1\| \\
&\quad + \sum_{j=1}^{k} \|\boldsymbol{x}_{j+1} - \boldsymbol{x}^*\| \|H_{j+1}\boldsymbol{g}_{j+1} - H_j \boldsymbol{g}_j\| \\
&\leq (\max_{j>k'} \|\boldsymbol{x}_j - \boldsymbol{x}^*\|) v_k + (\max_{j\leq k'} \|\boldsymbol{x}_j - \boldsymbol{x}^*\|) v_{k'}.
\end{aligned}$$

Since $u_k \to \infty$, we must have $v_k \to \infty$. Hence,

$$0 \leq \limsup_{k\to\infty} u_k/v_k \leq \max_{j>k'} \|\boldsymbol{x}_j - \boldsymbol{x}^*\|.$$

Letting $k' \to \infty$, we obtain $\lim_{k\to\infty} u_k/v_k = 0$. □

Corollary 10.2.7 *Let f be a twice continuously differentiable real function on R^n. If the sequence generated by DFP method with exact line search converges, then it converges to a stationary point.*

Proof. Since f is twice continuously differentiable, it's gradient satisfies the Lipschitz condition in a compact region containing the sequence. By Theorem 10.2.2, this sequence converges to a stationary point. □

Corollary 10.2.8 *Let f be a twice continuously differentiable strictly concave real function on R^n. Suppose f has a maximum. Then every infinite sequence generated by DFP method with exact line search converges to the maximum point.*

Proof. Since f is twice continuously differentiable, the second derivative of f in the level set $\{\boldsymbol{x} \mid f(\boldsymbol{x}) \geq f(\boldsymbol{x}_1)\}$ is bounded. It follows that there exists $L > 0$ such that for $\boldsymbol{x}$ and $\boldsymbol{y}$ in the level set

$$\|\boldsymbol{g}(\boldsymbol{x}) - \boldsymbol{g}(\boldsymbol{y})\| \leq L\|\boldsymbol{x} - \boldsymbol{y}\|.$$

By Theorem 10.2.2, for any sequence $\{\boldsymbol{x}_k\}$ generated by DFP method with exact line search we have

$$\liminf_{k\to\infty} \|\boldsymbol{g}_k\| = 0.$$

It follows that there exists a subsequence $\{\boldsymbol{x}_k\}_{k\in K}$ such that

$$\boldsymbol{g}_k \to \mathbf{0} \text{ as } k \to \infty, k \in K.$$

Since the level set $\{\boldsymbol{x} \mid f(\boldsymbol{x}) \geq f(\boldsymbol{x}_1)\}$ is bounded, we may assume that $\{\boldsymbol{x}_k\}_{k\in K}$ converges to a point $\boldsymbol{x}^*$. Clearly, $\boldsymbol{g}(\boldsymbol{x}^*) = \mathbf{0}$. Since f is concave, $\boldsymbol{x}^*$ is a maximum point. Since $\{f(\boldsymbol{x}_k)\}$ is increasing, it converges to $f(\boldsymbol{x}^*)$. □

10.3 Goldfarb's Method

Consider the linearly constrained problem (4.1) with the regularity condition. The following is a method of quasi-Newton type for it.

Goldfard's Method: *Choose an initial feasible point $\boldsymbol{x}_1$ and set $H_1 = P_{J_1}$. At iteration $k = 1, 2, \cdots$, we carry out the following steps:*

1. *Compute $\boldsymbol{g}_k$, $P_{J_k}\boldsymbol{g}_k$ and $(u_{kj}, j \in J_k)^T = (A_{J_k}^T A_{J_k})^{-1} A_{J_k}^T \boldsymbol{g}_k$. If $P_{J_k}\boldsymbol{g}_k = \mathbf{0}$ and $u_{kj} \leq 0$ for $j \in J_k \setminus M'$, then $\boldsymbol{x}_k$ is a Kuhn-Tucker point and the computation halts.*

2. *If $P_{J_k}\boldsymbol{g}_k \neq \mathbf{0}$ or $u_{kj} > 0$ for some $j \in J_k \setminus M'$, then we have either*

$$\|P_{J_k}\boldsymbol{g}_k\| > \frac{1}{2}u_{kh}b_{hh}^{-1/2} \tag{10.10}$$

or

$$\|P_{J_k}\boldsymbol{g}_k\| \leq \frac{1}{2}u_{kh}b_{hh}^{-1/2} \tag{10.11}$$

where b_{jj} is the jth diagonal element of $(A_{J_k}^T A_{J_k})^{-1}$ and

$$u_{kh}b_{hh}^{-1/2} = \max\{u_{kj}b_{jj}^{-1/2} \mid j \in J_k \setminus M'\}. \tag{10.12}$$

If (10.10) holds, then set $G_k = H_k$. Otherwise, (10.11) holds and we set

$$G_k = H_k + \frac{P_{J_k\setminus h}\boldsymbol{a}_h\boldsymbol{a}_h^T P_{J_k\setminus h}}{\boldsymbol{a}_h^T P_{J_k\setminus h}\boldsymbol{a}_h}. \tag{10.13}$$

3. *Let $\boldsymbol{d}_k = G_k\boldsymbol{g}_k$. Compute*

$$\bar{\lambda}_k = \begin{cases} \infty & \text{if } \boldsymbol{a}_j^T\boldsymbol{d}_k \geq 0 \text{ for all } j \notin J_k, \\ \min\left\{ \frac{b_j - \boldsymbol{a}_j^T\boldsymbol{x}_k}{\boldsymbol{a}_j^T\boldsymbol{d}_k} \,\middle|\, \boldsymbol{a}_j^T\boldsymbol{d}_k < 0 \right\} & \text{otherwise.} \end{cases}$$

Find a new point $\boldsymbol{x}_{k+1} = \boldsymbol{x}_k + \lambda_k\boldsymbol{d}_k$ $(0 < \lambda_k \leq \bar{\lambda}_k)$ by a line search procedure. If such a new point cannot be found, then halt; otherwise, go to the next step.

4. *If $J_{k+1} \setminus J_k = \{1, \cdots, t\} \neq \emptyset$, then compute H_{k+1} by the following recursive formula*

$$H^{(i+1)} = H^{(i)} - \frac{H^{(i)}\boldsymbol{a}_i\boldsymbol{a}_i^T H^{(i)}}{\boldsymbol{a}_i^T H^{(i)}\boldsymbol{a}_i} \tag{10.14}$$

where $H^{(0)} = G_k$ and $H_k = H^{(t)}$. If $J_{k+1} = J_k$, then set

$$H_{k+1} = G_k - \frac{\boldsymbol{\sigma}_k\boldsymbol{\sigma}_k^T}{\boldsymbol{\sigma}_k^T\boldsymbol{g}_k} - \frac{G_k\boldsymbol{y}_k\boldsymbol{y}_k^T G_k}{\boldsymbol{y}_k^T G_k\boldsymbol{y}_k} \tag{10.15}$$

where $\boldsymbol{\sigma}_k = \boldsymbol{x}_{k+1} - \boldsymbol{x}_k = \lambda_k\boldsymbol{d}_k$ and $\boldsymbol{y}_k = \boldsymbol{g}_{k+1} - \boldsymbol{g}_k$.

No result on the global convergence of Goldfarb's method has been found in the literature. However, if we make a little change in step 4, that is, set $H_{k+1} = P_{J_{k+1}}$ instead of employing a recursive formula to compute H_{k+1} when $J_{k+1} \neq J_k$, then the global convergence can be established under certain condition. We introduce one of such results in the following.

First of all, we show some basic facts.

Lemma 10.3.1 *If a matrix H is symmetric and positive semidefinite, then $H\boldsymbol{y} = \boldsymbol{0}$ if and only if $\boldsymbol{y}^T H\boldsymbol{y} = 0$.*

Proof. Since H is symmetric and positive semidefinite, there is a matrix A such that $H = A^T A$. Thus, $\boldsymbol{y}^T H\boldsymbol{y} = \boldsymbol{y}^T A^T A\boldsymbol{y} = \|A\boldsymbol{y}\|^2$. It follows immediately that $\boldsymbol{y}^T H\boldsymbol{y} = 0$ implies $H\boldsymbol{y} = A^T A\boldsymbol{y} = \boldsymbol{0}$. The inverse is obvious. □

Lemma 10.3.2 *All matrices H_k and G_k computed by the algorithm are positive semidefinite, and for any $\boldsymbol{y}$,*

(1) $H_k\boldsymbol{y} = \boldsymbol{0}$ *if and only if* $P_{J_k}\boldsymbol{y} = \boldsymbol{0}$,

(2) $G_k\boldsymbol{y} = \boldsymbol{0}$ *if and only if*
$P_{J_k}\boldsymbol{y} = \boldsymbol{0}$ *when (10.10) holds and*
$P_{J_k\backslash h}\boldsymbol{y} = \boldsymbol{0}$ *when (10.11) holds.*

Furthermore, $G_k\boldsymbol{g}_k = \boldsymbol{0}$ if and only if $\boldsymbol{x}_k$ is a Kuhn-Tucker point, and if $\boldsymbol{x}_k$ is not a Kuhn-Tucker point, then $\boldsymbol{d}_k$ is an ascending feasible direction.

Proof. We prove it by induction on k. For $k = 1$ it is trivial. Next, suppose that H_k is positive semidefinite and (1) holds for k. We show that G_k and H_{k+1} are positive semidefinite, (2) holds for k and (1) holds for $k+1$.

First, we consider G_k. If (10.10) holds, then $G_k = H_k$, so G_k is positive semidefinite and (2) holds for k. If (10.11) holds, then G_k is determined by (10.13). So, for any $\boldsymbol{y}$ we have

$$\boldsymbol{y}^T G_k\boldsymbol{y} = \boldsymbol{y}^T H_k\boldsymbol{y} + \frac{\boldsymbol{a}_h^T P_{J_k\backslash h}\boldsymbol{y}}{\boldsymbol{a}_h^T P_{J_k\backslash h}\boldsymbol{a}_h} \geq 0.$$

Thus, G_k is positive semidefinite and in addition, $\boldsymbol{y}^T G_k\boldsymbol{y} = 0$ if and only if $\boldsymbol{y}^T H_k\boldsymbol{y} = 0$ and $\boldsymbol{a}_h^T P_{J_k\backslash h}\boldsymbol{y} = 0$. Moreover, by (4.13),

$$\|P_{J_k\backslash h}\boldsymbol{y}\|^2 = \|P_{J_k}\|^2 + \frac{(\boldsymbol{a}_k^T P_{J_k\backslash h}\boldsymbol{y})^2}{\boldsymbol{a}_h^T P_{J_k\backslash h}\boldsymbol{a}_h}.$$

It follows that $P_{J_k\setminus h}\boldsymbol{y} = \mathbf{0}$ if and only if $P_{J_k}\boldsymbol{y} = \mathbf{0}$ and $\boldsymbol{a}_h^T P_{J_k\setminus h}\boldsymbol{y} = 0$. By Lemma 10.3.1 and the induction hypothesis, $\boldsymbol{y}^T H_k \boldsymbol{y} = 0$ if and only if $P_{J_k}\boldsymbol{y} = \mathbf{0}$. Therefore, $G_k\boldsymbol{y} = \mathbf{0}$ if and only if $P_{J_k\setminus h}\boldsymbol{y} = \mathbf{0}$. From (2), it is easy to see that $G_k\boldsymbol{g}_k = \mathbf{0}$ if and only if $\boldsymbol{x}_k$ is a Kuhn-Tucker point.

Now, we look at H_{k+1}. If $J_{k+1} \setminus J_k \neq \emptyset$, then $H_{k+1} = P_{J_{k+1}}$, so H_{k+1} is positive semidefinite and (1) holds for $k+1$. If $J_{k+1} \setminus J_k = \emptyset$, then

$$J_{k+1} = \begin{cases} J_k & \text{if (10.10) holds,} \\ J_k \setminus h & \text{if (10.11) holds.} \end{cases}$$

By (2), we have that $G_k\boldsymbol{y} = \mathbf{0}$ if and only if $P_{J_{k+1}}\boldsymbol{y} = \mathbf{0}$. Thus, in order to show (1) for $k+1$, it suffices to prove that $H_{k+1}\boldsymbol{y} = \mathbf{0}$ if and only if $G_k\boldsymbol{y} = \mathbf{0}$. Note that H_{k+1} is determined by (10.15). So for any $\boldsymbol{y}$,

$$\begin{aligned} \boldsymbol{y}^T H_{k+1}\boldsymbol{y} &= \boldsymbol{y}^T G_k \boldsymbol{y} - \frac{(\boldsymbol{\sigma}_k^T \boldsymbol{y})^2}{\boldsymbol{\sigma}_k^T \boldsymbol{y}_k} - \frac{(\boldsymbol{y}_k^T G_k \boldsymbol{y})^2}{\boldsymbol{y}_k^T G_k \boldsymbol{y}_k} \\ &= \frac{(\boldsymbol{y}^T G_k \boldsymbol{y})(\boldsymbol{y}_k^T G_k \boldsymbol{y}_k) - (\boldsymbol{y}_k^T G_k \boldsymbol{y})^2}{\boldsymbol{y}_k^T G_k \boldsymbol{y}_k} - \frac{(\boldsymbol{\sigma}_k^T \boldsymbol{y})^2}{\boldsymbol{\sigma}_k^T \boldsymbol{y}_k}. \end{aligned}$$

Since $\boldsymbol{g}_{k+1}^T \boldsymbol{d}_k = 0$, we have $\boldsymbol{\sigma}_k^T \boldsymbol{y}_k = -\lambda_k \boldsymbol{g}_k^T G_k \boldsymbol{g}_k \leq 0$. Note that when H_{k+1} is computed, $\boldsymbol{x}_k$ must not be a Kuhn-Tucker. (Otherwise, the computation halts before H_{k+1} is computed.) Thus, $G_k\boldsymbol{g}_k \neq \mathbf{0}$. By Lemma 10.3.1, $\boldsymbol{\sigma}_k^T \boldsymbol{y}_k \neq 0$. Now, from (10.15), we have $\boldsymbol{y}^T H_{k+1}\boldsymbol{y} \geq 0$. Hence H_{k+1} is positive semidefinite. In addition, from (10.15), we can also see that $\boldsymbol{y}^T H_{k+1}\boldsymbol{y} = 0$ if and only if $\boldsymbol{\sigma}_k^T \boldsymbol{y} = 0$ and $(\boldsymbol{y}^T G_k \boldsymbol{y})(\boldsymbol{y}_k^T G_k \boldsymbol{y}_k) = (\boldsymbol{y}_k^T G_k \boldsymbol{y})^2$. But the latter equality holds if and only if $G_k(\boldsymbol{y} - \beta \boldsymbol{y}_k) = \mathbf{0}$ for some real number β. Suppose that $H_{k+1}\boldsymbol{y} = \mathbf{0}$. Then $-\beta\lambda_k \boldsymbol{g}_k^T G_k \boldsymbol{g}_k = \beta\lambda \boldsymbol{g}_k^T G_k \boldsymbol{y}_k = \lambda_k \boldsymbol{g}_k^T G_k \boldsymbol{y} = \boldsymbol{\sigma}_k^T \boldsymbol{y} = 0$. So, $\beta = 0$. Hence $G_k\boldsymbol{y} = \mathbf{0}$. Conversely, suppose $G_k\boldsymbol{y} = \mathbf{0}$. Then $\boldsymbol{\sigma}_k^T \boldsymbol{y} = -\lambda_k \boldsymbol{g}_k^T G_k \boldsymbol{y} = 0$ and $\boldsymbol{y}^T G_k \boldsymbol{y} = 0 = \boldsymbol{y}_k^T G_k \boldsymbol{y}$. Thus $\boldsymbol{y}^T H_{k+1}\boldsymbol{y} = 0$. Hence $H_{k+1}\boldsymbol{y} = \mathbf{0}$.

Now, suppose that $\boldsymbol{x}_k$ is not a Kuhn-Tucker point. Then $G_k\boldsymbol{g}_k \neq \mathbf{0}$. By Lemma 10.3.1, $\boldsymbol{g}_k^T \boldsymbol{d}_k = \boldsymbol{g}_k^T G_k \boldsymbol{g}_k > 0$, i.e. $\boldsymbol{d}_k$ is an ascending direction. If (10.10) holds, then by (2), for any $j \in J_k$, $\boldsymbol{a}_j^T \boldsymbol{d}_k = (G_k \boldsymbol{a}_j)^T \boldsymbol{g}_k = 0$. So $\boldsymbol{d}_k$ is a feasible direction. If (10.11) holds, then by (2), for $j \in J_k \setminus h$, $\boldsymbol{a}_j^T \boldsymbol{d}_k = 0$. Moreover, by (10.13), $\boldsymbol{a}_h^T \boldsymbol{d}_k = \boldsymbol{a}_h^T P_{J_k\setminus h}\boldsymbol{g}_k = u_{kh}\|P_{J_k\setminus h}\boldsymbol{a}_h\|^2 \geq 0$. Thus, $\boldsymbol{d}_k$ is a feasible direction. □

Lemma 10.3.3 *Let $\{\boldsymbol{x}_k\}$ be an infinite sequence generated by G-KWL method with a normal line search procedure. If no cluster point of the sequence* *is a Kuhn-Tucker point, then there exists k' such that for $k > k'$, (10.10) holds and $J_k = J_{k+1}$.*

Proof. It suffices to prove that there exists k'' such that for $k > k''$ (10.10) holds. In fact, if (10.10) holds, then $J_k \subseteq J_{k+1}$. Since the number of indices is finite, the existence of k'' implies the existence of k'. Next, for contradiction, we suppose that such a k'' does not exist, i.e. there are infinitely many k such that (10.11) holds, i.e. $J_k \setminus J_{k+1} \neq \emptyset$. It follows that there are infinitely many k such that $J_{k+1} \setminus J_k \neq \emptyset$.

Case 1. For sufficiently large k, $J_{k+1} \setminus J_k \neq \emptyset$. It follows that for sufficiently large k, $|J_{k+1} \setminus J_k| = |J_k \setminus J_{k+1}| = 1$. Thus, for sufficiently large k, $H_k = P_{J_k}$. This means that for large k, the algorithm becomes Rosen's method with a difference in multiplers $b_{jj}^{-1/2}$. However, those multiplers do not affect the convergence proof. Thus, by Theorem 6.2.1, we see that every cluster point of $\{\boldsymbol{x}_k\}$ is a Kuhn-Tucker point, contradicting our assumption.

Case 2. There are infinitely many k such that $J_{k+1} \subseteq J_k$. It follows that there exists an infinite set K such that for $k \in K$, $J_{k+1} \subseteq J_k$ and $J_k \setminus J_{k-1} \neq \emptyset$. Note that for $k \in K$, $H_k = P_{J_k}$ and

$$\boldsymbol{d}_k = \begin{cases} P_{J_k}\boldsymbol{g}_k & \text{if (10.10) holds,} \\ P_{J_k \setminus h}\boldsymbol{g}_k & \text{if (10.11) holds.} \end{cases}$$

Let $\boldsymbol{x}^*$ be a cluster point of $\{\boldsymbol{x}_k\}_{k \in K}$. Let K' be an infinite subset of K such that $\boldsymbol{x}_k \to \boldsymbol{x}^*$ and $\boldsymbol{d}_k \to \boldsymbol{d}^*$ as $k \to \infty$, $k \in K'$. Since $J_{k+1} \subseteq J_k$, we have $\lambda_k < \bar{\lambda}_k$. By Theorem 5.2.3, $(\boldsymbol{g}^*)^T \boldsymbol{d}^* = 0$. Now, by an argument similar to the proof of Lemma 6.2.2, we can prove that $\boldsymbol{x}^*$ is a Kuhn-Tucker point, a contradiction. □

It is worth mentioning that in the proof of Lemma 10.3.3 we did not use (10.15). In fact, Lemma 10.3.3 holds for every variable metric formula in Huang's family.

When the objective function is concave or the generated sequence is convergent, every cluster point of the sequence is a Kuhn-Tucker point if and only if there is a cluster point being a Kuhn-Tucker point. In this case, Lemma 10.3.3 enables us to study the convergence under the condition that the active constraints are fixed. That is, the computation is landing in an affine subset, which is equivalent to the unconstrained case. Note that in such a case, the modified Goldfarb method is equivalent to a variable metric method. Thus, the convergence of the modified Goldfarb method depends on the same conditions as the convergence of the corresponding variable metric method. When we use (10.15), the corresponding variable metric method is the well-known DFP method.

Corresponding to the convergence theorems for DFP method, we have the following result for the modified Goldfarb method.

Theorem 10.3.4 *Suppose that the level set* $\{\boldsymbol{x} \in \Omega \mid f(\boldsymbol{x}) \geq f(\boldsymbol{x}_1)\}$ *is bounded, and f is concave function, having continuous bounded second derivative. Then the modified Goldfarb method with the exact line searches either stops at a maximum or generates an infinite sequence whose cluster points are all maximum points.*

Proof. Although $\bar{\lambda}_k$ may be equal to ∞, the exact line search is always successful under the assumption. Thus, if the algorithm stops, then it must stops at Step 1. Hence the current point $\boldsymbol{x}_k$ is a Kuhn-Tucker point. If the algorithm generates an infinite sequence, then by Lemma 10.3.3, either a cluster point of the generated sequence is a Kuhn-Tucker point or there exists k' such that for $k > k'$, (10.10) holds and $J_k = J$ for a certain J. In the former case, every cluster point of the generated sequence is already a maximum point since f is concave. In the latter case, by Theorem 10.3.4, we have $P_J\boldsymbol{g}^* = \mathbf{0}$ for any cluster point $\boldsymbol{x}^*$ of the sequence. Letting $k \to \infty$ in (10.10), we see that $u_j^* \leq 0$ for $j \in J \setminus M'$ where $(u_j^*, j \in J)^T = (A_J^T A_J)^{-1} A_J^T \boldsymbol{g}^*$. It follows that $\boldsymbol{x}^*$ is a Kuhn-Tucker point. Hence $\boldsymbol{x}^*$ is a maximum point. □

Exercises

1. Prove that $b_{jj}^{-1} = \|P_{J_k \setminus j} \boldsymbol{a}_j\|^2$.

Chapter 11

Minimax

Nothing takes place in the world
whose meaning is not that of some maximum or minimum.
L. EULER

Classical minimax theory initiated by Von Neumann, together with duality and saddle point analysis, has played a critical role in optimization and game theory. However, minimax problems and techniques appear in a very wide area of disciplines. For example, many combinatorial optimization problems, including scheduling, location, allocation, packing, searching, and triangulation, can be represented as a minimax problem.

Consider the problem $\min_{x \in X} \max_{i \in I} f_i(x)$ where X is a convex set, I is a finite set of indices and $f_i(x)$'s are continuous concave functions of x. In this chapter, we study a characterization of $x \in X$ at which the minimax value is achieved. We also study some interesting applications of this characterization.

11.1 Optimality Condition

Minimax is an important topic in optimization. There are two fundamental ideas used to solve minimax problems.

The first is the search for a basis, that is, for the problem

$$\min_{x \in X} \max_{y \in Y} f(x, y)$$

determine first a finite subset B of X such that

$$\min_{x \in X} \max_{y \in Y} f(x, y) = \min_{x \in B} \max_{y \in Y} f(x, y)$$

and then search for an optimal x^* from B in finitely many steps.

The second is the determination of a saddle point. A point (x^*, y^*) is called a *saddle point* of $f(x, y)$ on the set $X \times Y$ if for any $x \in X$ and $y \in Y$,

$$f(x^*, y) \leq f(x^*, y^*) \leq f(x, y^*).$$

It follows that for a saddle point (x^*, y^*),

$$\min_{x \in X} \max_{y \in Y} f(x, y) = f(x^*, y^*) = \max_{y \in Y} \min_{x \in X} f(x, y).$$

These two ideas have resulted two important mathematical branches. P. L. Chebyshev is probably the first person who made an important contribution to the first idea. He discovered the theory of best approximation. The second idea was extensively studied by Von Neumann. He initiated the game theory. Since Von Neumann, many efforts have been made to find various sufficient conditions for a point being a saddle point. This involves a great deal of mathematics including fixed point theory.

While there is a very large literature on minimax, we select only a small part for discussion in this chapter. In fact, including all results about minimax could be subject of a book instead of a chapter. In this chapter, we will focus only on recent developments of the first idea.

The original problem considered by Chebyshev is as follows: Given a list of values of some real function:

$$y_k = f(x_k), \quad k = 0, \ldots, m,$$

find a polynomial p of degree at most n $(n < m)$ which provides the best approximation at these m points, that is, p minimizes

$$\max_{k=0,\ldots,m} |y_k - p(x_k)|.$$

Chebyshev gave a beautiful result about the solution of this problem.

First, consider $m = n + 1$. In this case, the polynomial of the best approximation exists and is unique. Chebyshev proved that a polynomial p is the best approximation if and only if for some h,

$$(-1)^k h + p(x_k) = y_k, \qquad \text{for } k = 0, \ldots, n + 1.$$

Furthermore, h and p can be constructed explicitly. This p is called a *Chebyshev interpolating polynomial.*

For general m, a subset of $n+2$ x_k's is called a *basis.* Each basis σ determines a Chebyshev interpolating polynomial p_σ and a value

$$h(\sigma) = \max_{x_k \in \sigma} |y_k - p_{\sigma^*}(x_k)|.$$

A basis σ^* is called an *extremal basis* if

$$h(\sigma^*) = \max_{\sigma} h(\sigma)$$

where σ is over all bases. Chebyshev showed the following:

Theorem 11.1.1 *There exists a unique polynomial of best approximation. A polynomial p is the polynomial of best approximation if and only if p is a Chebyshev interpolating polynomial for some extremal basis.*

There are other ways to characterize the extremal basis. In fact, Chebyshev also proved that σ^* is an extremal basis if and only if

$$h(\sigma^*) = \max_{k=0,\ldots,m} |y_k - p_{\sigma^*}(x_k)|.$$

For each polynomial p, define

$$I(p) = \{i \mid |y_i - p(x_i)| = \max_{k=0,\ldots,m} |y_k - p(x_k)|\}.$$

$I(p)$ is *maximal* if no polynomial q exists such that $I(p) \neq I(q)$ and $I(p) \subset I(q)$. From the second characterization of the extremal basis, it is not hard to prove the following.

Proposition 11.1.2 *σ^* is an extremal basis if and only if $I(p_{\sigma^*})$ is maximal.*

The Chebyshev problem can be transformed into a linear programming problem as follows:

$$\begin{array}{ll} \min & z \\ \text{subject to} & -z \le a_0 + a_1 x_k + \cdots + a_n x_k^n - y_k \le z \\ & k = 0, \ldots, m. \end{array}$$

Note that this linear programming has $n+2$ variables and $2(m+1)$ constraints. For an extremal basis σ^*, p_{σ^*} would make $n+2$ constraints *active* (i.e., the equality sign holds for those constraints). This means that each extremal basis corresponds to a feasible basis of the above linear programming in the following standard form:

$$\begin{array}{ll} \min & z \\ \text{subject to} & u_k - z = a_0 + a_1 x_k + \cdots + a_n x_k^n - y_k = z - v_k \\ & u_k \geq 0, \quad v_k \geq 0 \\ & k = 0, \ldots, m. \end{array}$$

Linear programming problems are closely related to minimax problems. In fact, there are several ways to transform linear programming into a minimax problem. For example, consider a linear programming

$$\begin{array}{ll} \min & c^T x \\ \text{subject to} & Ax = b \\ & x \geq 0 \end{array}$$

and its dual

$$\begin{array}{ll} \max & b^T y \\ \text{subject to} & A^T y \leq c. \end{array}$$

For any feasible solution x of the original linear programming and any feasible solution y of the dual linear programming, $c^T x \geq b^T y$. The equality sign holds only if the two feasible solutions are actually optimal solutions for the two linear programming, respectively. This is equivalent to the following minimax problem which achieves the minimax value 0:

$$\min_{(x,y)} \max(c^T x - b^T y, -x, Ax - b, b - Ax, A^T y - c).$$

In the above, we see already the problem in the following form:

$$\min_{x \in X} \max_{i=1,\cdots,m} f_i(x).$$

Now, we consider it with a little general conditions. We assume that X is a polytope in $\mathbf{R}^n$ and $f_i(x)$'s are continuous concave functions of x. We will extend Chebyshev's idea to this problem.

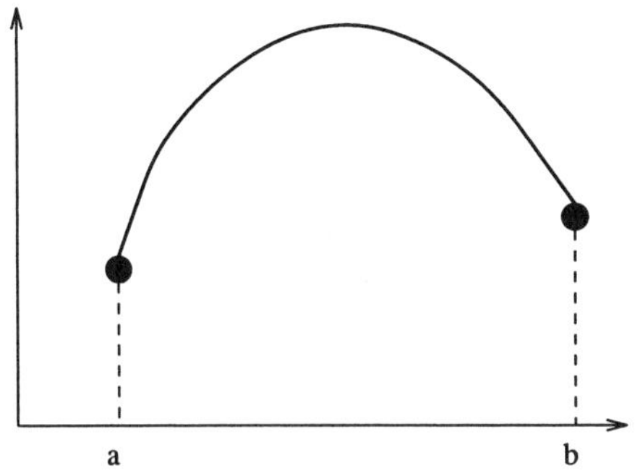

Figure 11.1: The minimum point of a concave function.

The simplest case is $m = n = 1$. As shown in Figure 11.1, the minimum value of a concave function $f_1(x)$ on the interval $[a, b]$ is achieved at a or b. For $m = 1$ and general n, it is well-known that the minimum value of $f_1(x)$ is achieved at a vertex of the polytope X. What we are interested in this chapter is the case $m > 1$. If $m > 1$ and $n = 1$, then as shown in Figure 11.2, $g(x) = \max_{i=1,\cdots,m} f_i(x)$ is a piecewise concave function. Thus, the minimum value of $g(x)$ on the interval $[a, b]$ is achieved at an endpoint of a concave piece.

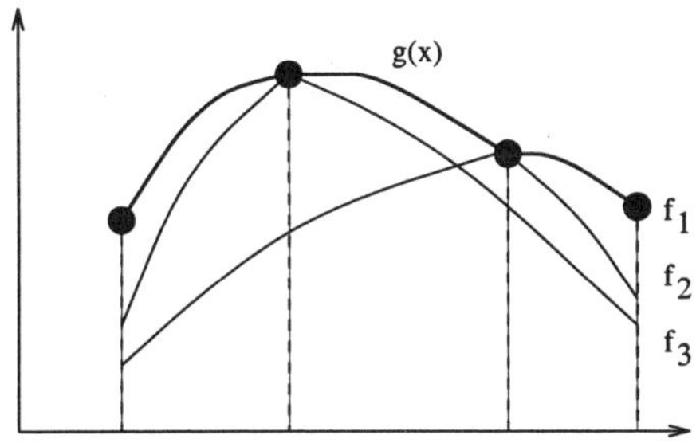

Figure 11.2: A piecewise concave function $g(x)$.

Similarly, for $m > 1$, the polytope P can be divided into small regions in each of which $g(x)$ is concave. These small regions can be defined by

$$X_i = \{x \in X \mid f_i(x) = g(x)\}.$$

However, they may not be convex. Thus, where the minimum value of $g(x)$ can be achieved is not so easy to see. Du and Hwang in 1990 found that the minimum value of $g(x)$ can still be achieved at a "vertex" of the small regions where the "vertex" is defined in the following way.

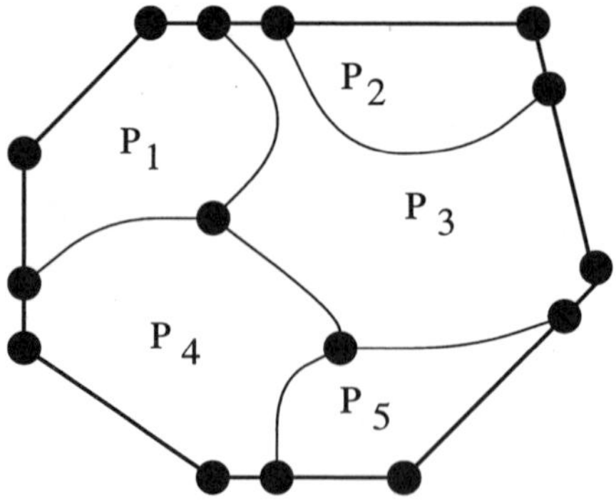

Figure 11.3: DH-point

Let us first give a new definition for the vertex of a polytope. Consider the polytope $X = \{x \mid a_j^T x \geq b_j, j = 1, \cdots, k\}$. Denote $J(x) = \{j \mid a_j^T x = b_j\}$. A point x in X is a vertex if $J(x)$ is maximal, i.e., there does not exist $y \in X$ such that $J(x)$ is a proper subset of $J(y)$. This definition is different from the traditional one: x is a vertex if $x = \frac{1}{2}y + \frac{1}{2}z$ for y, $z \in X$ implies $x = y = z$. However, they are equivalent for polytopes.

Now, a point x in X is called a *DH-point* if $J(x) \cup M(x)$ is maximal where

$$M(x) = \{i \mid f_i(x) = g(x)\}.$$

Theorem 11.1.3 (Du and Hwang, 1990) *The minimum value of $g(x)$ is achieved at a DH-point.*

Proof. Let x^* be a minimum point for $g(x)$. Since all $f_i(x)$ are continuous, there is a neighborhood V of x^* such that for any $x \in V$, $M(x) \subseteq M(x^*)$. Let $Y = \{x \in X \mid a_j^T x = b_j \text{ for } j \in J(x^*)\}$. Then x^* is a *relative interior point* of Y, that is, for any $x \in Y$ and for sufficiently small number λ, $x^* + \lambda(x^* - x) \in Y$. Consider a DH-point $\hat{x}$ such that $M(x^*) \cup J(x^*) \subseteq M(\hat{x}) \cup J(\hat{x})$, i.e., $M(x^*) \subseteq M(\hat{x})$ and $J(x^*) \subseteq J(\hat{x})$. The latter inclusion implies that $\hat{x} \in Y$. We will show that $\hat{x}$ is also a minimum point. Therefore, the theorem is proved.

For contradiction, suppose that $\hat{x}$ is not a minimum point. Choose a positive λ sufficiently small such that

$$x(\lambda) = x^* + \lambda(x^* - \hat{x}) \in V \cap Y.$$

Thus, $M(x(\lambda)) \subseteq M(x^*) \subseteq M(\hat{x})$. Consider an index $i \in M(x(\lambda))$. Since x^* is a minimum point of $g(x)$, we have

$$f_i(x^*) < f_i(\hat{x}), \quad \text{and} \quad f_i(x^*) \leq f_i(x(\lambda)).$$

Note that

$$x^* = \frac{\lambda}{1+\lambda}\hat{x} + \frac{1}{1+\lambda}x(\lambda).$$

By the concavity of $f_i(x)$,

$$f_i(x^*) \geq \frac{\lambda}{1+\lambda}f_i(\hat{x}) + \frac{1}{1+\lambda}f_i(x(\lambda)) > f_i(x^*),$$

a contradiction. □

From the proof of the theorem we can also conclude the following:

Corollary 11.1.4 *If $I(x) \subseteq I(y)$, then $g(x) \geq g(y)$.*

Let us make some remarks on this minimax theorem.

Remark 1. A function f is *pseudo-concave* in a region if for any x and y in the region and for any $\lambda \in [0, 1]$,

$$f(\lambda x + (1-\lambda)y) \geq \min(f(x), f(y)).$$

The pseudo-concavity is clearly weaker than the concavity. In the theorem, the concavity of f_i can be replaced by the pseudo-concavity. For this replacement, the proof needs to be modified as follows: Choose a minimum point x^* with maximal $J(x)$ and a point $\hat{x}$ in Y with $M(x^*) \subseteq M(\hat{x})$. Prove the existence of V as above. By the pseudo-concavity of $f_i(x)$,

$$f_i(x^*) \geq \min(f_i(\hat{x}), f_i(x(\lambda)) \geq f_i(x^*),$$

for $i \in M(x(\lambda))$, $x(\lambda) = x^* + \lambda(\hat{x} - x^*) \in Y \cap V$ and $\lambda > 0$. It follows that for $x(\lambda) \in Y \cap V$, $x(\lambda)$ is a minimum point. Note that all minimum points form a closed set. There exists the maximum value λ^* such that $x(\lambda^*)$ is a minimum point. Clearly, $x(\lambda^*)$ cannot be a relative interior point of Y. (Otherwise, by the above argument, we can obtain a larger λ such that $x(\lambda)$ is a minimum point.) Thus, $J(x^*)$ is a proper subset of $J(x(\lambda^*))$, contradicting the choice of x^*. We state the result of this remark in the following.

Theorem 11.1.5 *Let $g(x) = \max_{i \in I} f_i(x)$ where f_i's are continuous pseudo-concave functions and I is a finite set of indices. Then the minimum value of $g(x)$ over a polytope is achieved at a DH-point.*

Remark 2. A similar result holds for the following minimax problem:

$$\min_{x} \max_{x \in I(x)} f_i(x)$$

where $I(x)$ is a finite index set varying as x varies. The following result has been very useful. The proof is similar to the proof of Theorem 11.1.3 (Figure 11.4).

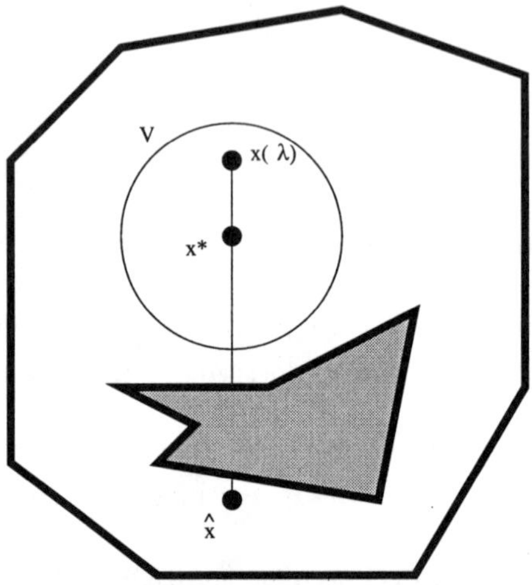

Figure 11.4: $I(x)$ is defined on a subset of X

Theorem 11.1.6 *Let $g(x) = \max_{i \in I} f_i(x)$ where the f_i's are continuous and pseudo-concave functions in a convex region X and $I(x)$ is a finite index set defined on a compact subset X' of P. Denote $M(x) = \{i \in I(x) \mid f_i(x) = g(x)\}$. Suppose that for any $x \in X$, there exists a neighborhood of x such that for any point y in the neighborhood, $M(y) \subseteq M(x)$. If the minimum value of $g(x)$ over X is achieved at an interior point of X', then this minimum value is achieved at a DH-point, i.e., a point with maximal $M(x)$ over X'. Moreover, if x is an interior minimum point in X' and $M(x) \subseteq M(y)$ for some $y \in X'$, then y is a minimum point.*

Remark 3. The finite index set I in above can be replaced by a compact set. The result can be stated as follows:

Theorem 11.1.7 (Du and Pardalos, 1994) *Let $f(x, y)$ be a continuous function on $X \times I$ where X is a polytope in $\mathbf{R}^m$ and I is a compact set in $\mathbf{R}^n$. Let $g(x) = \max_{y \in Y} f(x, y)$. If $f(x, y)$ is concave with respect to x, then the minimum value of $g(x)$ over X is achieved at some DH-point.*

The proof of this theorem is also the same as the proof of the previous theorem except that the existence of the neighborhood V needs to be derived from the compactness of I and the existence of $\hat{x}$ needs to be derived by Zorn's lemma.

11.2 Steiner Trees

Given three points on the Euclidean plane, let us study the problem of finding a point with the minimum total distance from this point to the three given points. This problem is called *Fermat problem.*

To solve this problem, let us consider a mechanical system as shown in Figure 11.5. Drill three holes on a table corresponding to the three points A,

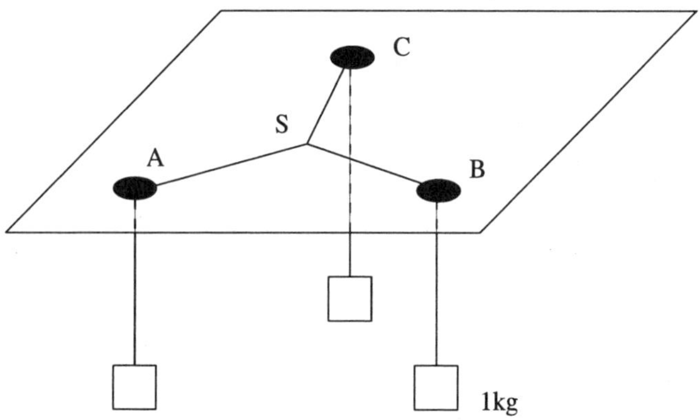

Figure 11.5: A mechanical system.

B, and C. Take three pieces of rope and put through three holes respectively. Tie them together above the table and hung a weight of 1kg on each piece of rope under the table. When the system is balanced, the common point S of the three rope pieces is the solution of Fermat problem. In fact, the potential energy of this system equals the product of 1 kg times the sum of heights of three weights which reaches the minimum when the system is balanced. The sum of heights of three weights equals a constant (three times the table height) minus the total length of rope pieces under the table which reaches the minimum when the total length of rope pieces above the table reaches the minimum, i.e. $SA + SB + SC$ reaches the minimum.

To describe S geometrically, let us consider two cases:

Case 1. S does not lie at A, B, and C. In this case, since the system of three forces of 1kg along SA, SB, and SC is balanced, we must have $\angle ASB = \angle BSC = \angle CSA = 120^o$. If follows that the given points A, B and C form a triangle with all inner angles less than 120°. Conversely, if the given points A, B and C form a triangle with all inner angles less than 120°, then S cannot lie at A, B, and C. For example, suppose S lies at A. Then the sum of the two forces along SB and SC is bigger than 1kg since $\angle BSC < 120^o$, which will pulls S away from A.

Case 2. S lies at A (or B or C). In this case, we must have $\angle BSC \geq 120^o$. Otherwise, the two forces along SB and SC will pull S away from A.

From the discussion in these two cases, we can see the solution of Fermat problem as follows.

If the given points A, B and C form a triangle with all inner angles less than 120°, then the solution S lies inside of the triangle ABC and satisfies the condition $\angle ASB = \angle BSC = \angle CSA = 120^o$.

If the triangle ABC has an angle not less than 120°, say $\angle A \geq 120^o$, then the solution is the point A.

There are two ways to generalize Fermat's problem. The first generalization is the problem of finding a point such that the sum of n distances from the point to n given points achieves minimum. This generalization is still called *Fermat problem.* The second one is the problem of finding a shortest network interconnecting n given points on the Euclidean plane. Clearly, the second problem is not a direct generalization. We see it as a generalization of Fermat problem just because, given three points, the shortest network interconnecting them is exactly the solution of Fermat problem. The second one is called the *Steiner tree problem.* In fact, such a shortest network must be a tree which is called a *Steiner minimum tree (SMT)* for the set of given points.

An SMT for a point set P, $SMT(P)$ may contain vertices not in P. Such vertices are called *Steiner points* while vertices in P are called *regular points.* It is clear that

(1) all leaves are regular points.

From the properties of SMT for three points, we can also see that

(2) any two edges meet at an angle of at least 120°,

(3) every Steiner point has degree exactly three, and every angle at a Steiner point equals 120°.

In fact, if two edges, say AB and BC, meet at an angle $\angle ABC$ of less than 120°, then by solution of the Fermat problem, we see that AB and BC cannot form a SMT for $\{A, B, C\}$. So, we can shorten the total length of the network by replacing AB and BC with an SMT for $\{A, B, C\}$. This proves (2). (3) follows immediately from (2).

A tree interconnecting the point set P and satisfying (1), (2) and (3) is called a *Steiner tree (ST)*. Its topology (the graph structure of the network) is called a *Steiner topology*.

A leaf in a ST must be a regular point. However, the inverse is false. If in a ST any regular point is a leaf, then the ST is called a *full* ST. An ST for n points can contain at most $n-2$ Steiner points. It contains exactly $n-2$ Steiner points if and only if it is a full ST. If a regular point is not a leaf, then we can split the ST at this regular point. In this way, the ST can be decomposed into edge-disjoint full sub-ST's. Such full sub-ST's are called *full components* of the ST. The topology of a full ST is also called a *full topology*.

Given a point set P and a full topology t, an ST with topology t on P can be constructed by Melzak's method as follows: The construction is a recursive procedure. First, find two regular points A and B which connect to the same Steiner point in the topology t. Then construct an equilateral triangle ABC (Figure 11.6). Using C instead of A and B, we get a new

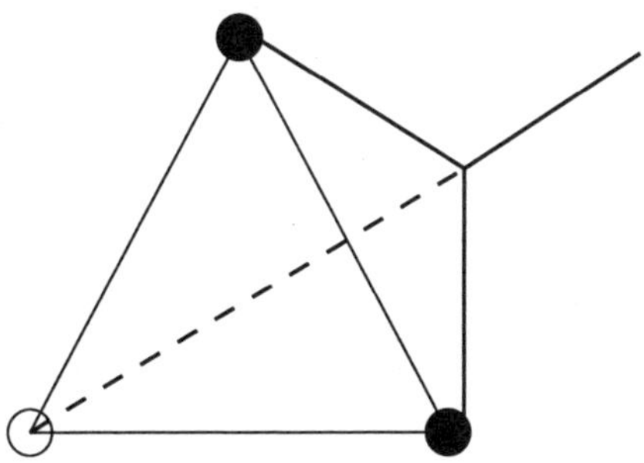

Figure 11.6: Melzak method.

point set P' $(= (P \setminus \{A, B\}) \cup \{C\})$ and a new topology on P'.

In Figure 11.7, there are two full ST's for four points A, B, C and D. Which one is shorter? There are several ways to answer this question.

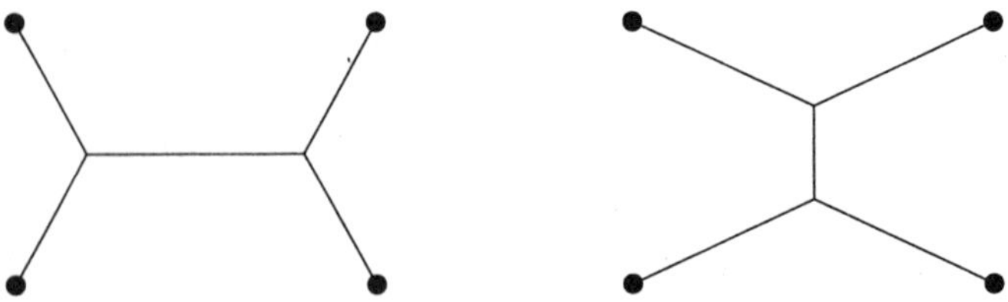

Figure 11.7: Full ST for four points.

Theorem 11.2.1 (Pollak) *Let E be the intersection point of two diagonals. If $\angle AEB \leq \angle BEC$, then the full ST $AB - CD$ is not longer than the full ST $BC - AD$.*

Theorem 11.2.2 (Ollerenshaw) *If two full Steiner trees exist for four points, then the one with longer segment between two Steiner points is shorter.*

In general, there are at least $\begin{bmatrix} 2n-4 \\ n-2 \end{bmatrix}$ Steiner topologies for n regular points. The large number of Steiner topologies is the main reason for the intractability of the Steiner tree problem.

11.3 Solution of Gilbert-Pollak's Conjecture

Computing SMT(P) has been shown to be an NP-hard problem, i.e. unlikely SMT has an algorithm running fast (in polynomial time). Therefore, it merits to study approximate solutions. A spanning tree on P is just a tree with vertex set P. A shortest spanning tree on P is also called a *minimum spanning tree* on P, denoted by MST(P). MST(P) can be computed quickly. However, it is usually longer than SMT(P). To compare SMT and MST, one studies the *Steiner ratio*

$$\rho = inf\{L_s(P)/L_m(P) \mid P\}$$

where $L_s(P)$ and $L_m(P)$ are lengths of SMT(P) and MST(P), respectively.

To get some feeling on the Steiner ratio, let us look at three points A, B, and C forming an equilateral triangle with unit edge length. Clearly, $L_s(A,B,C) = \sqrt{3}$ and $L_m(A,B,C) = 2$. So, $L_s(A,B,C)/L_m(A,B,C) = \sqrt{3}/2$. By the definition of the Steiner ratio, $\rho \leq \sqrt{3}/2$.

Gilbert and Pollak in 1968 conjectured $\rho = \frac{\sqrt{3}}{2}$. That is, the equilateral triangle gives the worse case for using MST to approximate SMT. To prove this conjecture, it suffices to prove

$$L_s(P) \geq (\sqrt{3}/2)L_m(P) \tag{11.1}$$

for any set P of n points. This conjecture for $n = 3$ can be easily verified in the following way.

Let A, B, and C be three points. If the triangle ABC has an inner angle not less than 120°, then $L_s(A,B,C) = L_m(A,B,C) > (\sqrt{3}/2)L_m(P)$. If the triangle ABC has no angle not less than 120°, then choose an angle not less than 60°, say $\angle A \geq 60^o$. Construct an equilateral triangle ABD. Then

$$\begin{aligned} L_s(A,B,C)^2 &= d(C,D)^2 \\ &= d(A,C)^2 + d(A,D)^2 - 2d(A,C)d(A,D)\cos\angle CAD \\ &\geq d(A,C)^2 + d(A,D)^2 - 2d(A,C)d(A,D)\cos 120^o \\ &\geq (3/4)(d(A,C)+d(A,D))^2 + (1/4)(d(A,C)-d(A,D))^2 \\ &\geq (3/4)L_m(A,B,C)^2. \end{aligned}$$

Thus, $L_s(A,B,C) \geq (\sqrt{3}/4)L_m(A,B,C)$.

Through many efforts, this conjecture was proved by Du and Hwang in 1990. The main idea of their proof is to transform Gilbert-Plollak conjecture to a minimax problem in section 11.1.

Given a Steiner topology t and a point set P, the ST of topology t on P is denoted by $t(P)$. Similarly, given a spanning tree topology s and a point set P, the spanning tree of topology s on P is denoted by $s(P)$. Let I be the set of spanning tree topologies. Since any ST can be decomposed into an edge-disjoint union of smaller full ST's, to prove Gilbert-Pollak's conjecture, it suffices to prove that for any full topology t, if $t(P)$ exists then $\ell(t(P)) \geq (\sqrt{3}/2)min_{s\in I}\ell(s(P))$ where $\ell(T)$ denotes the length of a tree T. It is equivalent to say that for any full topology t,

$$\min_{P\in E_t}\max_{s\in I} g_{t,s}(P) \geq 0 \tag{11.2}$$

where $E_t = \{P \mid t(P) \text{ exists}\}$ and $g_{t,s}(P) = \ell(t(P)) - (\sqrt{3}/2)\ell(s(P))$. Note that a point set can be determined by a full Steiner tree and a full Steiner tree can be determined by its topology and all edge lengths. Let $t(x)$ denote the Steiner tree with full topology t and $2n-3$-dimensional parameter vector x whose components are its edge lengths. Let $P(t;x)$ denote the set of regular points determined by $t(x)$. Then $E_t = \{P(t;x) \mid x > 0\}$. Denote $X = \{x = (x_1, ..., x_{2n-3}) \mid x \geq 0 \text{ and } x_1 + \cdots + x_{2n-3} = 1\}$. For simplicity of notations, we write $g_{t,s}(x)$ and $s(t;x)$ instead of $g_{t,s}(P(t;x))$ and $s(P(t;x))$, respectively. By similarity, it is easy to know that (11.2) holds iff for every t,

$$\min_{x \in X} \max_{s \in I} g_{t,s}(x) \geq 0.$$

Here, we have allowed the edge length to be zero; in this case, $t(x)$ can be seen as a limiting ST with topology t and a positive edge-length vector y as y goes to x. Note that, throughout this chapter, a parameter vector is said to exist for a point set and a topology if either an ST with the topology for the point set exists or its limit exists. The following lemma shows the concavity of the function $g_{t,s}(x)$.

Lemma 11.3.1 *$\ell(s(t;x))$ is a convex function with respect to x.*

Proof. Let A and B be two regular points. We show that the distance between A and B, $d(A,B)$ is a convex function of x. Find a path in $t(x)$ which connects points A and B. Suppose the path has k edges with lengths $x_{1'}, ..., x_{k'}$ and with directions $e_1, ..., e_k$, respectively, where $e_1, ..., e_k$ are unit vectors. Then $d(A,B) = \|x_{1'}e_1 + \cdots + x_{k'}e_k\|$ where $\|\cdot\|$ is the Euclidean norm. Note that a norm is a convex function and the part inside the norm is linear with respect to x. Thus, $d(A,B)$ is a convex function with respect to x. Finally, we note that a sum of convex functions is also a convex function. □

We have transformed the Gilbert-Pollak conjecture into a minimax problem (11.2). By the minimax theorem, we know that to prove (11.2), it suffices to prove that at every DH-point x, $f_t(x) = \max_{s \in I} g_{t,s}(x) \geq 0$. Note that $I(x) = \{s \mid s(t;x)$ is a minimum spanning tree on $P(t;x)\}$. Thus, a DH-point x in the interior of X corresponds to a point set $P(t;x)$ with maximal number of minimum spanning trees. According to this property, we can find a geometric character of $P(t;x)$. So, the problem reduces to checking the conjecture on some special point sets. To explain this intuitively, let us use this approach to verify the conjecture for $n = 4$.

Suppose that there is a set P of four points such that

$$L_s(P) < (\sqrt{3}/2)L_m(P).$$

We will find a contradiction in the following. First, we claim that every SMT(P) T must be a full ST. In fact, if T is not a full ST, then T can be decomposed into two trees T' and T'' which are ST's for point sets P' and P'', respectively. P' and P'' have 2 or 3 points. So, $L_s(P') \geq (\sqrt{3}/2)L_s(P')$ and $L_s(P'') \geq (\sqrt{3}/2)L_m(P'')$. Adding them together, we obtain that $L_s(P) \geq (\sqrt{3}/2)(L_m(P') + L_m(P'')) \geq (\sqrt{3}/2)L_m(P)$, a contradiction.

Now, suppose that t is the topology of T. Then t is full. Clearly, there exists $x \in X$ such that $t(x)$ is similar to T, so that $\max_{s\in I} g_{t,s}(x) = \ell(t(x)) - (\sqrt{3}/2)L_m(P(t;x)) < 0$. Denote $f_t(x) = \max_{s\in I} g_{t,s}(x)$ and $F(t) = \min_{x\in X} f_t(x)$. Then $F(t) < 0$. A point $x^* \in X$ is called a *minimum point* if $f_t(x^*) = F(t)$.

Lemma 11.3.2 *Every minimum point x^* is an interior point of X.*

Proof. Note that $t(x^*)$ has five edges. Four of them are incident to regular points. One of them is between two Steiner points. The point x^* is an interior point of X if and only if $t(x^*)$ has no edge of length 0. Since $f_t(x^*) = F(t) < 0$, $\ell(t(x^*)) < (\sqrt{3}/2)L_m(P(t;x^*))$. By the argument in the second paragraph of this section, the four edges incident to regular points cannot have length 0. So, we need only to prove that the edge between two Steiner points cannot be of length 0. Suppose to the contrary that this edge has length 0. Clearly, $t(x^*)$ must not be a SMT on $P(t;x^*)$. Suppose that T is a SMT on $P(t;x^*)$. Then $\ell(T) < \ell(t(x^*)) = 1 < (\sqrt{3}/2)L_m(P(t;x^*))$. So, T must be a full ST. Let t' be the topology of T. Let $x' \in X$ be such that $t'(x')$ is similar to T. Then

$$\begin{aligned} f_{t'}(x') &= 1 - (\sqrt{3}/2)L_m(P(t';x')) \\ &= 1 - (\sqrt{3}/2)L_m(P(t;x^*))/\ell(T) \\ &< 1 - (\sqrt{3}/2)L_m(P(t;x^*)) \\ &= f_t(x^*). \end{aligned}$$

Thus, $F(t') < F(t)$. However, there are only two full topologies for four points. They are homeomorphic to each other. So, $F(t) = F(t')$, a contradiction. □

By the minimax theorem, a minimum point x^* must be a DH-point. So, the set $I(x^*)$ of minimum spanning tree topologies is maximal. Now, we want to discover the geometric structure of the point set $P(t;x^*)$ by using this property. First, we prove some properties of minimum spanning trees.

Lemma 11.3.3 *Two minimum spanning trees can never cross, i.e., edges meet only at vertices.*

Proof. Suppose that AB and CD are two edges crossing at the point E (Figure 11.8) and they belong to two minimum spanning trees T and T', respectively.

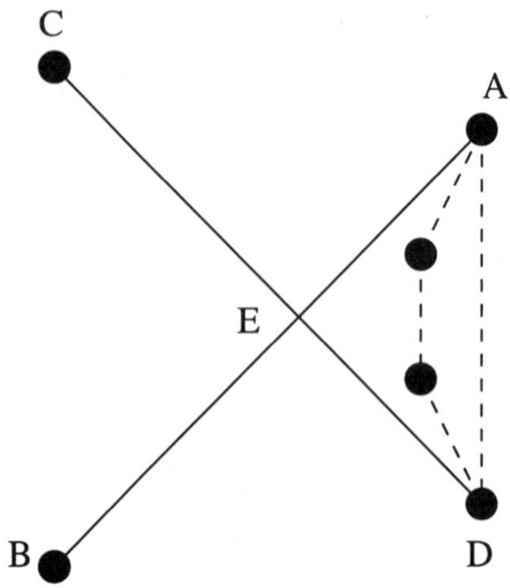

Figure 11.8: The proof of Lemma 11.3.3.

Without loss of generality, assume that EA has a smallest length among the four segments EA, EB, EC and ED. Removing the edge CD from the tree T', the remainder has two connected components containing C and D, respectively. Without loss of generality, assume that A and C are in the same component. Note that $\ell(AD) < \ell(EA) + \ell(ED) \leq \ell(CD)$. Connecting the two components by AD results in a shorter spanning tree, contradicting the minimality of T'. $\square$

Let

$$M(t;x) = \{m \mid m \text{ is the topology of a minimum spanning tree on } P(t;x)\}.$$

Let $\Gamma(t;x)$ be the union of all minimum spanning trees on $P(t;x)$. By Lemma 11.3.3, $\Gamma(t;x)$ is a planar graph.

Lemma 11.3.4 *Every polygon of* $\Gamma(t;x)$ *has at least 2 equal longest edges.*

Proof. Suppose to the contrary that $\Gamma(t;x)$ has a polygon Q with the unique longest edge e. Let m be the minimum spanning tree containing e. For every edge e' of Q not in m, the union of m and e' contains a cycle. If this cycle contains e, then $(m \setminus \{e\}) \cup \{e'\}$ is an spanning tree shorter than m, a contradiction. Thus, such a cycle does not contain e. Hence, for every e' in Q not in m, m has a path connecting two endpoints of e'. These paths and e form a cycle in m, a contradiction. □

Let $C(t;x^*)$ be the convex hull of the set $P(t;x^*)$ of four points. Now, we are ready to determine the geometric structure of $P(t;x^*)$.

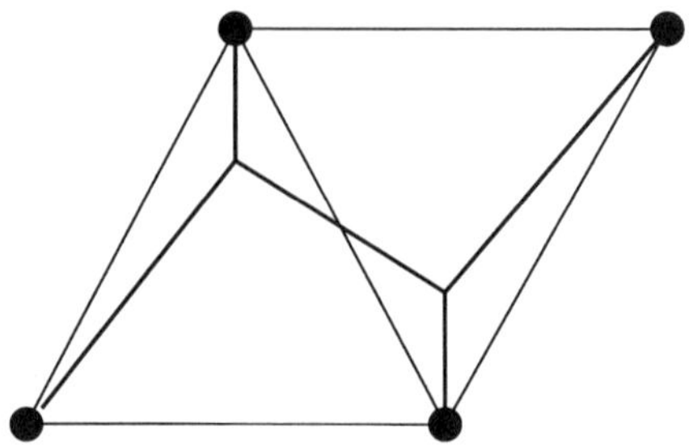

Figure 11.9: Two equilateral triangles.

Lemma 11.3.5 *If x^* is a minimum point, then $\Gamma(t;x^*)$ divides $C(t;x^*)$ into two equilateral triangles (Figure 11.9).*

Proof. If the lemma is false then one of the following must occur: (a) $\Gamma(t;x^*)$ has a free edge, an edge not on any polygon of $\Gamma(t;x^*)$. (b) $\Gamma(t;x^*)$ has a polygon of more than three edges. (c) Neither (a) nor (b) occurs and $\Gamma(t;x^*)$ has a nonequilateral triangle. We will show that in each case, the number of minimum spanning trees can be increased.

First, assume (a) or (b) occurs. Embedding $\Gamma(t;x^*)$ into a triangulation of $C(t;x^*)$, we can find a triangle Δ with an edge e in $\Gamma(t;x^*)$ and at least one edge not in $\Gamma(t;x^*)$. Let T be a minimum spanning tree containing e. Among edges not in $\Gamma(t;x^*)$, there is an edge e' of Δ such that $T' = (T \setminus \{e\}) \cup \{e'\}$ is a spanning tree. (Such an edge e' must exist for, if Δ has only one edge not in $\Gamma(t;x^*)$ then this edge must have the desired property; if Δ has two edges not in $\Gamma(t;x^*)$ then the one which lies between two connected components of $T \setminus \{e\}$ meets the requirement.) Clearly, $\ell(e) < \ell(e')$. Now, we decrease the length of e' and fix all other edge-lengths in the triangulation, so that every edge in $\Gamma(t;x^*)$ does not change its length. Let ℓ be the length of the shrinking e'. At the beginning, $\ell = \ell(e')$. At the end, $\ell = \ell(e) < \ell(e')$. For each ℓ, denote by $\bar{P}(\ell)$ the corresponding set of regular points. Then $P(\ell(e')) = P(t;x^*)$. Consider the set L of all $\ell \in [\ell(e), \ell(e')]$

satisfying the condition that there is a minimum point y in X such that $\bar{P}(\ell) = P(t; y)$. Since $\ell(e') \in L$, L is nonempty. Moreover, L is a closed set since all minimum points form a closed set. Now, we consider the minimal element ℓ^* of L. Suppose $\bar{P}(\ell^*) = P(t; y)$. Then for any $m \in M(t; x^*)$, $\ell(m(t; y)) = \ell(m(t; x^*))$. Since both x^* and y are minimum points, we have $f_t(x^*) = f_t(y)$, that is, the length of a minimum spanning tree for $P(t; x^*)$ equals that of $P(t; y)$. Hence $M(t; x^*) \subseteq M(t; y)$. However, x^* is a DH-point. Thus, $M(t; x^*) = M(t; y)$. It follows that $\ell > \ell(e)$. By Lemma 11.3.2, y is an interior point of X. This means that there exists a neighborhood of ℓ^* such that for ℓ in it, the ST of topology t exists for the point set $\bar{P}(\ell)$. Thus, there exists $\ell < \ell^*$ such that $\bar{P}(\ell) = P(t; z)$ for some vector z (not-necessarily in X). Since $\ell(m(t; x))$ is continuous with respect to x, there is a neighborhood of y such that for every point y' in the neighborhood, $M(t; y') \subseteq M(t; y)$. So, ℓ can be chosen to make z satisfy $M(t; z) \subseteq M(t; y)$. Note that $M(t; x^*) = M(t; y)$ and for every $m \in M(t; x^*)$, $\ell(m(t; z)) = \ell(m(t; x^*))$. It follows that for every $m \in M(t; x^*)$, $m(t; z)$ is a minimum spanning tree for $P(t; z)$. Thus, $M(t; z) = M(t; x^*)$ and $f_t(x^*) = f(z)$. Suppose $hz \in X$ where h is a positive number. By Corollary 11.1.4, $f_t(x^*) = f_t(hz) = hf_t(z)$. So, $h = 1$, i.e., $z \in X$. Hence, z is a minimum point, contradicting the minimality of ℓ^*.

Next, assume that neither (a) nor (b) occurs. In this case, we increase the length of all shortest edges and fix all other edge-lengths in $\Gamma(t; x^*)$. Let r be the ratio of lengths between the shortest edge and the longest edge. Suppose at the beginning, $r = r_0 < 1$. For each r, denote by $\bar{P}(r)$ the corresponding set of regular points. Then $P(r_0) = P(t; x^*)$. Consider the set R of all $r \in [r_0, 1]$ satisfying the condition that there is a minimum point y in X such that $\alpha\bar{P}(r) = P(t; y)$ for some constant α and $M(t; x^*) \subseteq M(t; y)$. Since $r_0 \in R$, R is nonempty. Moreover, R is a closed set since all minimum points form a closed set. Now, we consider the maximum number r^* in R. Suppose $\alpha\bar{P}(r^*) = P(t; y)$ and $M(t; x^*) \subseteq M(t; y)$. Since x^* is a DH-point, we must have $M(t; x^*) = M(t; y)$. It follows that $r^* < 1$. By Lemma 11.3.2, y is an interior point of X. This means that there exists a neighborhood of r^* such that for r in it, the ST of topology t exists for the point set $\bar{P}(r)$. Thus, there exists a neighborhood of r^* such that for r in the neighborhood $\beta\bar{P}(r) = P(t; z)$ for some vector z in X and a constant β. Note that all shortest edges between regular points in $P(t; y)$ must belong to $\Gamma(t; y)$. We can choose $r > r^*$ so close to r^* preserving the same set of shortest edges from $P(r^*)$ to $P(r)$. This means that we can choose $r > r^*$ to make z satisfy $M(t; z) = M(t; y)$. By Corollary 11.1.4, z is a minimum point. Therefore,

$r \in R$, contradicting the maximality of r^*. □

Now, we consider Figure 11.9. It is easy to see that

$$L_s(P(t;x^*))/L_m(P(t;x^*)) = \sqrt{7}/3 > \sqrt{3}/2.$$

So, $f_t(x^*) > 0$, contradicting to $F(t) < 0$.

Exercises

1. Let D, E and F be three points on three edges BC, CA and AB of a triangle ABC, respectively. Let $per(\triangle ABC)$ denote the perimeter of the triangle ABC. Prove that

$$per(\triangle DEF) \geq \min(per(\triangle AEF), per(\triangle BFD), per(\triangle CDE)).$$

2. Let D, E and F be three points on three edges BC, CA and AB of a triangle ABC, respectively. Prove that

$$\max(S_{\triangle AEF}, S_{\triangle BFD}, S_{\triangle CDE}) \geq S_{\triangle DEF} \geq \min(S_{\triangle AEF}, S_{\triangle BFD}, S_{\triangle CDE})$$

3. Let D, E and F be three points on three edges BC, CA and AB of a triangle ABC, respectively. Prove that

$$\max(r_{\triangle AEF}, r_{\triangle BFD}, r_{\triangle CDE}) \geq r_{\triangle DEF}.$$

4. Prove Theorems 11.2.1 and 11.2.2.

5. The circle packing problem is to determine the maximal radius of N equal circles that can be packed into a unit square. Formulate this problem as a minimax optimization problem.

6. How should n points be arranged into a unit square such that the minimal distance between them is maximized?

7. Given a set V of n points in the Euclidean plane and an integer $m \leq n$, find a subset $S \subset V$ of size m such that the minimal spanning tree for S achieves the maximum over all minimal spanning trees of all subsets of size m.

Chapter 12

Relaxation

Our nature consists in motion;
complete rest is death.
PASCAL

An optimization problem usually consists of an objective function and a feasible domain. "Relaxation" is an important technique in designs of approximation algorithms for intractable optimization problems. By relaxation, we mean to relax some constraints on feasible solutions, so that the feasible domain is enlarged. The purpose of relaxation on feasible domain is to change the intractability of the optimization problem. For example, the integer programming is NP-hard. We may relax integers to real numbers. However, the optimal solution obtained from relaxation is often not a feasible solution of the original problem. Thus, one has to modify it to construct a feasible solution for the original problem.

From the idea of relaxation, we see that an approximation using relaxation technique usually contains two steps. In the first step, one analyze the feasible solutions to get a new relaxed feasible domain and find an optimal solution in the new feasible domain. In the second step, one has to modify this optimal solution to obtain a feasible solution in the original domain.

12.1 Vertex Cover and MIN 2SAT

Many combinatorial optimization problems can be transformed to integer programming. By relaxing integers to real numbers, the integer programming can be relaxed to linear programming which can be solved in polynomial-time. Rounding optimal solution of linear programming to a feasible solution of original problem will give an approximation. This is a quite general approach to find an approximation. In this and next several sections, we present some examples using this approach.

Given a graph G, find its minimum vertex cover. The complexity of this problem has been studied in the first section. In this section, we want to discuss approximation solution of the vertex cover.

A simple approximation is as follows.

Algorithm VC1

First, find a maximal matching by a greedy algorithm.
Output the vertex set of the maximal matching.

This approximation has performance ratio two. In fact, every vertex cover has to have an endpoint of each edge in the maximal matching. Thus, the size of a minimum vertex cover is at least the size of the maximal matching. This implies that the number of vertices of the maximal matching is at most twice the size of the minimum vertex cover. This bound is also tight on input graphs consisting of only one maximal matching.

Next, we want to extend the vertex cover problem to weighted one. That is, given a graph $G(V, E)$ with vertex weight, find a minimum weight vertex cover. In weighted case, the above idea seems not to work well. At least, we cannot see immediately how to extend it to the weighted case. Now, we introduce an interesting and powerful technique to solve this case. First, transform the vertex cover to a 0-1 integer programming.

Suppose $V = \{v_1, v_2, \cdots, v_n\}$. Let x_i be a 0-1 variable representing vertex v_i being in a vertex cover or not. That is, if v_i is in the vertex cover, set $x_i = 1$; otherwise, set $x_i = 0$. Let w_i be the weight of vertex v_i. Then every vertex cover corresponds to a feasible solution in the following 0-1 integer programming and the minimum weight vertex cover corresponds to the optimal solution of the 0-1 integer programming.

$$\begin{array}{lll} \text{(IP1)} & \min & w_1x_1 + w_2x_2 + \cdots + w_nx_n \\ & \text{subject to} & x_i + x_j \geq 1 \text{ for } (v_i, v_j) \in E \end{array}$$

$$x_i = 0 \text{ or } 1 \text{ for } i = 1, 2, \cdots, n.$$

Now, they relax this 0-1 integer programming into a linear programming

$$\begin{array}{lll} \text{(LP1)} & \min & w_1x_1 + w_2x_2 + \cdots + w_nx_n \\ & \text{subject to} & x_i + x_j \geq 1 \text{ for } (v_i, v_j) \in E \\ & & 0 \leq x_i \leq 1 \text{ for } i = 1, 2, \cdots, n. \end{array}$$

and give the following approximation.

Algorithm VC2

First, find an optimal solution x^* for linear programming (LP1)
Output the following approximation x^A for (ILP1):
$x_i^A = 1$ *if* $x_i^* \geq 0.5$ *and* $x_i^A = 0$ *if* $x_i^* < 0.5$.

For each $(v_i, v_j) \in E$, since $x_i^* + x_j^* \geq 1$, we have at least one of x_i^* and x_j^* not smaller than 0.5. Therefore, at least one of x_i^A and x_j^A equals 1. This guarantees that x^A is a feasible solution of (IP1). Clearly,

$$\sum_{i=1}^{n} w_i x_i^A \leq 2 \sum_{i=1}^{n} w_i x_i^*$$

and the optimal solution of (ILP1) has objective function value not smaller than $\sum_{i=1}^{n} w_i x_i^*$. Therefore, the following theorem is proved.

Theorem 12.1.1 *The weighted vertex cover problem has a polynomial-time approximation with performance ratio two.*

In the above approximation, the construction of feasible solution of the integer programming from an optimal solution of linear programming is called *rounding*. There are many techniques about rounding.

Given a Boolean formula 2-CNF, find a satisfying assignment that contains a minimum number of true variables. This problem is called *MIN 2SAT*. MIN 2SAT can be seen as a generalization of vertex cover problem. In fact, for each graph $G(V, E)$, construct a 2-CNF $F(G)$ as follows: For each vertex v_i, give a Boolean variable x_i and for each edge (v_i, v_j), give a clause $(x_i + x_j)$. Then each vertex cover of G corresponds to a satisfying assignment of $F(G)$. G has a vertex cover of size k iff $F(G)$ has a satisfying assignment with k true variables.

It is similar to vertex cover problem that MIN 2SAT can also be transformed into a 0-1 integer programming. Consider a Boolean formula 2-CNF F. Suppose that F has n Boolean variables $x_1, x_2, \cdots, x_n$. We may use the same notation x_i to denote a corresponding 0-1 integer variable. Now, MIN 2SAT is equivalent to the following.

$$
\begin{array}{llcl}
(IP2) & \text{minimize} & x_1 + x_2 + \cdots + x_n & \\
& \text{subject to} & x_i + x_j \geq 1 & \text{for clause } (x_i \vee x_j) \text{ in } F \\
& & (1 - x_i) + x_j \geq 1 & \text{for clause } (\bar{x}_i \vee x_j) \text{ in } F \\
& & x_i + (1 - x_j) \geq 1 & \text{for clause } (x_i \vee \bar{x}_j) \text{ in } F \\
& & (1 - x_i) + (1 - x_j) \geq 1 & \text{for clause } (\bar{x}_i \vee \bar{x}_j) \text{ in } F \\
& & x_i = 0 \text{ or } 1 & \text{for } i = 1, 2, \cdots, n.
\end{array}
$$

The relaxation of (IP2) is obtained by replacing $x_i = 0,\ \text{or}1$ with $0 \leq x_i \leq 1$, that is,

$$
\begin{array}{llcl}
(\text{LP2}) & \text{minimize} & x_1 + x_2 + \cdots + x_n & \\
& \text{subject to} & x_i + x_j \geq 1 & \text{for clause } (x_i \vee x_j) \text{ in } F \\
& & (1 - x_i) + x_j \geq 1 & \text{for clause } (\bar{x}_i \vee x_j) \text{ in } F \\
& & x_i + (1 - x_j) \geq 1 & \text{for clause } (x_i \vee \bar{x}_j) \text{ in } F \\
& & (1 - x_i) + (1 - x_j) \geq 1 & \text{for clause } (\bar{x}_i \vee \bar{x}_j) \text{ in } F \\
& & 0 \leq x_i \leq 1 & \text{for } i = 1, 2, \cdots, n.
\end{array}
$$

Similarly, suppose x^* is an optimal solution of (LP2). We might consider to construct an approximation solution x^A of (IP2) by setting $x_i^A = 1$ if $x_i^* > 0.5$ and setting $x_i^A = 0$ if $x_i^* < 0.5$. This will satisfy all inequalities in which two terms has at least one > 0.5 in x^*. However, we have a trouble to determine x_i^A when $x_i^* = 0.5$. For example, if both $x_i \vee x_j$ and $\bar{x}_i \vee \bar{x}_j$ are clauses of F and $x_i^* = x_j^* = 0.5$, then setting $x_i = x_j = 0$ or $x_i = x_j = 1$ cannot be satisfied. What should we do in this case? The following is a solution.

Algorithm MIN-2SAT

First, find an optimal solution x^* for linear programming (LP2).
Set $x_i^A = 1$ if $x_i^* > 0.5$ and $x_i^A = 0$ if $x_i^* < 0.5$.
Consider all clauses of F whose two terms have value 0.5 with assignment x^*. Suppose those clause form a 2-CNF F'.
If F' is not satisfiable, then F is not satisfiable;
else find a satisfying assignment $(x_i^A, x_i^* = 0.5)$ of F' and output x^A.

First, let us explain that the solution x^A generated by algorithm MIN-2AT feasible solution of (IP2). Clearly, it satisfies F'. So, we need to consider only those clauses of F which are not clauses of F'. If $x_i \vee x_j$ is such a clause, then we must have $x_i^* > 0.5$ or $x_j^* > 0.5$ since $x_i^* + x_j^* \geq 1$. Thus, either $x_i^A = 1$ or $x_j^A = 1$ and hence $x_i \vee x_j$ is satisfied. A similar argument can be applied to $x_i \vee \bar{x}_j$, $\bar{x}_i \vee x_j$, and $\bar{x}_i \vee \bar{x}_j$ when they are clauses in F but not in F'.

Next, we want to show that algorithm MIN-2SAT in polynomial time. To see this, it suffices to explain that 2SAT is polynomial-time solvable. That is, given a Boolean formula 2-CNF F', we can determine, in polynomial time, whether F' is satisfiable or not, and if F' is satisfiable, a satisfying assignment would be found. Let us start with constructing a directed graph $G(F')$. The vertex set of $G(F')$ consists of all literals x_i and $\bar{x}_i$ for variables x_i appearing in F'. For each clause $y_i \vee y_j$ of F', construct two edges $\bar{y}_i \to y_j$ and $\bar{y}_j \to y_i$ for $G(F')$. Thus, each edge $y \to z$ means that $y = 1$ implies $z = 1$ if $F' = 1$. This graph has an important property that if there is a path from y to z, then there is a path from $\bar{z}$ to $\bar{y}$.

Now, if x_i and $\bar{x}_i$ for some variable x_i are strongly connected in $G(F')$ (i.e., there is a directed cycle passing through both x_i and $\bar{x}_i$), then it is clear that F' is not satisfiable. If for any variable x_i, x_i and $\bar{x}_i$ are not strongly connected, then we can find a satisfying assignment for F' as follows.

If there is a path from x_i to $\bar{x}_i$, then set $x_i = 0$ and assign 1 to all literals to which there is a path from $\bar{x}_i$. (Of course, if a literal is assigned with 1, then its negation is assigned with 0 simultaneously without mentioning.) If there is a path from $\bar{x}_i$ to x_i, then set $x_i = 1$ and assign 1 to all literals to which there is a path from x_i. This assignment would not generate any contradiction. To show this, suppose that a variable w is assigned with contradicting values 0 and 1. Then it must happen in the way that w is assigned with 1 due to setting $u = 1$ and $\bar{w}$ is assigned with 1 due to setting

$v = 1$ for some literals u and v satisfying that there are two paths from $\bar{u}$ to u and $\bar{v}$ to v respectively (see Figure 12.1). From the mentioned property of

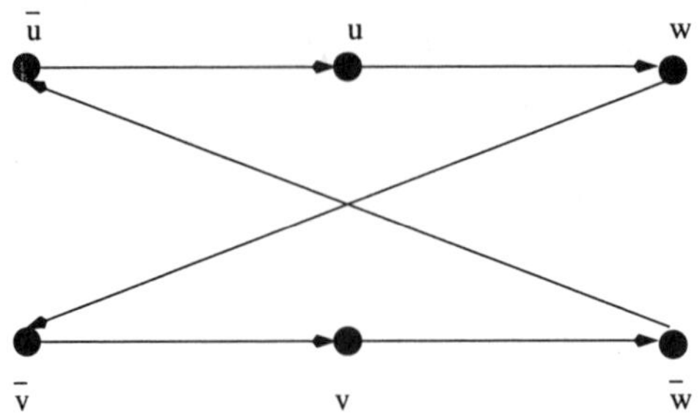

Figure 12.1: A cycle passing through both u and $\bar{u}$.

$G(F')$, there exists a path from w to $\bar{v}$ and there also exists a path from $\bar{w}$ to $\bar{u}$. This implies that u and $\bar{u}$ are strongly connected, a contradiction.

Now, delete all assigned literals from graph $G(F)$. The remainder will not contain any path from a literal to its negation. Pick up any unassigned variable x_i from the graph, set $x_i = 1$ and assign 1 to all literals to which there is a path from x_i, then delete all assigned literals. Repeat this process until the graph becomes empty. This completes the construction for a satisfying assignment of F'.

A example is provided in Figure 12.2. Since there is a path from x_1 to

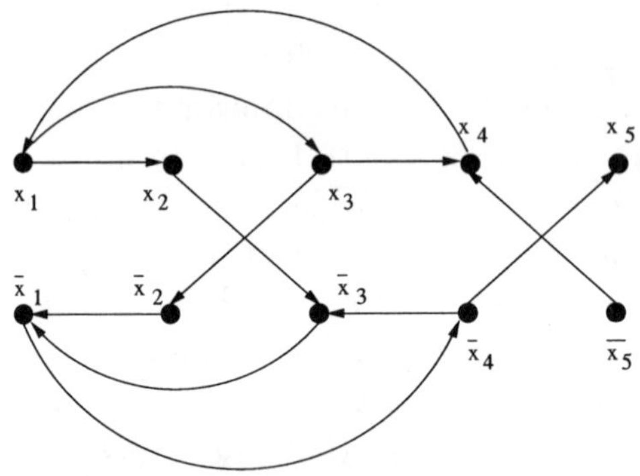

Figure 12.2: $G((\bar{x}_1 \vee x_2) \wedge (\bar{x}_2 \vee x_3) \wedge (\bar{x}_1 \vee x_3) \wedge (\bar{x}_3 \vee x_4) \wedge (x_4 \vee x_5) \wedge (x_1 \vee \bar{x}_4))$.

$\bar{x}_1$, we have to set $x_1 = 0$ and consequently assign 1 to $\bar{x}_4$. Deleting variables x_1 and x_4 from the graph, we obtain a graph in Figure 12.3. Now, set $x_2 = 1$ and consequently assign 1 to $\bar{x_3}$. Finally, set $x_5 = 1$. We obtain a satisfying

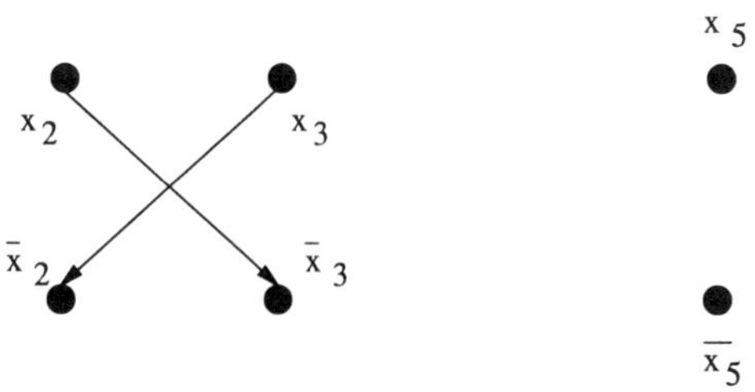

Figure 12.3: The remainder after deleting x_1 and x_4.

assignment $x_1 = 0, x_2 = 1, x_3 = 0, x_4 = 0, x_5 = 1$.

From the above analysis, we can conclude the following.

Theorem 12.1.2 *MIN 2SAT has a polynomial-time approximation with performance ratio two.*

12.2 General Cover Problem

Consider the set cover problem: Given a hypergraph $H(V, E)$, find a cover of minimum cardinality. This problem is equivalent to the following integer programming.

$$\begin{array}{ll} \text{minimize} & \sum_{v \in V} x_v \\ \text{subject to} & \sum_{u \in e} x_u \geq 1, \text{ for all } e \in E \\ & x_v = 0 \text{ or } 1 \text{ for all } v \in V \end{array}$$

Now, we consider a general problem

$$\begin{array}{lll} (IP3) & \text{minimize} & \sum_{j=1}^{n} c_j x_j \\ & \text{subject to} & \sum_{j=1}^{n} a_{ij} x_j \geq b_i, i = 1, 2, \cdots, m \\ & & x_j \in \{0, 1\} \text{ for } j = 1, 2, \cdots, n, \end{array}$$

where a_{ij}, b_i, and c_j are nonnegative integers. This problem is called *general cover problem*.

First, we note that b_i and c_j may be assumed to be positive. In fact, if $b_i = 0$, then we can remove the ith constraint; if $c_j = 0$, then we may set $x_j = 1$ and remove corresponding column.

Denote $f = \max_{1 \le i \le m} \sum_{j=1}^{n} a_{ij}$. Consider the following algorithm.

Algorithm GVC1

Let x^* be an optimal solution of the LP relaxation of (IP3).
If $x_j^* \ge 1/f$ then set $x_j^A = 1$.
If $x_j^* < 1/f$ then set $x_j^A = 0$.
Output x^A.

Theorem 12.2.1 *Algorithm GVC1 produces an approximation within a factor of f from optimal.*

Proof. Clearly, $x_j^A \le f x_j^*$ for every $j = 1, 2, \cdots, n$. Thus,

$$\sum_{j=1}^{n} c_j x_j^A \le f \sum_{j=1}^{n} c_j x_j^*.$$

To prove the feasibility of x^A, let $J_0 = \{j \mid x_j^* < 1/f\}$ and $J_1 = \{j \mid x_j^* \ge 1/f\}$. First, we show that

$$\sum_{j \in J_0} a_{ij} x_j^* < 1.$$

If $a_{ij} = 0$ for all $j \in J_0$, then $\sum_{j \in J_0} a_{ij} x_j^* = 0 < 1$. If $a_{ij} > 0$ for some $j \in J_0$, then $\sum_{j \in J_0} a_{ij} x_j^* < \sum_{j \in J_0} a_{ij} \frac{1}{f} \le 1$.

Now, for every $i = 1, 2, \cdots, m$, we have

$$\begin{aligned}
\sum_{j=1}^{n} a_{ij} x_j^A &\ge \sum_{j \in J_1} a_{ij} x_j^* \\
&\ge b_i - \sum_{j \in J_0} a_{ij} x_j^* \\
&> b_i - 1
\end{aligned}$$

Since both $\sum_{j=1}^{n} a_{ij}x_j^A$ and $b_i - 1$ are integers, we have

$$\sum_{j=1}^{n} a_{ij}x_j^A \geq b_i.$$

$\square$

12.3 Rounding with Duality

The linear-programming-based approximation is good for large scale problems since there exist many softwares for solving linear programming. However, linear programming cannot be solved very fast in general. This limits the application of those approximations. In this section, we introduce a technique to speed up running time of approximation.

An interesting technique about rounding is to use the duality theory of linear programming. From the duality theory, we know that every *dual feasible solution* (i.e., a feasible solution of dual linear programming) provides a lower bound for minimum value of primal linear programming. Thus, we can use this lower bound to establish the performance ratio of approximation. This means that a 'good enough' dual feasible solution may do the same job as the optimal solution of the linear programming. This would save running time.

Now, let us start to find the dual linear programming of the relaxation (LP3) of general cover problem. To do it, we first transform (LP3) to the standard form as follows:

$$\begin{array}{llll}
\text{(LP4)} & Z_4 = \min & c^T x & \\
& \text{subject to} & Ax - w = b & \\
& & x + v = \mathbf{1} & \\
& & x \geq 0, w \geq 0, v \geq 0, &
\end{array}$$

where w in R^m and v in R^n are vectors of introduced variables and $\mathbf{1} = (1, \cdots, 1)^T$. The dual of (LP4) is

$$\begin{array}{lll}
\text{(DLP4)} & Z_4 = \max & y^T b - z^T \mathbf{1} \\
& & y^T A - z^T \leq c^T \\
& & y \geq 0, z \geq 0.
\end{array}$$

The complementary-slackness condition between this two linear programming is

$$\begin{aligned}(c^T + z^T - y^T A)x &= 0\\ z^T w &= 0\\ y^T v &= 0.\end{aligned}$$

Note that $w = Ax - b$ and $v = \mathbf{1} - x$. This condition can also be written as

$$\begin{aligned}(c^T + z^T - y^T A)x &= 0\\ z^T(Ax - b) &= 0\\ y^T(1 - x) &= 0.\end{aligned}$$

A more explicit writting may help the reader's vision.

$$(c_j + z_j - \sum_{i=1}^{m} a_{ij}y_i)x_j = 0 \quad \text{for } j = 1, 2, \cdots, n. \tag{12.1}$$

$$z_j(1 - x_j) = 0 \quad \text{for } j = 1, 2, \cdots, n, \tag{12.2}$$

$$y_i(\sum_{j=1}^{n} a_{ij}x_j - b_i) = 0 \quad \text{for } i = 1, 2, \cdots, m \tag{12.3}$$

Note that if x is a feasible solution of (LP3) and (y, z) is a feasible solution of (DLP4), then

$$c^T x - y^T b + z^T \mathbf{1} = (c^T + z^T - y^T A)x + z^T(Ax - b) + y^T(1 - x).$$

Thus, if more equations in the complementary-slackness condition are satisfied, then they are closer to each other. Of course, we cannot expect all equations satisfied when we obtain x from (y, z) through rounding. Here, we would like to have (12.1) satisfied and hence rounding (y, z) in following way

$$x_j^A = \begin{cases} 1 & \text{if } \sum_{i=1}^{m} a_{ij}y_i - z_j = c_j,\\ 0 & \text{if } \sum_{i=1}^{m} a_{ij}y_i - z_j < c_j.\end{cases} \tag{12.4}$$

Two important properties of this rounding is proved in the following.

Lemma 12.3.1 *Suppose (y, z) is a feasible solution of (DLP4) and there exists a feasible solution x of (LP3) such that*

$$\sum_{i=1}^{m} a_{ij}y_i - z_j < c_j \Rightarrow x_j = 0. \tag{12.5}$$

Then x^A is a feasible solution of (LP3).

Proof. Note that $x^A \geq x$. Thus, the feasibility of x implies the feasibility of x^A since all coefficients in constraints are non-negative. □

Lemma 12.3.2 *Suppose (y, z) is a feasible solution of (DLP4), satisfying*

$$\sum_{j=1}^{n} z_j \leq \sum_{i=1}^{m} (b_i - 1) y_i. \tag{12.6}$$

Then $c^T x^A \leq f Z_3$ where $f = \max_{1 \leq i \leq m} \sum_{j=1}^{n} a_{ij}$.

Proof. Note that $0 \leq x_j^A \leq 1$. Thus,

$$\begin{aligned}
c^T x^A &= \sum_{i=1}^{m} b_i y_i - \sum_{j=1}^{n} z_j + \sum_{j=1}^{n} z_j (1 - x_j^A) + \sum_{i=1}^{m} y_i (\sum_{j=1}^{n} a_{ij} x_j^A - b_i) \\
&= \sum_{j=1}^{n} z_j (-x_j^A) + \sum_{i=1}^{m} y_i (\sum_{j=1}^{n} a_{ij} x_j^A) \\
&\leq f \sum_{i=1}^{m} y_i \\
&\leq f (\sum_{i=1}^{m} b_i y_i - \sum_{j=1}^{n} z_j) \\
&\leq f Z_3.
\end{aligned}$$

□

The following approximation is designed according to the above two lemmas. The algorithm keeps computing a pair of vectors x^k and (y^k, z^k) satisfying conditions (12.5) and (12.6). (y^k, z^k) is always a feasible solution of (DLP4) and x^k is a candidate for (and finally becomes) a feasible solution of (LP3).

Algorithm GVC2

Initially, set $x^0 = 0$, $(y^0, z^0) = (0, 0)$, $k = 0$.
while x^k *is not prime feasible* **do begin**
  set $J(k) = \{j \mid x_j^k = 0\}$;

set $I(k) = \{i \mid \sum_{j=1}^{n} a_{ij}x_j^k \leq b_i - 1\}$;
choose $r \in J(k)$ *such that*
$\frac{c_r - \sum_{i=1}^{m} a_{ir}y_i^k}{\sum_{i \in I} a_{ir}} = \alpha = \min_{j \in J(k)} \frac{c_j - \sum_{i=1}^{m} a_{ij}y_i^k}{\sum_{i \in I} a_{ij}}$;
set $x_j^{k+1} = x_j^k$ *if* $j \neq r$ *and* $x_r^{k+1} = 1$;
set $y_i^{k+1} = y_i^k + \alpha$ *if* $i \in I(k)$ *and* $y_i^{k+1} = y_i^k$ *if* $i \notin I(k)$
set $z_j^{k+1} = \max(\sum_{i=1}^{m} a_{ij}y_i^{k+1} - c_j, 0)$;
$k \leftarrow k + 1$;

end-while

Output x^k *with* $Z_A = \sum_{j=1}^{n} c_j x_j^k$.

Algorithm GVC2 runs in $O(n^2)$ time. Next, we show that it has the same performance ratio as that of previous approximation.

Lemma 12.3.3 *For any* k*, we have the following:*
(a) (y^k, z^k) *is dual-feasible,*
(b) $\sum_{i=1}^{m} a_{ij}y_i^k < c_j \Rightarrow x_j^k = 0$*, and*
(c) $\sum_{j=1}^{n} z_j^k \leq \sum_{i=1}^{m}(b_i - 1)y_i^k$.

Proof. We prove it by induction on k. Initially, all (a), (b), and (c) are true trivially for x^0 and (y^0, z^0). Suppose they hold for $k \geq 0$. Now, we consider $k + 1$.

To see (a) for $k + 1$, we note that $y_i^{k+1} \geq y_j^k \geq 0$ for all i and $z_j^{k+1} = \max(\sum_{i=1}^{m} a_{ij}y_i^{k+1} - c_j, 0)$ which implies that $z_j^{k+1} \geq 0$ and $z_j^{k+1} \geq \sum_{i=1}^{m} a_{ij}y_i^{k+1} - c_j$ for all j.

To see (b) for $k+1$, we note that $\sum_{i=1}^{m} a_{ij}y_i^k \leq \sum_{i=1}^{m} a_{ij}y_i^{k+1}$ for all j. If $\sum_{i=1}^{m} a_{ij}y_i^{k+1} < c_j$, then $\sum_{i=1}^{m} a_{ij}y_i^k < c_j$. By induction hypothesis, $x_j^k = 0$. Since $\sum_{i=1}^{m} a_{ir}y_i^{k+1} = c_r$, we know $j \neq r$. Therefore, $x_j^{k+1} = 0$.

To see (c) for $k + 1$, we first note that by the choice of α, $z_j^{k+1} = 0$ for $j \in J(k)$ and $z_j^{k+1} = \sum_{i=1}^{m} a_{ij}y_i^{k+1} - c$ for $j \notin J(k)$. Therefore,

$$
\begin{aligned}
\sum_{j=1}^{n} z_j^{k+1} &= \sum_{j \notin J(k)} \sum_{i=1}^{m} a_{ij}y_i^{k+1} - c_j \\
&= \sum_{j \notin J(k)}^{n} \sum_{i \in I(k)} a_{ij}\alpha + \sum_{i=1}^{m} a_{ij}y_i^k - c_j
\end{aligned}
$$

$$
\begin{aligned}
&\leq \sum_{j \notin J(k)} \Big(\sum_{i \in I(k)} a_{ij}\alpha + z_j^k \Big) \\
&\leq \sum_{i \in I(k)} \sum_{j \notin J(k)} a_{ij}\alpha + \sum_{i=1}^{m} (b_i - 1) y_i^k \quad \text{(by induction hypothesis)} \\
&\leq \sum_{i \in I(k)} (b_i - 1)\alpha + \sum_{i=1}^{m} (b_i - 1) y_i^{k+1} \\
&= \sum_{i=1}^{m} (b_i - 1) y_i^{k+1}.
\end{aligned}
$$

□

The following is an immediate consequence of Lemma 12.3.3.

Theorem 12.3.4 *Algorithm GVC2 produces an approximation within a factor of f from optimal where $f = \max_{1 \leq i \leq m} \sum_{j=1}^{n} a_{ij}$.*

Exercises

1. Consider the following general form of linear programming.

$$
\begin{array}{rl}
\min & c^T x \\
\text{subject to} & Ax = b \\
 & Bx \geq d,
\end{array}
$$

   where c and x are in R^n, b is in R^m, d is in $R^{m'}$, A is an $m \times n$ matrix, and B is an $m' \times n$ matrix. Prove that its dual is as follows

$$
\begin{array}{rl}
\max & y^T b + z^T d \\
\text{subject to} & y^T A + z^T B = c^T \\
 & z \geq 0,
\end{array}
$$

   where $y \in R^m$ and $z \in R^{m'}$.

2. Define

$$
\begin{aligned}
J(z) &= \{ j \mid z_j > 0 \}, \\
I_0(z) &= \{ i \mid \sum_{j \in J(y,z)} a_{ij} = b_i \},
\end{aligned}
$$

$$\begin{aligned} I_+(z) &= \{i \mid \sum_{j \in J(y,z)} a_{ij} > b_i\}, \\ I_-(z) &= \{i \mid \sum_{j \in J(y,z)} a_{ij} < b_i\}. \end{aligned}$$

Show that

(a) for any optimal solution (y^*, z^*) of (DLP4), we have $y_i^* = 0$ for all $i \in I_+(z^*)$;

(b) there exists an optimal solution (y^*, z^*) of (DLP4) such that $y_i^* = 0$ for all $i \in I_0(z^*)$.

3. Show that there exists an optimal solution (y^*, z^*) of (DLP4) such that

$$\sum_{j=1}^{n} z_j^* \leq \sum_{i=1}^{m} (b_i - 1) y_i^*.$$

Chapter 13

Semidefinite Programming

Any given program will expand
to fill all available memory.
ARTHUR BLOCH

Semidefinite programming is a nonlinear programming with linear objective function and semidefinite constraints, which has a lot of properties similar to linear programming. It is an important tool to design polynomial-time approximation for NP-hard problems.

13.1 Spectrahedron

Let $\mathcal{S}_n$ be the family of symmetric matrices of order n. For any two matrices $A, B \in \mathcal{S}_n$, denote by $A \cdot B$ the inner product of A and B, i.e., $A \bullet B = \mathrm{Tr}(AB)$, the trace of AB. If $A - B$ is positive semidefinite (definite), then we write $A \succeq B$ ($A \succ B$).

A standard form of the semidefinite programming is as follows:

$$\begin{aligned} \min \quad & U \bullet Q_0 \\ s.t. \quad & U \bullet Q_i = c_i, \quad i = 1, 2, \cdots, m \\ & U \succeq 0 \end{aligned} \tag{13.1}$$

where $Q_0, Q_1, \cdots, Q_m$ are given linearly independent symmetric matrices of order n and $c_1, \cdots, c_m$ are given constants. Its feasible domain

$$\Omega = \{U \mid U \bullet Q_i = c_i, 1 \leq i \leq n, U \succeq 0\}$$

is a spectrahedron. In general, a *spectrahedron* is defined to be a closed convex set in the form

$$\{\boldsymbol{x} \mid Q(\boldsymbol{x}) \preceq Q_0\}$$

where $Q(\boldsymbol{x}) = \sum_{i=1}^n x_i Q_i$.

The spectrahedron is a generalization of polyhedron

$$\mathcal{P} = \{\boldsymbol{x} \mid A\boldsymbol{x} \leq \boldsymbol{b}\}$$

where A is an $m \times n$ matrix and $\boldsymbol{b}$ is an m-dimensional vector. In fact, suppose $A = (\boldsymbol{a}_1, \boldsymbol{a}_2, \cdots, \boldsymbol{a}_n)$. Then $\mathcal{P}$ can have a spectrahedron representation as follows:

$$\{\boldsymbol{x} \mid \sum_{i=1}^n x_i Diag(\boldsymbol{a}_i) \preceq Diag(\boldsymbol{b})$$

where

$$Diag(\boldsymbol{b}) = \begin{pmatrix} b_1 & 0 & \cdots & 0 & \\ 0 & b_2 & \cdots & 0 & \\ \vdots & \vdots & \ddots & \vdots & \\ 0 & 0 & \cdots & \cdots & b_m \end{pmatrix}.$$

The spectrahedron has an important property.

Proposition 13.1.1 *The intersection of two spectrahedra is still a spectrahedron.*

Proof. Consider two spectrahedra

$$\begin{aligned} \mathcal{G} &= \{\boldsymbol{x} \mid \sum_{i=1}^m x_i G_i \preceq G_0\}, \\ \mathcal{H} &= \{\boldsymbol{x} \mid \sum_{i=1}^m x_i H_i \preceq H_0\}. \end{aligned}$$

Define

$$Q_i = \begin{pmatrix} G_i & \\ & H_i \end{pmatrix}.$$

Note that two symmetric matrices A and B both are positive semidefinite if and only if

$$\begin{pmatrix} A & \\ & B \end{pmatrix}$$

is positive semidefinite. Therefore,

$$\mathcal{G} \cap \mathcal{H} = \{\boldsymbol{x} \mid \sum_{i=1}^{m} x_i Q_i \preceq Q_0\}$$

which is a spectrahedron. □

An immediate consequence of this proposition is that the feasible region Ω in the standard form of semidefinite programming is a spectrahedron because Ω is the intersection of a polyhedron

$$\{U \mid U \bullet Q_i = c_i, 1 \leq i \leq m\}$$

and a spectrahrdron

$$\{U \mid U \succeq 0\}.$$

13.2 Duality

The semidefinite programming (13.1) has a dual programming

$$\begin{aligned} \max \quad & \mathbf{c}^T \boldsymbol{x} \\ & \sum_{i=1}^{m} x_i Q_i + Z = Q_0 \\ & Z \succeq O. \end{aligned} \tag{13.2}$$

Two programmings (13.1) and (13.2) have the following relations:

Lemma 13.2.1 *Suppose U is a feasible solution of (13.1) and $(\boldsymbol{x}, Z)$ is feasible solution of (13.2). Then $\mathbf{c}^T \boldsymbol{x} \leq U \bullet Q_0$.*

Proof. Since

$$\begin{aligned} \mathbf{c}^T \boldsymbol{x} &= \sum_{i=1}^{m} c_i x_i \\ &= \sum_{i=1}^{m} (U \bullet Q_i) x_i \\ &= U \bullet (\sum_{i=1}^{m} x_i Q_i), \end{aligned}$$

we have

$$\begin{aligned} U \bullet Q_0 - \mathbf{c}^T \mathbf{x} &= U \bullet (Q_0 - \sum_{i=1}^{m} x_i Q_i) \\ &= \mathrm{Tr}(U(Q_0 - \sum_{i=1}^{m} x_i Q_i)) \\ &= \mathrm{Tr}(UZ) \\ &\geq 0. \end{aligned}$$

Here, we note that the product of two positive semidefinite matrices is positive semidefinite. □

Clearly, if $\mathrm{Tr}(UZ) = 0$ for primal feasible solution U and dual feasible solution $(\mathbf{x}, Z)$, then U and $(\mathbf{x}, Z)$ are respectively primal and dual optimal solutions. Note that since $Q_1, \cdots, Q_m$ are linearly independent, $\mathbf{x}$ can be uniquely determined by Z through the equality

$$\sum_{i=1} x_i Q_i + Z = Q_0.$$

Therefore, we may use either $\mathbf{x}$ or Z to represent a feasible dual solution $(\mathbf{x}, Z)$. Meanwhile, we call (U, Z) a feasible primal-dual pair if U and Z are feasible solutions of (13.1) and (13.2), respectively. The *standard duality gap* of semidefinite programming is defined to be

$$\inf_{U \succeq 0} \sup_{\mathbf{x} \in R^m} [\mathbf{c}^T \mathbf{x} + U \bullet (Q_0 - Q(\mathbf{x}))] - \sup_{\mathbf{x} \in R^m} \inf_{U \succeq 0} [\mathbf{c}^T \mathbf{x} + U \bullet (Q_0 - Q(\mathbf{x}))].$$

It is equal to $\mathrm{Tr}(U^* Z^*)$ when an optimal feasible primal-dual pair (U^*, Z^*) exists.

When does the standard duality gap equals zero? There are many sufficient conditions. The following is one of them.

A feasible solution U of (13.1) is said to be *strictly feasible* if U is positive definite. A feasible solution Z of (13.2) is said to be *strictly feasible* if Z is positive definite. The following theorem can be found in [128].

Theorem 13.2.2 *If (13.1) has a strictly feasible solution, then the standard duality gap equals zero.*

Theorem 13.2.3 *For any semidefinite programming (13.1), there is one with the same optimal value and with strictly feasible solution.*

Proof. Define

$$Q'_i = \begin{pmatrix} Q_i & 0 \\ 0 & c_i - Q_i \bullet I \end{pmatrix} \text{ for } i = 1, 2, \cdots, m$$

and

$$Q'_0 = \begin{pmatrix} Q_0 & 0 \\ 0 & M \end{pmatrix}$$

where M is a large positive constant. Then, the semidefinite program

$$\begin{array}{ll} \min & Q'_0 \bullet U' \\ s.t. & Q'_i \bullet U' = c_i \text{ for } i = 1, 2, \cdots, m \\ & U' \succeq 0 \end{array}$$

has the same optimal value as that of (13.1) and a strictly feasible solution, the identity matrix. □

13.3 Semidefinite Relaxation

The semidefinite programming can be solved efficiently. (This will be discussed in detail in the next chapter.) Like the linear programming, we can use the semidefinite programming to design approximation algorithms with relaxation technique.

First, let us study a standard relaxation from the *multiquadratic programming*:

$$\begin{array}{ll} \min & \boldsymbol{x}^T Q_0 \boldsymbol{x} + 2\boldsymbol{b}_0^T \boldsymbol{x} + c_0 \\ s.t. & \boldsymbol{x}^T Q_i \boldsymbol{x} + 2\boldsymbol{b}_i^T \boldsymbol{x} + c_i \quad \forall i = 1, 2, \cdots, m. \end{array} \tag{13.3}$$

It can be rewritten as follows:

$$\begin{array}{ll} \min & U \bullet Q_0 + 2\boldsymbol{b}_0^T \boldsymbol{x} + c_0 \\ s.t. & U \bullet Q_i + 2\boldsymbol{b}_i^T \boldsymbol{x} + c_i = 0 \quad \forall i = 1, 2, \cdots, m \\ & U - \boldsymbol{x}\boldsymbol{x}^T = 0 \end{array}$$

By relaxing the constraint $U - \boldsymbol{x}\boldsymbol{x}^T = 0$ to $U - \boldsymbol{x}\boldsymbol{x}^T \succeq 0$, we obtain

$$\begin{array}{ll} \min & U \bullet Q_0 + 2\boldsymbol{b}_0^T \boldsymbol{x} + c_0 \\ s.t. & U \bullet Q_i + 2\boldsymbol{b}_i^T \boldsymbol{x} + c_i = 0 \quad \forall i = 1, 2, \cdots, m \\ & U - \boldsymbol{x}\boldsymbol{x}^T \succeq 0. \end{array} \tag{13.4}$$

This is a semidefinite programming. Indeed, we have

Lemma 13.3.1 $U - \boldsymbol{x}\boldsymbol{x}^T \succeq 0$ *if and only if*

$$\begin{pmatrix} U & x^T \\ x & 1 \end{pmatrix} \succeq 0.$$

Proof. First, assume $U - \boldsymbol{x}\boldsymbol{x}^T \succeq 0$, that is, for any n-dimensional vector $\boldsymbol{y}$,

$$\boldsymbol{y}^T(U - \boldsymbol{x}\boldsymbol{x}^T)\boldsymbol{y} = \boldsymbol{y}^T U \boldsymbol{y} - (\boldsymbol{y}^T\boldsymbol{x})^2 \geq 0.$$

Therefore, for any n-dimensional vector $\boldsymbol{y}$ and scalar α,

$$\begin{aligned} & (\boldsymbol{y}^T, \alpha)\begin{pmatrix} U & \boldsymbol{x}^T \\ \boldsymbol{x} & 1 \end{pmatrix}\begin{pmatrix} \boldsymbol{y} \\ \alpha \end{pmatrix} \\ = \; & \boldsymbol{y}^T U \boldsymbol{y} + 2\alpha \boldsymbol{y}^T\boldsymbol{x} + \alpha^2 \\ \geq \; & (\boldsymbol{y}^T\boldsymbol{x})^2 + 2\alpha\boldsymbol{y}^T\boldsymbol{x} + \alpha^2 \\ = \; & (\boldsymbol{y}^T\boldsymbol{x} + \alpha)^2 \geq 0. \end{aligned}$$

Conversely, assume

$$\begin{pmatrix} U & \boldsymbol{x}^T \\ \boldsymbol{x} & 1 \end{pmatrix} \succeq 0.$$

Choose $\alpha = -\boldsymbol{y}^T\boldsymbol{x}$. Then,

$$\begin{aligned} 0 \; \leq \; & (\boldsymbol{y}^T, \alpha)\begin{pmatrix} U & \boldsymbol{x}^T \\ \boldsymbol{x} & 1 \end{pmatrix}\begin{pmatrix} \boldsymbol{y} \\ \alpha \end{pmatrix} \\ = \; & \boldsymbol{y}^T(U - \boldsymbol{x}\boldsymbol{x}^T)\boldsymbol{y} + (\boldsymbol{y}^T\boldsymbol{x} + \alpha)^2 \\ = \; & \boldsymbol{y}^T(U - \boldsymbol{x}\boldsymbol{x}^T)\boldsymbol{y}. \end{aligned}$$

This means that $U - \boldsymbol{x}\boldsymbol{x}^T \succeq 0$. □

By Lemma 13.3.1, (13.4) can be rewritten as the following primal standard form of semidefinite programming:

$$\begin{aligned} \min \quad & \begin{pmatrix} U & \boldsymbol{x}^T \\ \boldsymbol{x} & 1 \end{pmatrix} \bullet \begin{pmatrix} Q_0 & \boldsymbol{b}_0^T \\ \boldsymbol{b}_0 & c_0 \end{pmatrix} \\ s.t. \quad & \begin{pmatrix} U & \boldsymbol{x}^T \\ \boldsymbol{x} & 1 \end{pmatrix} \bullet \begin{pmatrix} Q_i & \boldsymbol{b}_i^T \\ \boldsymbol{b}_i & c_i \end{pmatrix} = 0 \quad \forall i = 1, 2, \cdots, m \end{aligned}$$

$$\begin{pmatrix} U & \boldsymbol{x}^T \\ \boldsymbol{x} & 1 \end{pmatrix} \succeq 0.$$

The above relaxation is called the *convexification relaxation* of multiquadratic programming. We next study an example:

MAX-BISECTION: Given a graph $G = (V, E)$ with non-negative weight w_{ij} for each edge in E, find a partition (V_1, V_2) of vertex set V to maximize the number of edges between V_1 and V_2 under condition that $|V_1| = |V_2|$.

This problem can be formulated as

$$\begin{aligned} \max \quad & \frac{1}{4} \sum_{1 \le i,j \le n} w_{ij}(1 - x_i x_j) \\ s.t. \quad & \sum_{i=1}^{n} x_i = 0 \\ & x_i^2 = 1, i = 1, 2, \cdots, n. \end{aligned} \tag{13.5}$$

Note that $\sum_{i=1}^{n} x_i = 0$ is equivalent to $\sum_{1 \le i,j \le n} x_i x_j = 0$. Thus, its convexification relaxation can be expressed as

$$\begin{aligned} \max \quad & \frac{1}{4} W \bullet (J - U) \\ s.t. \quad & J \bullet U = 0 \\ & U_{ii} = 1, i = 1, 2, \cdots, n \\ & U \in \mathcal{S}_n \\ & U \succeq 0. \end{aligned} \tag{13.6}$$

where $W = (w_{ij})$ and J is the $n \times n$ matrix of all ones.

Suppose $\boldsymbol{x}^*$ is an optimal solution of the Max-Bisection problem (13.5) and U^* is an optimal solution of its relaxation (13.6). Clearly,

$$w^* = \frac{1}{4} \sum_{1 \le i,j \le n} w_{ij}(1 - \boldsymbol{x}_i^* \boldsymbol{x}_j^*) \le w_R = \frac{1}{4} W \bullet (J - U^*).$$

There is an interesting random rounding technique for obtaining an approximation solution of Max-Bisection from U^*.

Random Rounding

- *Generate a random vector u from a multivariate normal distribution with 0 mean and covariance matrix U^*, i.e., $u \in N(0, U^*)$.*

- *Choose*

$$S = \begin{cases} \{i \mid u_i \geq 0\} & \text{if } |\{i \mid u_i \geq 0\}| \geq n/2 \\ \{i \mid u_i < 0\} & \text{otherwise.} \end{cases}$$

- *For each* $i \in S$*, denote* $\zeta(i) = \sum_{j \notin S} w_{ij}$*. Assume* $S = \{i_1, i_2, \cdots, i_{|S|}\}$ *with* $\zeta(i_1) \geq \zeta(i_2) \geq \cdots \geq \zeta(i_{|S|})$*. Assign* $S_A = \{i_1, i_2, \cdots, i_{n/2}\}$ *and* $\bar{S}_A = \{1, 2, \cdots, n\} - S$*.*

To evaluate the quality of bisection $(S_A, \bar{S}_A)$, let us define two random variables:

$$\begin{aligned} w &= w(S) = \sum_{i \in S, j \notin S} w_{ij} \\ m &= |S|(n - |S|). \end{aligned}$$

Set

$$z = \frac{w}{w^*} + \frac{m}{m^*}$$

where $m^* = n^2/4$.

Lemma 13.3.2 *In the random rounding procedure, if* S *satisfies* $z \geq \alpha$*, then*

$$w(S_A) = \sum_{i \in S_A, j \notin S_A} w_{ij} \geq 2(\sqrt{2\alpha} - 1)w^*.$$

Proof. Assume $w(S) = \lambda w^*$ and $|S| = \beta n$. Then $m/m^* = 4\beta(1-\beta)$.

$$z = \lambda + 4\beta(1-\beta).$$

From the definition of S_A, it is easy to see that

$$w(S_A) \geq \frac{n \cdot w(S)}{2|S|}.$$

Therefore,

$$\begin{aligned} w(S_A) &\geq \frac{w(S)}{2\beta} \\ &= \frac{\lambda w^*}{2\beta} \\ &= \frac{z - 4\beta(1-\beta))w^*}{2\beta}. \end{aligned}$$

Let us study the function

$$g(\beta) = \frac{z - 4\beta(1-\beta)}{2\beta}.$$

Rewrite it as

$$4\beta^2 - \beta(4 - 2g(\beta)) + z = 0.$$

Then we know that

$$(2 - g(\beta))^2 - 4z \geq 0.$$

Note that $0 < g(\beta) \leq 1$. Thus,

$$g(\beta) \geq 2(\sqrt{z} - 1) \geq 2(\sqrt{\alpha} - 1).$$

□

Next, we want to establish $E(z)$. Before doing so, let us first show two lemmas.

Lemma 13.3.3 $1 - \frac{2}{\pi}\arcsin x \geq \rho(1-x)$ *where* $\rho = 0.878567\cdots$.

Proof. Denote $\phi = \arcsin x$. The inequality is equivalent to

$$1 - \frac{2}{\pi}\phi \geq 0.878567(1 - \sin\phi).$$

As shown in Fig. 13.1, draw tangent line of curve $x = \sin\phi$ through point

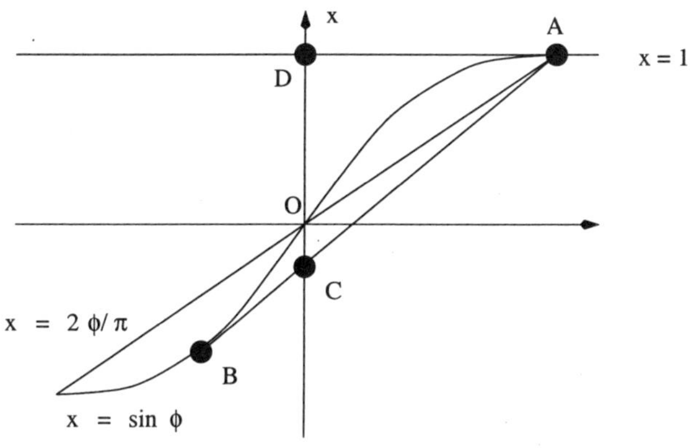

Figure 13.1: Lemma 13.3.3

$A(\pi/2, 1)$, which intersects the line $\phi = 0$ at point C. Suppose O is the original point and D is the intersection point of lines $x = 1$ and $\phi = 0$. Define

$$\rho = \frac{|CD|}{|OD|} = |CD|.$$

Then

$$1 - \frac{2}{\pi}\phi \geq \rho(1 - \sin\phi).$$

By calculation, we know $\rho = 0.878567\cdots$. □

Define

$$\hat{u}_i = \begin{cases} 1 & \text{if } u_i \geq 0 \\ -1 & \text{otherwise.} \end{cases}$$

Lemma 13.3.4 $E[\hat{u}_i\hat{u}_j] = \frac{2}{\pi}\arcsin(U^*_{ij})$.

Proof. See [11]. □

Lemma 13.3.5 $E[z] \geq 2\rho$ *where* $\rho = 0.878567\cdots$.

Proof. Note that

$$\begin{aligned} w(S) &= \frac{1}{4}\sum_{i,j} w_{ij}(1 - \hat{u}_i\hat{u}_j) \\ |S|(n - |S|) &= \frac{1}{4}\sum_{i,j}(1 - \hat{u}_i\hat{u}_j. \end{aligned}$$

Therefore, by Lemmas 13.3.4 and 13.3.3,

$$\begin{aligned} E[w] &= \frac{1}{4}\sum_{i,j} w_{i,j}(1 - \frac{2}{\pi}\arcsin U^*_{ij}) \\ &\geq \frac{1}{4}w_{i,j}\rho(1 - U^*_{ij}) \\ &\geq \rho w^* \end{aligned}$$

and

$$\begin{aligned} E[m] &= \frac{1}{4}\sum_{i,j}(1 - \frac{2}{\pi}\arcsin U^*_{ij}) \\ &\geq \frac{1}{4}\rho(1 - U^*_{ij}) \\ &\geq \rho m^*. \end{aligned}$$

Thus, $E[z] \geq 2\rho$. □

When $\alpha = 2\rho$, $2(\sqrt{\rho} - 1) = 0.651$. Therefore, the following is obtained.

Theorem 13.3.6 *There is a polynomial-time random algorithm for Max-Bisection, which produces cut with expected total weight at least 0.651 times the optimal.*

Exercises

1. Show that ellipsoid $\{\boldsymbol{x} \mid \boldsymbol{x}^T Q\boldsymbol{x} + \boldsymbol{b}^T\boldsymbol{x} + c \leq 0\}$ is a spectrahedron.

2. Show that $\mathcal{E}_n = \{U \in \mathcal{S}_n \mid U_{ii} = 1, U \succeq 0\}$, called *elliptope*, is a spectrahedron with 2^n vertices where a vertex is a matrix in form $\boldsymbol{v}\boldsymbol{v}^T$.

3. A *face* of a spectrahedron is the intersection of a hyperplane and the spectrahedron.

   (a) Show that the smallest face of a spectrahedron G containing point $\bar{\boldsymbol{x}}$ is

   $$F_G(\bar{\boldsymbol{x}}) = \{\boldsymbol{x} \in G \mid \mathrm{Null}(Q_0 - Q(\bar{\boldsymbol{x}})) \subseteq \mathrm{Null}(Q_0 - Q(\boldsymbol{x}))\}$$

   where $G = \{\boldsymbol{x} \mid Q(\boldsymbol{x}) \succeq Q_0\}$ and for a matrix A, $\mathrm{Null}(A) = \{\boldsymbol{y} \mid A\boldsymbol{y} = 0\}$.

   (b) Construct a spectrahedron such that the dimensions of its faces are triangular integers $k(k+1)/2$ for $k = 0, 1, \cdots, n$.

4. Consider a spectrahedron $G = \{\boldsymbol{x} \mid Q(\boldsymbol{x}) \succeq Q_0\}$. A *plate* of G of order k is defined to be the closure of a connected component of $\{\boldsymbol{x} \in G \mid \mathrm{rank}(Q_0 - Q(\boldsymbol{x})) = k\}$.

   (a) Find all plates of spectrahedron

   $$\{\boldsymbol{x} \in R^3 \mid x_1^2 + (x_2 - 2)^2/4 + x_3^2/4 \leq 1, x^T x \leq 1\}.$$

   (b) Show that the relative interior of any face is contained in exactly one plate.

   (c) Show that every spectrahedron has finitely many plates.

(d) Show that every plate of a polyhedron is a face.

5. Show the following:

(a) If U^* is an optimal solution of primal semidefinite programming (13.1), then

$$U^* \bullet Q_0 = \max_{\boldsymbol{x} \in R^m} \min_{U \succeq 0} \{\boldsymbol{c}^T \boldsymbol{x} + U \bullet (Q_0 - Q(\boldsymbol{x}))\}$$

where $Q(\boldsymbol{x}) = \sum_{i=1}^m x_i Q_i$.

(b) If $\boldsymbol{x}^*$ is an optimal solution of dual semidefinite programming (13.2), then

$$\boldsymbol{c}^T \boldsymbol{x}^* = \min_{U \succeq 0} \max_{\boldsymbol{x} \in R^m} \{\boldsymbol{c}^T \boldsymbol{x} + U \bullet (Q_0 - Q(\boldsymbol{x}^*))\}.$$

6. Suppose $(\bar{\boldsymbol{x}}, \bar{Z})$ is a strictly feasible solution of (13.2). Show that for any $\boldsymbol{x}$ satisfying

$$\text{Tr}(Z\bar{Z}^{-2}Z) \le 1$$

$(\boldsymbol{x}, Z)$ is a feasible solution of (13.2) where $Z = Q_0 - \sum_{i=1}^m x_i Q_i$.

7. Show that if (13.2) has a strictly feasible solution, then the standard duality gap equals zero.

8. Given a graph $G = (V, E)$, a *clique* (*independent set*) of G is a vertex subset such that every two vertices in it are adjacent (not adjacent) each other. Assume $V = \{1, 2, \cdots, n\}$. The *characteristic vector* $\boldsymbol{x}$ of a vertex subset V' is defined by $x_i = 1$ if $i \in V'$ and 0 if $i \notin V'$. Show the following:

(a) If $\boldsymbol{u}$ and $\boldsymbol{v}$ are characteristic vectors of a clique and an independent set, respectively, then $\boldsymbol{u}^T \boldsymbol{v} \le 1$.

(b) Let INDEP(G) be the convex hull of characteristic vectors of all independent sets in G. Then INDEP(G) is a subset of the following polyhedron:

$$\begin{aligned}\text{QINDEP}(G) = \{\boldsymbol{x} \ge 0 \mid \boldsymbol{x}^T \boldsymbol{u} \le 1 \ \forall \text{ characteristic vectors } \boldsymbol{u} \\ \text{of cliques of } G\}.\end{aligned}$$

(c) Consider maximum independent set problem:

$$\begin{array}{lll} \max & \sum_{i=1}^n x_i & \\ s.t. & x_i x_j = 0 & \forall (i,j) \in E \\ & x_i(x_i - 1) = 0 & \forall i \in V. \end{array}$$

Its convexification relaxation is maximizing $\sum_{i=1}^n x_i$ over

$$\begin{aligned} S(G) &= \{(U, \boldsymbol{x}) \mid \boldsymbol{x} \geq 0, U \succeq \boldsymbol{x}\boldsymbol{x}^T, U_{ii} = x_i \ \ \forall i \in V, \\ & \qquad U_{ij} = 0 \forall (i,j) \in E\}. \end{aligned}$$

(d) Define $\mathrm{TH}(G) = \{\boldsymbol{x} \mid \exists U : (U, \boldsymbol{x}) \in S(G)\}$. Then

$$\mathrm{INDEP}(G) \subseteq \mathrm{TH}(G) \subseteq QINDEP(G).$$

9. Let $\mathcal{C}_n$ denote the convex hull of all matrices $\boldsymbol{v}\boldsymbol{v}^T$ for $\boldsymbol{v} \in \{-1, +1\}^n$. For a matrix A, denote by $f_o(A)$ the matrix whose (i,j)th entry is $f(A_{ij})$ where A_{ij} is the (i,j)th entry of A. Show that

$$\mathcal{C}_n \subseteq \mathcal{E}_n \subseteq \{\sin_o(\frac{\pi}{2}U) \mid U \in \mathcal{E}_n\}.$$

10. Show that the following nonlinear semidefinite program has the same objective function value as MAXCUT:

$$\frac{1}{\pi} \max\{W \bullet (\arccos_o(U)) \mid U \in \mathcal{E}_n\}$$

where W is a weight matrix and $\mathcal{E}_n$ is an elliptope.

11. Consider a positive semidefine symmetric matrix $A = (a_{ij})$. Show that if $a_{ii} = 1$ for all i, then $|a_{ij}| \leq 1$ for all i and j.

Chapter 14

Interior Point Methods

Small rooms or dwelling discipline the mind,
large one weaken it.
LEONARDO DA VINCI

Since Karmarkar proposed the first interior-point algorithm for linear programming, many variations have been discovered. In particular, they can be extended to semidefinite programming. We will study some of them in this chapter.

14.1 Affine Scaling

A key idea of Karmarkar's algorithm is that at each iteration, a affine mapping is employed to transform the current point to the center of a ball inscribed in the feasible region. This enable us to fix search step length, so that searching along the steepest descent direction of the potential function would obtain more benefit.

Actually, this idea can also be used in linear programming in the standard form:

$$\begin{aligned} \min \quad & \boldsymbol{c}^T \boldsymbol{x} \\ s.t. \quad & A\boldsymbol{x} = \boldsymbol{b} \\ & \boldsymbol{x} \geq 0 \end{aligned}$$

where A is an $m \times n$ matrix with rank m. We may assume the existence of a known relative interior feasible point, that is, a feasible point $\boldsymbol{x} > 0$. In fact, if we do not know such a feasible point, then we consider the following instead.

$$\begin{array}{ll} \min & \boldsymbol{c}^T\boldsymbol{x} + Mz \\ s.t. & A\boldsymbol{x} + (\boldsymbol{b} - A\mathbf{1})z = \boldsymbol{b} \\ & \boldsymbol{x} \geq 0, z \geq 0 \end{array}$$

where z is a new variable and M is a sufficiently large constant such that the minimum point of this problem must have $z = 0$. Clearly, $\boldsymbol{x} = \mathbf{1}, z = 1$ is a relative interior feasible point of this new problem.

Now, assume a relative interior feasible point $\boldsymbol{x}_k > 0$ is found. We describe how to search the next feasible point with the similar idea.

Let $D_k = Diag(\boldsymbol{x}_k)$, the matrix with components of vector $\boldsymbol{x}_k$ as diagonal elements. Consider affine mapping

$$\boldsymbol{y} = D_k^{-1}\boldsymbol{x}.$$

It transforms the linear programming into the following

$$\begin{array}{ll} \min & (c^T D_k)\boldsymbol{y} \\ s.t. & (AD_k)\boldsymbol{y} = \boldsymbol{b} \\ & \boldsymbol{y} \geq 0. \end{array}$$

Clearly, $\mathbf{1} = D_k^{-1}\boldsymbol{x}_k$, that is, this affine mapping transforms current point $\boldsymbol{x}_k$ to a new position which has the same distance to every coordinate plane. Next, we choose the steepest descent direction as search direction, that is, the opposite direction of the gradient projection on affine space $(AD_k)\boldsymbol{y} = \boldsymbol{b}$,

$$\boldsymbol{d} = -(I - D_k A^T (AD_k^2 A^T)^{-1} AD_k)(c^T D_k),$$

and the new point to be

$$\boldsymbol{y}_{k+1} = \mathbf{1} + \boldsymbol{d}/\|\boldsymbol{d}\|.$$

With the inverse of the affine mapping, we can obtain a new point in the original space

$$\boldsymbol{x}_{k+1} = D_k(\mathbf{1} + \boldsymbol{d}/\|\boldsymbol{d}\|.$$

If $\boldsymbol{x}_{k+1} > 0$, then we can continue our search in the same way. If $\boldsymbol{x}_{k+1}$ has a 0-component, then $\boldsymbol{x}_{k+1}$ is an optimal solution. In fact, the existence of 0-component of $\boldsymbol{x}_{k+1}$ implies the existence of 0-component of $\boldsymbol{y}_{k+1}$. Thus, the ball $\{\boldsymbol{y} \mid (AD_k)\boldsymbol{y} = \boldsymbol{b}, \|\boldsymbol{y} - \mathbf{1}\| \leq 1\}$ touches the boundary in $\boldsymbol{y} \geq 0$. This means that the ball must tangents the boundary, a hyperplane H in the affine space $(AD_k)\boldsymbol{y} = \boldsymbol{b}$; $\boldsymbol{y}_{k+1}$ is the tangent point. Thus, the search direction $\boldsymbol{d}$ must be perpenticular to the hyperplane H. Since $-\boldsymbol{d}$ is also the gradient projection direction at $\boldsymbol{y}_{k+1}$, $\boldsymbol{y}_{k+1}$ is a Kuhn-Tucker point. Hence, $\boldsymbol{y}_{k+1}$ is a minimum point.

This affine scaling method for linear programming was first proposed by Dikin in 1967. Therefore, it is called Dikin affine scaling method.

14.2 Primal-Dual Affine Scaling

The idea of affine scaling can also be combined with primal-dual methods. Let us introduce this method based on semidefinite programming. Actually, the most important fact is that the semidefinite programming can be solved efficiently. There are many efficient algorithms. We introduce one of them in this section.

Let us start with a pair of primal and dual strictly feasible solutions U and $(\boldsymbol{x}, Z)$. In each iteration, we want to compute a new pair of $U + \Delta U$ and $(\boldsymbol{x} + \Delta\boldsymbol{x}, Z + \Delta Z)$ to reduce the duality gap, i.e., to minimize $(U + \Delta U) \bullet (Z + \Delta Z)$ with respect to $(\Delta U, \Delta\boldsymbol{x}, \Delta Z)$ under constraints

$$\Delta U \bullet Q_i = 0, \tag{14.1}$$

$$U + \Delta U \succeq 0, \tag{14.2}$$

$$\sum_{i=1}^{m} \Delta x_i Q_i + \Delta Z = 0 \tag{14.3}$$

$$Z + \Delta Z \succeq 0. \tag{14.4}$$

Since Q_i, $i = 1, 2, \cdots, m$ are linearly independent, $\Delta\boldsymbol{x}$ is uniquely determined by ΔZ through (14.3). Thus, we may focus our interest in $(\Delta U, \Delta Z)$.

First, we transform the current point (U, Z) to the "center" of the primal-dual space. To do so, we employ the *Nesterov-Todd scaling matrix*

$$D = Z^{-1/2}(Z^{1/2}UZ^{1/2})^{1/2}Z^{-1/2}.$$

It satisfies

$$D^{-1/2}UD^{-1/2} = D^{1/2}ZD^{1/2} = V. \tag{14.5}$$

Define

$$D_V = D_U + D_Z$$

where

$$
\begin{aligned}
D_U &= D^{-1/2}\Delta U D^{-1/2}, \\
D_Z &= D^{1/2}\Delta Z D^{1/2}.
\end{aligned}
$$

Then

$$
\begin{aligned}
D_U \bullet D_Z &= \mathrm{Tr}(D_U D_Z) \\
&= \mathrm{Tr}(\Delta U \Delta Z) \\
&= -\sum_{i=1}^{m} x_i U \bullet Q_i \\
&= 0.
\end{aligned}
$$

Actually, D_Z is in the linear subspace L generated by $D^{-1/2}Q_iD^{-1/2}$, $i = 1, 2, \cdots, m$, and D_U is in its orthgonal subspace $L^{\perp}$. Therefore, D_U and D_Z, can be determined by D_V; so are ΔU and ΔZ. Therefore, we may want to study D_V instead of ΔU and ΔZ.

Can we represent the duality gap $(U + \Delta U) \bullet (Z + \Delta Z)$ through D_V instead of ΔU and ΔZ ? The answer is YES. In fact,

$$
\begin{aligned}
D^{-1/2}(U + \Delta U)(Z + \Delta Z)D^{1/2} &= (V + D_U)(V + D_Z) \\
&= V^2 + VD_Z + D_UV + D_UD_Z.
\end{aligned}
$$

Since

$$\mathrm{Tr}(D_U D_Z) = 0$$

and

$$\mathrm{Tr}(D_U V) = \mathrm{Tr}(V D_U),$$

we have

$$(U + \Delta U) \bullet (Z + \Delta Z) = \mathrm{Tr}(V^2 + VD_V).$$

What are constraints on D_V? There are four constraints (14.1) – (14.4) for ΔU and ΔZ. For any $D_V \in \mathcal{S}_n$, we can decompose it into $D_V = D_U + D_Z$ where D_U and D_Z are in L and $L^{\perp}$, respectively. Then we can obtain ΔU and ΔZ satisfying (14.1) and (14.3). Thus, (14.1) and (14.3) do not produce any constraint to D_V. The remaining constraints (14.2) and (14.4) are equivalent to

$$
\begin{aligned}
V + D_U &= D^{-1/2}(U + \Delta U)D^{-1/2} \succeq 0 \\
V + D_Z &= D^{1/2}(Z + \Delta Z)D^{1/2} \succeq 0.
\end{aligned}
$$

To satisfy these two constraints, we may ask

$$\|V^{-1/2}D_UV^{-1/2}\|_2 + \|V^{-1/2}D_ZV^{-1/2}\|_2 \leq 1$$

where $\|\cdot\|_2$ is the spectral norm, that is, $\|A\|_2 = (\lambda_{\max}(A^TA))^{1/2}$. This implies

$$\|V^{-1/2}D_VV^{-1/2}\|_2 \leq 1.$$

This constraint is not sufficient to make (14.2) and (14.4) hold. But, we can still consider to obtain D_V by minimizing $\mathrm{Tr}(V^2 + VD_V)$ over an ellipsoid

$$\mathcal{CE} = \{D_V \mid \|V^{-1/2}D_VV^{-1/2}\|_2 \leq 1\}.$$

In fact, we may consider $(\Delta U, \Delta Z)$ obtained from such a D_V as a search direction instead of difference to the new point.

Lemma 14.2.1 *The optimal solution of minimizing $Tr(V^2 + VD_V)$ over ellipsoid $\mathcal{CE}$ is*

$$D_V = D_U + D_Z = -V.$$

Proof. Note that $I \succeq V^{-1/2}D_VV^{-1/2} \succeq -I$. It follows that $V + D_V \succeq 0$ and $V - D_V \succeq 0$. Hence,

$$0 \leq \mathrm{Tr}((V + D_V)(V - D_V)) = \|V\|^2 - \|D_V\|^2$$

where $\|A\| = (\mathrm{Tr}(AA^T))^{1/2}$ is the Frobenius norm of A. Therefore, $\|D_V\| \leq \|V\|$.

$$\begin{aligned}\mathrm{Tr}(V^2 + VD_V) &= \mathrm{Tr}(V^2) + \mathrm{Tr}(VD_V)\\ &\geq \mathrm{Tr}(V^2) - \|V\| \cdot \|D_V\|\\ &\geq \mathrm{Tr}(V^2) - \|V^2\| \cdot \|V^2\|.\end{aligned}$$

The equality sign holds if and only if

$$D_V = -aV,$$

for some positive scalar a and

$$\|D_V\| = \|V\|.$$

Hence, $D_V = -V$. □

By Lemma 14.2.1, we choose $D_V = -V$, that is,

$$\Delta U + D\Delta Z D = -V. \tag{14.6}$$

Above discussion suggests the following primal-dual affine scaling algorithm for semidefinite programming.

Primal-Dual Affine Scaling Algorithm *Initially, we assume that a strictly feasible primal-dual pair* (U, Z) *exists and is known. At each iteration, we carry out the following steps until* $Tr(UZ) \leq \varepsilon$*:*

- *Compute* $(\Delta U, \Delta Z)$ *from (14.6), (14.1), and (14.3).*
- *Choose a search step length* α *and set*

$$\begin{aligned} U &:= U + \alpha \Delta U, \\ Z &:= Z + \alpha \Delta Z. \end{aligned}$$

How to determine the step length α? We will discuss it in the next section. Before ending this section, it may be worth mentioning the following: The central idea of this algorithm is that the search direction $(\Delta U, \Delta Z)$ is chosen to minimize the duality gap $(U + \Delta U) \bullet (Z + \Delta Z)$ over an inscribed ellipsoid in the primal-dual space. However, there are many ways to define the inscribed ellipsoid. Different ellipsoids will give different primal-dual affine scaling methods. What we introduced is just one of them.

14.3 Central Path

If U and $(\boldsymbol{x}, Z)$ satisfy the following conditions,

$$\begin{aligned} Q_i \bullet U &= c_i, i = 1, 2, \cdots, m \\ \sum_{i=1}^{m} x_i Q_i + Z &= Q_0 \\ U, Z &\succeq 0 \\ UZ &= 0, \end{aligned}$$

then U and (x, Z) are optimal solutions of (13.1) and (13.2), respectively. We relax $UZ = 0$ to $UZ = \mu I$. Then we would obtained a system of relaxed optimality conditions

$$\begin{aligned} Q_i \bullet U &= c_i, i = 1, 2, \cdots, m \\ \sum_{i=1}^{m} x_i Q_i + Z &= Q_0 \\ U, Z &\succeq 0 \\ UZ &= \mu I. \end{aligned}$$

This system has unique solution $(U(\mu), \boldsymbol{x}(\mu), Z(\mu))$ which forms a path when μ varies. This path is called the *central path.*

How to measure the centrality of a pair of primal and dual feasible solutions, U and $(\boldsymbol{x}, Z)$? There are several ways. One of them is defined by

$$\kappa(UZ) = \frac{\lambda_{\max}(UZ)}{\lambda_{\min}(UZ}$$

where $\lambda_{\max}(UZ)$ and $\lambda_{\min}$ are respectively the maximum and the minimum eigenvalues of UZ.

This measure is important in determination of step length α in the Primal-Dual Affine Scaling Algorithm.

Theorem 14.3.1 *Let $\tau_0 = \lceil \kappa(UZ) \rceil$ where (U, Z) is an initial strictly feasible primal-dual pair. For any $\varepsilon > 0$, if choose $\alpha = \frac{1}{nL\tau_0}$ where $L = \lceil \ln \frac{Tr(UZ)}{\varepsilon} \rceil$, then the Primal-Dual Affine Scaling Algorithm runs at most $\tau_0 n L^2$ iterations to obtain a feasible primal-dual pair (U^*, Z^*) such that $\kappa(U^* Z^*) \leq 3\tau_0$ and $Tr(U^* Z^*) \leq \varepsilon$.*

We will divide the proof of this theorem into several lemmas.

Denote $U(\alpha) = U + \alpha \Delta U$ and $Z(\alpha) = Z + \alpha \Delta Z$.

Lemma 14.3.2 *If $\det(U(\alpha)Z(\alpha)) > 0$ for $0 \leq \alpha \leq \bar{\alpha}$, then $U(\alpha) \succ 0$ and $Z(\alpha) \succ 0$.*

Proof. It is not hard to know that the minimum eigenvalue of $U(\alpha)$, $\lambda_{\min}(U(\alpha))$, is a continuous function with respect to α and so is $\lambda_{\min}(Z(\alpha))$. Note that $U(0) = U \succ 0$ and $Z(0) = Z \succ 0$. If $U(\alpha) \succ 0$ (or $Z(\alpha)$) does not hold for some α, then there exists $\alpha' \in (0, \alpha]$ such that $\lambda_{\min}(U(\alpha')) = 0$ (or

$\lambda_{\min}(Z(\alpha')) = 0)$. Therefore, $\det(U(\alpha')Z(\alpha')) = 0$, a contradiction. □

To study the spectral of UZ, we note that every square matrix can be uniquely represented as sum of a symmetric matrix and a skew-symmetric matrix as follows

$$A = \frac{1}{2}(A + A^T) + \frac{1}{2}(A - A^T).$$

The following lemma is very helpful.

Lemma 14.3.3 *Let $A = Q + S$ where Q is a symmetric matrix of order n and S is a skew-symmetric matrix of order n. If $Q \succ 0$, then* $\det[A] > 0$. *Moreover, if all eigenvalues of A are real numbers, then*

$$\kappa(A) \le \kappa(Q).$$

Proof. Define $Q(t) = Q + tS$. Then for any nonzero vector $x \in R^n$,

$$x^T Q(t) x = x^T Q x > 0$$

since $x^T S x = 0$. This implies that 0 is not an eigenvalue of $Q(t)$. Note that $\det[Q(t)]$ is the product of all eigenvalues of $Q(t)$. Hence, $\det[Q(t)] \neq 0$ for any $t \in R$. Moreover, $Q(0) = Q \succ 0$. Therefore, $\det[Q(t)] > 0$ for any $t \in R$. In particular, $\det[Q(1)] = \det[A] > 0$.

For $\lambda < \lambda_{\min}(Q)$, we have $Q - \lambda I \succ 0$ and hence $\det[A - \lambda I] > 0$, that is, λ is not an eigenvalue of A. Similarly, for $\lambda > \lambda_{\max}(Q)$, λ is not an eigenvalue of A, neither. It follows that

$$0 < \lambda_{\min}(Q) \le \lambda_{\min}(A) \le \lambda_{\max}(A) \le \lambda_{\max}(Q).$$

This proved the second part of the lemma. □

Note that

$$\det[U(\alpha)Z(\alpha)] = \det[(V + \alpha D_U)(V + \alpha D_Z)]$$

and

$$\begin{aligned} M(\alpha) &= \frac{1}{2}[(V + \alpha D_U)(V + \alpha D_Z) + [(V + \alpha D_U)(V + \alpha D_Z)]^T] \\ &= V^2 + \frac{1}{2}\alpha[(D_U + D_Z)V + V(D_Z + D_U)] + \frac{1}{2}\alpha^2(D_U D_Z + D_Z D_U) \\ &= (1-\alpha)V^2 + \frac{1}{2}\alpha^2(D_U D_Z + D_Z D_U). \end{aligned}$$

By Lemma 14.3.3, to show $\det[U(\alpha)Z(\alpha)] >$, it suffices to show $M(\alpha) \succ 0$. To study when $M(\alpha) \succ 0$, the following lemma is very useful.

Lemma 14.3.4 *$r(D_U D_Z + D_Z D_U) \le \frac{1}{2}\|V\|^2$ where $r(A) = \max_i |\lambda_i(A)| = \max(|\lambda_{\min}(A)|, |\lambda_{\max}(A)|)$, is called the* spectral radius *of A.*

Proof. Note that

$$D_U D_Z + D_Z D_U = \frac{1}{2}[(D_U + D_Z)^2 - (D_U - D_Z)^2].$$

Thus

$$-\frac{1}{2}(D_U - D_Z)^2 \preceq D_U D_Z + D_Z D_U \preceq \frac{1}{2}(D_U + D_Z)^2.$$

It follows that

$$-\frac{1}{2}\|D_U - D_Z\|^2 I \preceq D_U D_Z + D_Z D_U \preceq \frac{1}{2}\|D_U + D_Z\|^2.$$

Since $D_U \bullet D_Z = 0$, we have

$$\|D_U + D_Z\| = \|D_U - D_Z\|.$$

Therefore,

$$\begin{aligned} r(D_U D_Z + D_Z D_U) &\le \frac{1}{2}\|D_U + D_Z\|^2 \\ &= \frac{1}{2}\|V\|^2. \end{aligned}$$

□

Lemma 14.3.5 *Suppose $\kappa(V^2) = \tau$. Choose*

$$\bar{\alpha} = \min(\frac{4}{\tau n}, \sqrt{\frac{4(t-1)}{(t\tau + 1)n}})$$

where $t > 1$. Then for $0 \le \alpha \le \bar{\alpha}$,

$$\det[U(\alpha) Z(\alpha)] > 0$$

and

$$\kappa(U(\alpha) Z(\alpha)) \le t\tau.$$

Proof. By Lemma 14.3.3, it suffices to show that for $0 \le \alpha \le \bar{\alpha}$,

$$M(\alpha) \succ 0$$

and

$$\kappa(M(\alpha)) \le t\tau.$$

Denote $\tau_1 = \lambda_{\min}(V^2)$ and $\tau_2 = \lambda_{\max}(V^2)$. By Lemma 14.3.4,

$$r(D_U D_Z + D_Z D_U) \le \frac{1}{2}\|V\|^2 \le \frac{1}{2}\tau_2 n.$$

Therefore,

$$\lambda_{\min}(M(\alpha)) \le (1-\alpha)\tau_1 - \frac{1}{4}\alpha^2\tau_2 n > 0$$

since

$$0 \le \alpha \le \frac{4}{\tau n} < \frac{1+\sqrt{1+\tau n}}{\frac{1}{2}\tau n}.$$

Hence, $M(\alpha) \succ 0$.

Note that

$$\kappa(M(\alpha)) \le \frac{(1-\alpha)\tau_2 + \frac{1}{4}\alpha^2\tau_2 n}{(1-\alpha)\tau_1 - -\frac{1}{4}\alpha^2\tau_2 n}.$$

To show $\kappa(M(\alpha)) \le t\tau$, it suffices to prove

$$(1-\alpha)\tau_2(t-1) - \frac{1}{4}\alpha^2\tau_2 n(t\tau+1) \ge 0,$$

that is,

$$(1-\alpha)(t-1) - \frac{1}{4}\alpha^2(t\tau+1)n \ge 0.$$

This can follow from

$$0 \le \alpha \le \sqrt{\frac{4(t-1)}{(t\tau+1)n}}.$$

□

Choose $t = 1 + \frac{1}{nL^2\tau_0}$. Then for $1 \le k \le nL^2\tau_0$, $\tau_0 t^k < 3\tau_0$. It follows that

$$sqrt\frac{4(t-1)}{(t\tau+1)n} \ge \frac{1}{nL\tau_0}$$

and

$$\frac{4}{\tau n} \ge \frac{1}{nL\tau_0}.$$

Therefore, if we choose $\alpha = \frac{1}{nL\tau_0}$, then we have

$$\begin{aligned}
\mathrm{Tr}(U(\alpha)Z(\alpha)) &= \mathrm{Tr}(V + \alpha D_U)(V + \alpha D_Z)) \\
&= \mathrm{Tr}(V^2 + \alpha V(D_U + D_Z)) \\
&= (1-\alpha)\mathrm{Tr}(V^2) = (1-\alpha)\mathrm{Tr}(UZ).
\end{aligned}$$

Hence, with $nL^2\tau_0$ iterations, the Primal-Dual Affine Scaling Algorithm will find a feasible primal-dual pair (U^*, Z^*) satisfying

$$\begin{aligned}
\mathrm{Tr}(U^*Z^*) &\le (1 - \frac{1}{nL\tau_0})^{nL^2\tau_0}\mathrm{Tr}(UZ) \\
&\le (\frac{1}{e})^L \mathrm{Tr}(UZ) \\
&= \varepsilon
\end{aligned}$$

and

$$\kappa(U^*Z^*) \le 3\tau_0.$$

Finally, it is worth mentioning that a general primal-dual pair of semidefinite programming may be embeded in a self-dual semidefinite programming with nonempty interior and known strictly feasible solution [94].

Exercises

1. Show that under the condition that both primal and dual linear programmings are nondegenerate, Dinkin's affine scaling method either stops at a minimum point or generates an infinite sequence in which every cluster point is a minimum point.

2. Apply the primal-dual affine scaling algorithm studied in sections 14.2-3 to the linear programming. What is the time-complexity?

3. Show that for any $D_V \in \mathcal{DE}$, $V + D_V \preceq 0$.

4. Show that the system of relaxed optimality conditions

$$\begin{aligned}
Q_i \bullet U &= c_i, i = 1, 2, \cdots, m \\
\sum_{i=1}^m x_i Q_i + Z &= Q_0 \\
U, Z &\preceq 0 \\
UZ &= 0,
\end{aligned}$$

has unique solution $(U(\mu), \boldsymbol{x}(\mu), Z(\mu))$.

5. Define $\delta(UZ) = \|I - V/\mu\|$ where $\mu = \text{Tr}(UZ)/n$ and V is defined in (14.5). Show that $\delta(UZ) \leq \sqrt{n}(1 - 1/\kappa(UZ))$.

6. Show that if $aI \preceq A \preceq bI$, then $\kappa(A) \leq b/a$. Give a counterexample to show that $aB \preceq A \preceq bB$ does not necessarily implies $\kappa(A) \leq b/a$ if $B \neq I$.

7. Show that $\kappa(A + B) \leq \max(\kappa(A), \kappa(B))$.

8. Consider the Frobenius norm $\|A\| = (\text{Tr}(AA^T))^{1/2}$. Show that the optimal solution of minimizing $\text{Tr}(V^2 + VD_V)$ over ellipsoid $\{D_V \mid \|V^{-1/2}D_V V^{-1/2}\| \leq 1\}$ is

$$D_V = D_U + D_Z = -V^3/\|V^2\|.$$

Chapter 15

From Local to Global

As we have seen in the previous chapters most of the existing methods in optimization focus on the very important problem of computing feasible points that satisfy necessary optimality conditions (Kuhn-Tucker points). In the previous chapter, we studied a special class of minimax problems for which a characterization of its global solutions was obtained.

In many important results regarding optimization algorithms, the underlying theory is based on the assumption that the objective function (or the feasible domain) is convex. Unless the function has constant Hessian (i.e. is quadratic) or has a very special structure, convexity is not easily recognizable. Even for multivariable polynomials there is no known computable procedure to decide convexity. Therefore, from the practical point of view, a general objective function can be assumed to be neither convex nor concave, having multiple local optima.

During the past decade several works have appeared on the subject of global optimization. Given a real valued function $f(x)$ defined on a set X in R^n, the global optimization problem to be considered is to find the function value f^* and a point $x^* \in X$ such that

$$f^* = f(x^*) = \text{global} \min_{x \in X} f(x) \ (\text{or} \ \max_{x \in X} f(x))$$

if such a point x^* exists. If the objective function and/or the feasible domain are nonconvex, then we may have exponentially many local minima (and stationary points) which are not global. From the complexity point of view such problems are NP-hard, i.e. no polynomial time algorithms are known for their solution.

15.1 Convex Envelopes

The first question that comes to mind when someone first encounters an optimization problem with a nonconvex objective function is "Can we reduce it to a convex problem?" The concept of the convex envelope of a function has been motivated by the efforts to answer this question.

The convex envelope of a function is the uniformly best convex underestimation of this function over a given domain. More precisely, let $f : S \to R$ be a lower semi-continuous function, where S is a nonempty convex set in R^n. Then the convex envelope of $f(x)$ taken over S is a function $F(x)$ such that

i) $F(x)$ is convex on S

ii) $F(x) \leq f(x)$ for all $x \in S$

iii) If $h(x)$ is any convex function defined on S such that $h(x) \leq f(x)$ for all $x \in S$, then $h(x) \leq F(x)$ for all $x \in S$.

Geometrically, the convex envelope $F(x)$ is precisely the function whose epigraph coincides with the convex hull of the epigraph of f.

In a similar way, the (upper) concave envelope of a function over a convex set S is defined to be the best concave overestimator of this function over S. Each nonconvex optimization problem with convex feasible set is associated with a convex problem whose optimal value is the same as that of the original problem. More precisely we have the following:

Theorem 15.1.1 *Consider the problem*

$$\textit{global} \min_{x \in S} f(x)$$

where S is a convex compact set in R^n. Let $F(x)$ be the convex envelope of $f(x)$ over S. Then

$$f^* := \min\{f(x) : x \in S\} = \min\{F(x) : x \in S\}$$

and

$$\{y \in S : f(y) = f^*\} \subseteq \{y \in S : F(y) = f^*\}.$$

Proof. Since, by definition, $F(x) \leq f(x)$ for all $x \in S$ we have that

$$\min\{F(x) : x \in S\} \leq f^*.$$

The constant function $h(x) = f^*$ is a convex underestimating function of f. Again by the definition of the convex envelope we have that $F(x) \geq f^*$ for all x and so

$$\min\{F(x) : x \in S\} \geq f(x^*).$$

We prove the second part by contradiction. Let x^* be a global minimum of $f(x)$ over S, and suppose that x^* is not a global minimum point of $F(x)$ over S. Let y^* be a global minimum point of $F(x)$ over S. Then

$$F(y^*) < F(x^*) \leq f^*.$$

The function $H(x) = \max\{f^*, F(x)\}$ is convex and $H(x) \geq F(x)$ for all $x \in S$, i.e., $H(y^*) > F(y^*)$, which contradicts the fact that $F(x)$, the convex envelope, is the largest convex underestimating function. Hence, $F(y^*) = F(x^*)$ and x^* globally minimizes $F(x)$ over S. □

The above result suggests that we may attempt to solve a nonconvex minimization problem by solving the corresponding convex minimization problem where the objective function is the convex envelope of the original problem. In general, finding the convex envelope of a function is as difficult as computing its global minimum. In several cases the convex envelope can be computed analytically. Moreover, many algorithms are based on the concept of the convex envelope.

15.2 Global Optimization Approaches to Discrete Problems

In graph theory many approaches have been developed that link the discrete universe to the continuous universe through geometric, analytic, and algebraic techniques. Such techniques include global optimization formulations, semidefinite programming, and spectral theory. For example, $z \in \{0,1\} \Leftrightarrow z+w=1, z \geq 0, w \geq 0, zw = 0$. That is an integer constraint is equivalent to continuous nonconvex constraints (complementarity!) In another approach, $z \in \{0,1\} \Leftrightarrow z - z^2 = z(1-z) = 0$. From these simple observations it is clear that an integer constraint is a very special nonconvex constraint. It is the view of many researchers that global optimization is an extension of integer programming. In fact, the first attempts to compute global solutions can be found in discrete optimization algorithms such as enumeration and branch and bound methods.

In many cases the formulation of a discrete problem as an equivalent continuous global optimization problem is not that obvious. As an example, consider the maximum clique problem. Let $G = G(V, E)$ be a graph where $V = \{1, \ldots, n\}$ denotes the set of vertices (nodes), and E denotes the set of edges. Denote by (i, j) an edge joining vertex i and vertex j. A clique of G is a subset C of vertices with the property that every pair of vertices in C is joined by an edge. In other words, C is a clique if the subgraph $G(C)$ induced by C is complete. The maximum clique problem is the problem of finding a clique set C of maximal cardinality. In the following theorem we provide an interesting formulation of the maximum clique problem as a maximization problem of a nonconvex quadratic function.

Theorem 15.2.1 (Motzkin and Strauss) *Consider the continuous indefinite quadratic programming problem*

$$\begin{aligned} \max \quad & f_G(x) = \sum_{(i,j)\in E} x_i x_j = \tfrac{1}{2} x^T A_G x \\ s.t. \quad & x \in S = \{x = (x_1, \ldots, x_n)^T : \sum_{i=1}^{n} x_i = 1, \\ & x_i \geq 0 \quad (i = 1, \ldots, n)\}, \end{aligned} \tag{15.1}$$

where A_G is the adjacency matrix of the graph G. If $\alpha = \max\{f_G(x) : x \in S\}$, then G has a maximum clique C of size $\omega(G) = 1/(1 - 2\alpha)$. This maximum can be attained by setting $x_i = 1/k$ if $i \in C$ and $x_i = 0$ if $i \notin C$.

Proof. The proof can be done by induction on the number of vertices of the graph (Exercise). □

The Motzkin-Strauss formulation of the maximum clique problem has been used to develop some very efficient heuristics. Moreover, a similar formulation can be obtained for the weighted maximum clique problem (for details see [66, 67]).

15.3 Nonconvex Quadratic Programming

After linear programming, quadratic programming is probably the most fundamental and most important optimization problem. Algorithms for the solution of nonconvex quadratic problems, are often basic subroutines to general nonlinear programming algorithms.

15.3.1 Concave Quadratic Programming

Next we consider a constrained concave quadratic programming problem of the form:

$$\begin{array}{rl} \min & \Psi(z,y) = \phi(z) + d^T y \\ \text{s.t.} & (z,y) \in \Omega = \{A_1 z + A_2 y = b,\ z \geq 0,\ y \geq 0\} \subseteq R^{n+k} \end{array} \qquad \text{(P)}$$

where $\phi(z)$ is a concave quadratic function, given by

$$\phi(z) = c^T z - \frac{1}{2} z^T Q z.$$

The symmetric $n \times n$ matrix Q is positive semi-definite. We assume that $A_1 \in R^{m\times n}, A_2 \in R^{m\times k}, b \in R^m, d \in R^k, c \in R^n$, and that the set Ω is a nonempty polytope. We denote by Ψ^* the global minimum objective function value.

Problems of the form (P) reflect the fact that in many applications the objective function is the sum of a concave function in $z \in R^n$ and of a linear function in $y \in R^k$, and usually, k is much larger than n.

First we show how to reduce the problem to an equivalent separable form and how to use piecewise linear approximations to obtain an approximate solution. This reduction is using the eigenstructure of the quadratic form and linear programming.

The symmetric real matrix Q has n real eigenvalues $\lambda_1, \ldots, \lambda_n$ and corresponding orthonormal eigenvectors $u_1, \ldots, u_n$. Moreover, it is well–known (spectral decomposition theorem) that $Q = UDU^T$, where U is the matrix with columns $u_1, \ldots, u_n$ and $D = \text{diag}(\lambda_1, \ldots, \lambda_n)$ (the $(n \times n)$–matrix (d_{ij}) with $d_{ii} = \lambda_i$ and $d_{ij} = 0$ for $i \neq j$). Furthermore, $U^T U = I$ (where I is the $(n \times n)$–unit matrix), since the eigenvectors are orthonormal.

For each $1 \leq i \leq n$ define

$$\begin{array}{rcl} \underline{\zeta}_i & = & \min\{u_i^T z : (z,y) \in \Omega\}, \\ \overline{\zeta}_i & = & \max\{u_i^T z : (z,y) \in \Omega.\}, \end{array} \qquad \text{(MCR)}$$

These $2n$ bounds are obtained by solving the multiple-cost row (MCR) linear programming (which can be solved efficiently by simplex type methods). Then apply the linear transformation

$$z = U(x + \underline{\zeta}) \tag{15.2}$$

where $\underline{\zeta} = (\underline{\zeta}_1, \ldots, \underline{\zeta}_n)^T$. This is equivalent to $x_i = u_i^T z - \underline{\zeta}_i$ $(1 \leq i \leq n)$. Further, let $\beta_i = \overline{\zeta}_i - \underline{\zeta}_i$ $(1 \leq i \leq n)$.

Then problem (P) can be formulated in the following separable form

$$\begin{aligned} \min \quad & \sum_{i=1}^{n} q_i(x_i) + d^T y \\ \text{s.t.} \quad & (A_1 U)x + A_2 y = b - A_1 U \underline{\zeta} \\ & Ux \geq U\underline{\zeta} \\ & 0 \leq x_i \leq \beta_i \quad (i = 1, \ldots, n) \\ & y_i \geq 0 \quad (i = 1, \ldots, k) \end{aligned}$$

where

$$q_i(x_i) = (c^T u_i - \lambda_i \underline{\zeta}_i)x_i - \frac{1}{2}\lambda_i x_i^2 + q_i \quad (i = 1, \ldots, n)$$

with q_i constant. By adding slack variables we can write the inequality $Ux \geq U\underline{\zeta}$ in equivalent equality form.

For simplicity of notation, we consider that the problem has the following separable quadratic programming problem:

$$\min\{\Psi(x, y) = \sum_{i=1}^{n} q_i(x_i) + d^T y : (x, y) \in \Omega,\ x \in R_x\}$$

where

$$\begin{aligned} q_i(x_i) &= c_i x_i - \tfrac{1}{2}\lambda_i x_i^2 \quad (i = 1, \ldots, n), \\ \Omega &= \{(x, y) : A_1 x + A_2 y = b,\ y \geq 0\}, \end{aligned} \tag{SP}$$

$$R_x = \{x : 0 \leq x_i \leq \beta_i \quad (i = 1, \ldots, n)\},$$

and each $\lambda_i > 0 \quad (i = 1, \ldots, n)$.

15.3.2 Pardalos-Rosen Algorithm

Let

$$\phi(x) = \sum_{i=1}^{n} q_i(x_i)$$

in the above separable quadratic formulation (SP), and let $\Gamma(x)$ be the convex envelope of $\phi(x)$ over the rectangle R_x.

Recall from the definition of the convex envelope that $\Gamma(x)$ is the uniformly best convex underestimating function of ϕ over R_x. Moreover, since R_x is a rectangle and $q_i(x_i)$ is concave for $i = 1, \ldots, n$,

$$\Gamma(x) = \sum_{i=1}^{n} \gamma_i(x_i) \text{ where } \gamma_i(x_i) = (c_i - \frac{1}{2}\lambda_i\beta_i)x_i \quad (1 \leq i \leq n). \qquad (15.3)$$

The convex envelopes $\gamma_i(x_i)$ of $q_i(x_i)$ over $[0, \beta_i]$ are the affine functions that coincide with q_i at $x_i = 0$ and $x_i = \beta_i$ (linear interpolation of q_i at $0, \beta_i$). It follows that a lower bound of the optimal objective function value of Problem (SP) is given by the optimal value of the linear programming problem:

$$\begin{aligned} \min \quad & \Gamma(x) + d^T y \\ \text{s.t.} \quad & (x, y) \in \Omega, \; x \in R_x. \end{aligned} \qquad \text{(LU)}$$

Let Ψ^* denote the optimal value of Problem (SP), and let $(\hat{x}, \hat{y})$ be an optimal vertex solution of Problem (LU). Then we have

$$\Gamma(\hat{x}) + d^T \hat{y} \leq \Psi^* \leq \phi(\hat{x}) + d^T \hat{y} = \Psi(\hat{x}, \hat{y}). \qquad (15.4)$$

Defining the error function

$$E(x) = \phi(x) - \Gamma(x) \qquad (15.5)$$

we see that

$$\Psi(\hat{x}, \hat{y}) - \Psi^* \leq E(\hat{x}). \qquad (15.6)$$

If $E(\hat{x})$ is sufficiently small, then we may consider $\Psi(\hat{x}, \hat{y})$ to be an acceptable approximation to the global minimum Ψ^* with $(\hat{x}, \hat{y})$ the corresponding approximate global minimum vertex.

We now derive a bound on the error $E(\hat{x})$ relative to the range of $\phi(x)$ over R_x. Let

$$\begin{aligned} \phi_{\max} &= \max\{\phi(x) : x \in R_x\}, \\ \phi_{\min} &= \min\{\phi(x) : x \in R_x\}. \end{aligned}$$

Hence, the range of ϕ over R_x is $[\phi_{\min}, \phi_{\max}]$. Next, we obtain a lower bound for

$$\Delta\phi = \phi_{\max} - \phi_{\min}.$$

Without loss of generality we may assume that

$$\lambda_1\beta_1^2 \geq \lambda_i\beta_i^2 \qquad i = 2, \ldots, n. \tag{15.7}$$

Define the ratios

$$\rho_i = \frac{\lambda_i\beta_i^2}{\lambda_1\beta_1^2} \leq 1 \qquad i = 1, 2, \ldots, n. \tag{15.8}$$

The unconstrained maximum of the quadratic function $q_i(x_i)$ is attained at the point $\bar{x}_i = c_i/\lambda_i$, for $i = 1, \ldots, n$. When $\bar{x}_i \in [0, \beta_i]$ the lower bound for $\Delta\phi$ depends on the distance $| \bar{x}_i - \beta_i/2 |$ between $\bar{x}_i$ and the midpoint of $[0, \beta_i]$. By scaling such that $| \beta_i | \to 1$, we express this dependence in terms of

$$\eta_i = \min\{1, | \frac{2\bar{x}_i}{\beta_i} - 1 |\} \qquad i = 1, \ldots, n. \tag{15.9}$$

Note that $0 \leq \eta_i \leq 1$, and $\eta_i = |\frac{2\bar{x}_i}{\beta_i} - 1|$ if and only if $\bar{x}_i \in [0, \beta]$.

Lemma 15.3.1 *We have that*

$$\Delta\phi \geq \frac{1}{8}\lambda_1\beta_1^2 \sum_{i=1}^{n} \rho_i(1 + \eta_i)^2 =: \overline{\Delta}\phi.$$

Proof. Let for $1 \leq i \leq n$

$$\begin{aligned} \bar{q}_i &= \max\{q_i(x_i) : 0 \leq x_i \leq \beta_i\}, \\ \underline{q}_i &= \min\{q_i(x_i) : 0 \leq x_i \leq \beta_i\} \end{aligned}$$

and

$$\Delta q_i = \bar{q}_i - \underline{q}_i.$$

There are four cases to consider:

1. $\bar{x}_i \in [0, \beta_i/2]$: We have $q_i(\beta_i) = \lambda_i\beta_i(\bar{x}_i - \frac{1}{2}\beta_i) \leq 0 = q_i(0)$ so that $\underline{q}_i = q_i(\beta_i)$ and $\overline{q}_i = q_i(\overline{x}_i) = \frac{1}{2}\lambda_i\bar{x}_i^2$. From (15.9) we have $\bar{x}_i = \frac{1}{2}\beta_i(1 - \eta_i)$, and hence

$$\begin{aligned} \Delta q_i &= \tfrac{1}{2}\lambda_i\bar{x}_i^2 - \lambda_i\beta_i\bar{x}_i + \tfrac{1}{2}\lambda_i\beta_i^2 = \tfrac{1}{2}\lambda_i(\beta_i - \bar{x}_i)^2 \\ &= \tfrac{1}{8}\lambda_i\beta_i^2(1 + \eta_i)^2. \end{aligned}$$

2. $\bar{x}_i \in [\beta_i/2, \beta_i]$: Now $q_i(\beta_i) \geq 0$, so that $\underline{q}_i = q_i(0) = 0$. From (15.9) we have $\bar{x}_i = \frac{1}{2}\beta_i(1+\eta_i)$, so that

$$\Delta q_i = \frac{1}{2}\lambda_i \bar{x}_i^2 = \frac{1}{8}\lambda_i \beta_i^2 (1+\eta_i)^2.$$

3. $\bar{x}_i \leq 0$: When $\bar{x}_i \leq 0$, we must have $c_i \leq 0$, so that $\overline{q_i} = 0$ and $\underline{q_i} \leq -\frac{1}{2}\lambda_i\beta_i^2$.
   Therefore, since $\eta_i = 1$,

$$\Delta q_i \geq \frac{1}{2}\lambda_i \beta_i^2 = \frac{1}{8}\lambda_i \beta_i^2 (1+\eta_i)^2.$$

4. $\bar{x}_i \geq \beta_i$: Now $\underline{q_i} = q_i(0) = 0$, and since $\eta_i = 1$,

$$\Delta q_i = \overline{q_i} = q_i(\beta_i) = \lambda_i\beta_i(\bar{x}_i - \beta_i/2) \geq \frac{1}{2}\lambda_i \beta_i^2 = \frac{1}{8}\lambda_i \beta_i^2 (1+\eta_i)^2.$$

   Finally, we have

$$\Delta\phi = \sum_{i=1}^{n} \Delta q_i \geq \frac{1}{8}\sum_{i=1}^{n} \lambda_i \beta_i^2 (1+\eta_i)^2 = \frac{1}{8}\lambda_1\beta_1^2 \sum_{i=1}^{n} \rho_i(1+\eta_i)^2.$$

An upper bound on $E(x)$ is easily obtained. We have

$$\begin{aligned} E(x) &= \phi(x) - \Gamma(x) = \sum_{i=1}^{n}[q_i(x_i) - \gamma_i(x_i)] \\ &= \tfrac{1}{2}\sum_{i=1}^{n} \lambda_i(\beta_i - x_i)x_i. \end{aligned}$$

This attains its maximum at $x_i = \beta_i/2$ $(1 \leq i \leq n)$, so that

$$E(x) \leq \tfrac{1}{8}\sum_{i=1}^{n} \lambda_i\beta_i^2 = \frac{1}{8}\lambda_1\beta_1^2 \sum_{i=1}^{n} \rho_i. \tag{15.10}$$

□

Theorem 15.3.2 *Let $\Gamma(x)$ be the convex envelope of $\phi(x)$ over R_x, and let $(\hat{x}, \hat{y})$ be an optimal solution of* $\min\{\Gamma(x) + d^T y : (x,y) \in \Omega,\ x \in R_x\}$. *Then an a-priori bound on the relative error in $\Psi(\hat{x}, \hat{y})$ is given by*

$$\frac{\psi(\hat{x},\hat{y}) - \psi^*}{\Delta\phi} \leq \frac{\sum_{i=1}^{n} \rho_i}{\sum_{i=1}^{n} \rho_i(1+\eta_i)^2} =: \sigma(\rho, \eta). \tag{15.11}$$

Proof. The proof follows from the previous lemma and the fact that $\psi(\hat{x},\hat{y}) - \psi^* \leq E(\hat{x})$. □

Note that $\sigma(\rho,\eta) \in [1/4,1]$ and, furthermore, $\sigma(\rho,\eta) < 1$ unless $\bar{x}_i = \beta_i/2$ for every i. In particular, if $\bar{x}_i \notin (0,\beta_i)$ for all $1 \leq i \leq n$, then $\sigma(\rho,\eta) = 1/4$.

Next we discuss an approach to obtain a guaranteed ε-approximate solution to the original problem. The error bound obtained above can improve the bound if we subdivide each interval $[0,\beta_i]$ into k_i subintervals of (for the sake of simplicity) equal length $h_i = \beta_i/k_i$. The result is a partition of the original rectangle R_x into $\prod_{i=1}^{n} k_i$ subrectangles. If we do with respect to each of these subrectangles what we did in Section 2.6.2 with respect to R_x, then the overall upper bound of the relative error will improve because of the monotonicity of convex envelopes. For $1 \leq i \leq n$ we interpolate $q_i(x_i)$ at the points $x_i = jh_i$ $(0 \leq j \leq k_i)$ to obtain the piecewise linear function $\Gamma_i(x_i)$ on $[0,\beta_i]$ and the underestimator $\Gamma(x) = \sum_{i=1}^{n} \Gamma_i(x_i)$. The upper bound (15.10) on $E(x)$ becomes

$$E(x) \leq \frac{1}{8}\sum_{i=1}^{n} \lambda_i \frac{\beta_i^2}{k_i^2} = \frac{1}{8}\lambda_1\beta_1^2 \sum_{i=1}^{n} \frac{\rho_i}{k_i^2} \tag{15.12}$$

since β_i is replaced by β_i/k_i. The lower bound for $\Delta\phi$ of Lemma 2.4 remains the same.

Corollary 15.3.3 *The above interpolation at $x_i = jh_i$ $(0 \leq j \leq k_i,\ 1 \leq i \leq n)$ yields*

$$\frac{\Psi(\hat{x},\hat{y}) - \Psi^*)}{\Delta\phi} \leq \frac{\sum_{i=1}^{n} (\rho_i/k_i^2)}{\sum_{i=1}^{n} \rho_i(1+\eta_i)^2}.$$

It follows easily that we can choose the k_i so that for any tolerance $\varepsilon > 0$ the relative error in the piecewise linear approximation is bounded by ε.

Corollary 15.3.4 *If for each $1 \leq i \leq n$*

$$k_i \geq (\frac{n}{\alpha}\rho_i)^{\frac{1}{2}},\ \textit{where}\ \alpha = \varepsilon \sum_{i=1}^{n} \rho_i\ (1+\eta_i)^2, \tag{15.13}$$

then an optimal solution $(\hat{x}, \hat{y})$ *of*

$$\begin{array}{ll} \min & \sum_{i=1}^{n} \Gamma_i(x_i) + d^T y \\ s.t. & (x, y) \in \Omega,\ x \in R_x \end{array}$$

satisfies

$$\frac{\Psi(\hat{x}, \hat{y}) - \Psi^*}{\Delta\phi} \leq \frac{E(\hat{x})}{\Delta\phi} \leq \varepsilon.$$

Proof. Exercise. □

To determine an ε– approximate solution to problem (P), we first have to compute the eigenvalues and eigenvectors of the matrix Q and to solve the "multiple–cost–row" linear program (MCR) to obtain the separable formulation (SP). Then we compute the linear underestimating approximation based on the enclosing rectangle R_x by solving the linear program (LU). With good luck this will give an ε–approximate solution and we are done. If so, the concave global minimization problem is solved by the time required to obtain the eigenvalues and eigenvectors and to solve a multiple–cost–row linear program with $(2n + 1)$ cost rows, m constraints and $n + k$ variables.

In the process of solving the multiple–cost–row linear program (MCR) with $2n$ cost rows, we evaluate $\Psi(x, y)$ at every vertex (x_i, y_i) of Ω encountered. The vertex (x_i^*, y_i^*) with minimum $\Psi(x_i, y_i)$ is chosen as the (first) incumbent and $IFV := \Psi(x_i^*, y_i^*)$ as incumbent function value which might give a better bound than $\Psi(\hat{x}, \hat{y})$, where $(\hat{x}, \hat{y})$ is an optimal solution of problem (LU). Thus, an implementation of the above ideas consists in the following steps:

1. Compute the eigenvalues λ_i of Q and the corresponding orthonormal eigenvectors u_i of Q.

2. Solve the multiple-cost-row linear programming problem (MCR) with $2n$ cost rows. Evaluate $\Psi(x, y)$ at every vertex (x_i, y_i) encountered.

3. Choose (x_i^*, y_i^*) satisfying $\Psi(x_i^*, y_i^*) = \min \Psi(x_i, y_i)$ as the incumbent, and set IFV $= \Psi(x_i^*, y_i^*)$. Construct $\Gamma(x)$. Compute the solution $(\hat{x}, \hat{y})$ of (LU), and the bound $\overline{\Delta}\phi$ of $\Delta\phi$ given in Lemma 2.4.

4. If $\Psi(\hat{x}, \hat{y}) < IFV$, set $IFV := \Psi(\hat{x}, \hat{y})$.

5. If $IFV - \Gamma(\hat{x}) - d^T\hat{y} \leq \epsilon\overline{\Delta}\phi$ then stop. The incumbent is an ϵ-approximate solution.

6. Otherwise, continue with one of the procedures given below.

Branch and Bound Algorithm One way to continue in Step 6 is a *branch and bound* procedure. Choose one of the intervals $[0, \beta_i]$, for example $[0, \beta_j]$ with $\beta_j = \max_{i=1,\dots,n} \beta_i$ or the interval that corresponds to $\rho_j = \max_{i=1,\dots,n} \rho_i$. Bisect the interval $[0, \beta_j]$ into $[0, \beta_j/2]$ and $[\beta_j/2, \beta_j]$. This results in a partition of R_x into two rectangles of equal volume. Replace the convex envelope $\gamma_j(x)$ of $q_j(x)$ by the two convex envelopes Γ_{j_1} and Γ_{j_2} that correspond to a linear interpolation of $q_j(x)$ at $0, \beta_j/2$ and $\beta_j/2$, β_j, respectively. Solve (LU) for the two intervals, and update IFV whenever a vertex of Ω is encountered which improves IFV. Set $(\hat{x}, \hat{y})$ to the argument of the smallest of the two optimal objective function values of the two linear programs (LU) which result, and check according to Step 5 whether we can stop or not. If we cannot terminate, then bisect one of the two rectangles and continue with refined interpolations, etc. A branch of the corresponding binary branch and bound tree can be pruned whenever the corresponding lower bound exceeds the incumbent function value IFV.

The resulting algorithm is a special case of a general branch and bound algorithm. Such an algorithm can be used only to solve small size problems.

Mixed Integer Programming Approach Another way to continue in Step 6 is to compute the numbers $k_i = \lceil (\frac{n}{\alpha}\rho_i)^{\frac{1}{2}} \rceil$ according to Corollary 2.5 (recall that, for $a \in R$, $\lceil a \rceil$ is the smallest integer greater or equal a) The corresponding piecewise linear underestimator

$$\Gamma(x) = \sum_{i=1}^{n} \Gamma_i(x_i)$$

can then be constructed by means of a standard *zero–one mixed integer formulation.* Minimizing $\Gamma(x)$ over the feasible set of problem (SP) amounts to solving a linear zero–one mixed integer program which gives the required ε approximation.

Recall that we subdivide each interval $[0, \beta_i]$ into k_i subintervals of length $h_i = \beta_i/k_i$. By introducing new variables w_{ij} $(1 \leq j \leq k_i,\ 1 \leq i \leq n)$

satisfying $0 \leq w_{ij} \leq 1$, one can express

$$x_i = h_i \sum_{j=1}^{k_i} w_{ij} \quad (1 \leq i \leq n). \tag{15.14}$$

Of course, in the representation (15.14) the w_{ij} are not unique. But if we introduce the vectors

$$\bar{w}_i = (w_{i1}, w_{i2}, \ldots, w_{ik_i}) \tag{15.15}$$

and restrict each $\bar{w}_i$ to have the form

$$\bar{w}_i = (1, \ldots, 1, \; w_{i\ell}, 0, \ldots, 0) \tag{15.16}$$

then any $x_i \in [0, \beta_i]$ will be represented by a unique vector $\bar{w}_i$. A standard way to enforce (15.16) is by introducing additional *zero–one variables* $z_{ij} \in \{0, 1\}$ satisfying

$$w_{ij+1} \leq z_{ij} \leq w_{ij} \quad (j = 1, \ldots, k_i - 1, i = 1, \ldots, n). \tag{15.17}$$

Notice that (15.17) implies

$$z_{ij+1} \leq z_{ij} \quad (j = 1, \ldots, k_i - 1),$$

and therefore the only possibilities for z_{ij} are

$$\begin{aligned} z_{ij} &= 1, \quad j = 1, \ldots, \ell, \\ z_{ij} &= 0, \quad j = \ell + 1, \ldots, k_i - 1 \end{aligned} \tag{15.18}$$

for some $\ell, 1 \leq \ell \leq k_i$. The constraints (15.17) then guarantee that $\bar{w}_i$ has the form (15.16).

It is easy to see that the piecewise linear function that interpolates $q_i(x_i)$ at the points jh_i $(j = 0, \ldots, k_i)$ is given by

$$\Gamma_i(x_i) = \sum_{j=1}^{k_i} \Delta q_{ij} w_{ij} \tag{15.19}$$

where

$$\Delta q_{ij} = q_i(jh_i) - q_i((j-1)h_i) \quad (i = 1, \ldots, n, \; j = 1, \ldots, k_i). \tag{15.20}$$

The piecewise linear approximation of the objective function $\phi(x,y)$ in (SP) as discussed in Section 2.6.3 is then given by

$$\sum_{i=1}^{n} \Gamma_i(x_i) + d^T y = \sum_{i=1}^{n} \sum_{j=1}^{k_i} \Delta q_{ij} w_{ij} + d^T y, \tag{15.21}$$

and the corresponding approximate optimal solution to the global minimum problem (SP) can be computed by solving the following linear zero–one mixed integer problem in the continuous variables w_{ij} and the zero–one variables z_{ij}:

$$\begin{aligned} \min \quad & \sum_{i=1}^{n} \sum_{j=1}^{k_i} \Delta q_{ij} w_{ij} + d^T y \\ \text{s.t.} \quad & \sum_{i=1}^{n} h_i a_i \sum_{j=1}^{k_i} w_{ij} + A_2 y = b \\ & 0 \le w_{ij} \le 1, \; y \ge 0 \\ & w_{ij+1} \le z_{ij} \le w_{ij} \quad (j = 1, \ldots, k_i - 1; i = 1, \ldots, n) \\ & z_{ij} \in \{0, 1\} \end{aligned} \tag{MI}$$

where a_i is the column number i of A_1.

An important factor that determines the worst case computational effort in a branch and bound method for zero–one mixed integer programs is obviously the number of possible nodes of the binary branch and bound tree (which is determined by the number of $0-1$ variables. The integer variables introduced above are of very special structure. Note that the number of zero–one variables z_{ij} in problem (MI) is given by

$$N = \sum_{i=1}^{n} k_i - n. \tag{15.22}$$

If all possible combinations of $z_{ij} \in \{0, 1\}$ were allowed, this problem would have a maximum of 2^N possibilities to be evaluated (which is the worst case number of nodes in the tree). However, in the justification of the above zero–one formulation we have seen that each $\bar{z}_i = (\bar{z}_{i1}, \ldots, \bar{z}_{ik_i-1})$ is of the form $(1, \ldots, 1, 0, \ldots, 0)$. Therefore, the maximum number of possible combinations is given by

$$\prod_{i=1}^{n} k_i$$

This number is bounded from above by k_1^n since we assumed that $\rho_1 \geq \rho_i$, and hence $k_1 \geq k_i \quad (i = 2, \ldots, n)$. Note that $k_1^n = 2^r$ where $r = n \log_2 k_1$ and $2^r \ll 2^N$.

In addition, note that, by the choice of $k_i = k_i(\varepsilon)$ we have the following upper bound:

$$\begin{aligned} N &= \sum_{i=1}^{n} k_i - n \leq \sum_{i=1}^{n}((\tfrac{n}{\alpha}\rho_i)^{\frac{1}{2}} + 1) - n \\ &= (\tfrac{n}{\alpha})^{\frac{1}{2}} \sum_{i=1}^{n} (\rho_i)^{1/2} \\ &\leq (\tfrac{n}{\alpha})^{\frac{1}{2}} n^{\frac{1}{2}} \left(\sum_{i=1}^{n} \rho_i\right)^{\frac{1}{2}} \quad \text{(by the Cauchy-Schwarz inequality)} \\ &= n\left((\sum_{i=1}^{n} \rho_i)/\alpha\right)^{\frac{1}{2}} = n\left(\tfrac{1}{\varepsilon}\sigma(\rho, \eta)\right)^{\frac{1}{2}}. \end{aligned}$$

15.3.3 Indefinite Quadratic Programming

The above ideas and methods for solving concave quadratic minimization problems can easily be extended to indefinite quadratic problems. Consider again problem (P):

$$\begin{aligned} &\min \quad \Psi(z, y) \\ \text{s.t.} \quad &(z, y) \in \Omega = \{A_1 z + A_2 y = b, \ z \geq 0, \ y \geq 0\} \end{aligned}$$

where

$$\begin{aligned} \Psi(z, y) &= \Phi(z) + d^T y, \\ \Phi(z) &= c^T z - \tfrac{1}{2} z^T Q z. \end{aligned}$$

Since we are concerned with indefinite quadratic problems, Q is a real symmetric indefinite $(n \times n)$–matrix. Assume that Q has r positive eigenvalues $\lambda_1, \ldots, \lambda_r$, and $(n - r)$ nonpositive eigenvalues $-\lambda_{r+1}, \ldots, -\lambda_n$. The reduction to separable form does not depend on the sign of the eigenvalues, and (after changing variables) we obtain exactly in the same way as for the concave case the equivalent separable formulation (SP):

$$\min\{\Psi(x, y) = \Phi(x) + d^T y : (x, y) \in \Omega, \ x \in R_x\}$$

where

$$\begin{aligned} \Omega &= \{(x, y) : A_1 x + A_2 y = b, \ y \geq 0\} \\ R_x &= \{x : 0 \leq x_i \leq \beta_i \quad (i = 1, \ldots, n)\} \end{aligned}$$

and

$$\Phi(x) = \sum_{i=1}^{n} q_i(x_i).$$

The nonlinear part of the objective function is now of the form

$$\Phi(x) = \Phi_1(x) + \Phi_2(x),$$

where

$$\Phi_1(x) = \sum_{i=1}^{r} q_i(x_i) = \sum_{i=1}^{r} c_i x_i - \frac{1}{2}\lambda_i x_i^2, \text{ the concave part,}$$

and

$$\Phi_2(x) = \sum_{i=r+1}^{n} q_i(x_i) = \sum_{i=r+1}^{n} c_i x_i + \frac{1}{2}\lambda_i x_i^2, \text{ the convex part.}$$

Note that $\lambda_i \geq 0$ for $i = 1, \ldots, n$ since, for $i = r+1, \ldots, n$, we denoted the (nonpositive) eigenvalues by $-\lambda_i$.

Next we discuss a straightforward approach to continue along the lines of the previous sections. Let $\Phi_2(x)$ as it is and construct the (piecewise) linear underestimation by linear interpolation for $\Phi_1(x)$ over the rectangle $[0, \beta_1] \times [0, \beta_2] \times \ldots \times [0, \beta_r]$ exactly as we did before for the function Φ (formally, replace Φ by Φ_1 and n by r.)

Setting $\Gamma_1(x) = \sum_{i=1}^{r} \gamma_i(x_i)$, where $\gamma_i(x_i) = (c_i - \frac{1}{2}\lambda_i\beta_i)x_i$ $(1 \leq i \leq r)$, problem (LU) must be replaced by

$$\begin{aligned} \min \quad & \Gamma_1(x) + \Phi_2(x) + d^T y \\ \text{s.t.} \quad & (x, y) \in \Omega, \; x \in R_x \end{aligned} \tag{CU}$$

which is a linearly constrained convex quadratic minimization problem. The approach of the preceding sections carries over in a similar way, where now convex quadratic programs of the type (CU) have to be solved in the branch–and–bound approach, whereas the zero–one mixed integer approach leads to a convex quadratic zero–one mixed integer program.

Exercises

1. Let $G = G(V, E)$ be a graph where $V = \{1, \ldots, n\}$ denotes the set of vertices (nodes), and E denotes the set of edges. If x^* is the solution of the following indefinite quadratic programming problem

$$\max f(x) = \textstyle\sum_{i=1}^{n} x_i - \sum_{(i,j)\in E} x_i x_j = e^T x - 1/2 x^T A_G x$$
$$\text{subject to } 0 \leq x_i \leq 1 \text{ for all } 1 \leq i \leq n$$

then, $f(x^*)$ equals the size of the maximum independent set of G.

2. Find a "finite" algorithm for the general quadratic programming problem

$$\begin{aligned} \min \quad & f(x) = \tfrac{1}{2} x^T Q x + c^T x \\ \text{s.t.} \quad & Ax \leq b \end{aligned}$$

where A is a real $(m \times n)$–matrix and Q is an indefinite symmetric real $(n \times n)$–matrix (hint: enumerate all Kuhn-Tucker points and use the fact that if the solution exists, then the solution occurs at a boundary point of the feasible domain).

Historical Notes

Who proposed the Steiner tree problem? It was an unsolved puzzle until 1986. The story included in section 1.1 was discovered by [158]. The stories about linear and nonlinear programming (section 1.1) can be found in [28, 100].

The floor planning problem (section 1.2) is an important problem in VLSI design. Many efforts have been made for solving floor planning problem with various optimization techniques [162].

The satisfiability problem (section 1.3) plays an important role in computational complexity theory [64]. The work introduced in section 1.3 belongs to Gu, Gu, and Du [77]. The logic puzzles in Exercise 1 come from [172] and [133].

The formulations for the network Steiner tree and the directed network Steiner tree problems in Exercises of Chapter 1 were found in [95].

Karmarkar [92] proposed the first interior-point method (section 2.3) which is the second polynomial time algorithm for linear programming since Klee and Minty [97] discovered an example that requires $\Omega(2^n)$ iterations for the simplex method or the dual-simplex method. In the last decade several interior-point methods for linear (and quadratic) programming problems have been proposed [9, 161, 163, 74].

Hua [86] initiated the idea of describing the steepest ascend method as blind man's method (section 3.1). He is a great Chinese mathematician who made a lot of contributions to the popularity of mathematics. This idea is extended in chapters 3-5 to introduce a convergence theory discovered initially for solving a long-standing open problem about Rosen's gradient projection method.

The gradient projection method proposed by J. B. Rosen in 1960 [152, 153] is one of earliest methods in the history of mathematical programming for solving constrained nonlinear programming problems (section 4.3). The importance of this method in the literature also stems from the fact that many more efficient algorithms, in linear and nonlinear programming, developed later (e.g. by Goldfarb [71], by Martagh and Sargent [118] and by Karmarkar [92]) incorporated the basic ideas propounded by Rosen.

The global convergence of Rosen's method was a long-standing open problem. Since Rosen's method is included in many textbooks, the convergence problem has become quite well-known. In fact, almost all books (such as [8, 10, 105, 135]) that have a chapter or a section to introduce Rosen's method recognize the non-existence of either a convergence proof or a counterexample.

The global convergence of Rosen's method relies on a parameter chosen at each iteration. To avoid the zigzag phenomena, Rosen chose the parameter to be a positive number with an upper bound. He also gave a convergence proof in his paper. However, it did not take very long for someone to point out that there is a serious mistake in the proof. Since then, substantial efforts have been made on the problem. Maybe, it is due to a natural expectation for seeking the simplicity. Most books wrongly state Rosen's method by setting the parameter to be zero which is exactly what Rosen tried to avoid. Du, Sun and Song [48] presented a counterexample for the method described in those books (section 4.3). It was indicated by Du and Zhang [50] that this counterexample also works for the case that the parameter varies and converges to zero as the computation runs from the first iteration to the infinity. It follows that Rosen's restriction on the parameter is not sufficient for the convergence.

The first convergent version was found by Polak [134]. Compared with the original one, Polak's version is too complicated. First of all, he used the ϵ-active set strategy. Secondly, the version contains a special procedure, which involves computing the gradient projection several times. Polak proved the convergence of his version under a condition, named by ϵ-hypothesis. Zhang [179] showed that ϵ-hypothesis is equivalent to the regularity condition (section 7.1). Zhang's work motivated from a theorem of Yue and Han [175].

Yue and Han utilized Polak's procedure to study the reduced gradient method. The reduced gradient method was first proposed by Wolfe [167]. An important step in the reduced gradient method is to compute a reduced basis. There are several ways for finding a reducible basis [175, 176, 49, 164, 165]. The one introduced in section 7.3 was initially given in [175] and improved in [49]. It has been known to be quite efficient in practice. Wolfe's original version of reduced gradient method is not globally convergent. A counterexample has been given by Wolfe [170] himself. Several convergent versions [105, 164, 165, 175, 176] appeared later. Among them, Wang [164] (section

7.3) derived his version from the Levitin-Polyak's method [104]. Levitin-Polyak's method is a gradient projection method different from Rosen's method. In each iteration, they chose the search direction to be the projection of gradient on the feasible region. In general, we cannot find a closed form formula for such a projection. However, theoretically, the method can be put in very general setting [5, 80, 91, 114].

Yue and Han [175] also obtained a new property of linearly constrained set in the format of the standard form of linear programming. This type of properties are very useful for the application of ϵ-active set strategy. Du and Sun [47] gave a generalization. Du [42] also found a simpler convergent version for Rosen's method by deleting Polak's special procedure (section 7.1).

To give a unified convergence theorem, Zangwill [177, 178] introduced a point-to-set mapping; each iteration that finds a new point from the current point is viewed as taking an element from the image set of the point-to-set mapping. He showed an abstract convergence theorem under the closeness of the mapping (section 3.3). However, for a constrained optimization problem, the closeness is lost when the line search procedure is stopped by a constraint. Du [45] proved a result to tell when an algorithm has the global convergence provable by Zangwill's theorem (section 3.3). McCormick suggested to search along a broken line, that is, if the search is stopped by reaching a new constraint then do not stop there and, keeping the constraint active, find a new direction to continue the search. Ritter [147, 149, 150] decomposed the broken line search into several line searches and applied it to a family of feasible direction method including the gradient projection method. Since the discovery of Zangwill's theorem, many abstract convergence results based on the point-to-set mapping have been established. The reader may find some of them in [89, 90, 116, 117]. Some sufficient conditions for global convergence, such as Bazaraa-Shetty's condition [10] (section 5.2), also appeared. However, none of them is powerful enough to show the convergence of Rosen's gradient projection method.

Although several convergent versions have been found, one still wanted to know what would happen when the parameter is also bounded below by a positive number. In fact, all the above convergent versions use some ideas different from Rosen's one for anti-zigzaging. Thus, it was still open whether Rosen's original idea that keeps the parameter away from zero and infinity works or not. (Allowing the parameter to approach zero as the iteration varies was considered as an oversight in [152].)

Zhang [181] made the first breakthrough in this direction; he showed that if the parameter equals a positive constant at every iteration, then Rosen's method is convergent in the three dimensional space. Soon later, Du and Zhang [50] established a more general result that if the parameter is chosen to be a positive constant with an upper bound, then Rosen's method is convergent in the n-dimensional space. This is the first solution for the convergence problem. In the same paper, Du and Zhang also conjecture that the upper bound on the constant can be deleted, that is, as long as the parameter does not vary as the iteration varies, it can be chosen to be any positive number. This conjecture was settled through several efforts including Du [51], He [81], and Du and Zhang [53] (section 6.2). Finally, Du, Wu and Zhang [55] showed a convergence theorem with a more general rule for selecting the parameter. Gui and Du [78] gave a method to deal with this degenerative case (section 7.2). Their method uses the lexicographic simplex procedure to deal with degenerate line constraints. It is worth mentioning that Bland's rule [12] can also be used here instead of the lexicographic simplex procedure.

The global convergence of Rosen's method also depends on the choice of line search. Du and Zhang [53] assume that the line search is normal (see section 5.2), including exact line search, Curry test, Goldstein test, Wolfe test, and Armijo rule [7]. If the line search is not normal, Hu [85] showed that the global convergence of Rosen's method may fail (section 6.2). But, with Hu's counterexample, the method in [42] still has the global convergence property. If the step length in the line search is uniformly bounded, the choice of parameter can be further relaxed [46] (section 6.3), which gives an improvement of Ritter's result [150].

When the objective function is not differentiable, most of the previously mentioned results on global convergence do not hold. A counterexample in section 6.2 showed that although the objective function is continuously differentiable at every point generated by Rosen's method, if it is not continuously differentiable in the entire feasible region, the global convergence is still lost. This tells us that in order to determine a search direction a convergent algorithm for nonsmooth optimization problems should not only consider the present point but also consider some points elsewhere. Thus, a new approach for determining the search direction is needed. Many algorithms for nonsmooth optimization have been discovered during the last twenty years. The reader may refer to [25, 92, 121, 125] for them. It is possible to develop an analogous convergence theory in nonsmooth optimization.

When the feasible region is constrained nonlinearly, the gradient projection on tangent plane could lose the feasibility. Thus, Rosen [153] suggested to choose the search direction in the following way: First, project the gradient to the tangent plane, and then pull the projection back to the feasible region. How is it pulled? Polak [135], Zhang [180], Lai [101] and Xie [173] gave different answers. Du [44] introduced an alternative idea: Rotate the tangent plane before projecting the gradient on it (section 7.2).

As the global convergence problem is resolved, a number of new techniques for studying the global convergence of "non-closed" algorithms have been discovered. In particular, a result derived from the Du and Zhang's proof provided a surprising fact: A large class of feasible direction methods (such as Zoutendijk method [184, 10] in section 5.2) share a convergence property that if the generated sequence of points has a cluster point but does not converge to it, then every cluster point of the sequence is a Kuhn-Tucker point. It is interesting to point out that Wolfe [170] has showed by a counterexample that Zoutendijk's method can generate a sequence of points converging to a point which is not a Fritz John point. This means that this property is very popular.

Let us explain it briefly. Consider an iterative algorithm for a maximization problem. Suppose that at each point, the algorithm has a certain way to find a search direction and then compute a new point along the search direction. The search direction is always a feasible ascendant direction. For some points, no feasible ascendant direction exists. Those points will be called stationary points. Our goal is to show that if the algorithm generates an infinite sequence of points with at least two cluster points, then every cluster point of the sequence is a stationary point. (The similar thing happens for the sequence with a cluster point which is not its limit.)

The process of an algorithm generating points can be seen as a person who climbs a mountain along a road. To decrease its slope, the road is usually constructed with a shape like a snake lying on the mountain. The longer is its length, the smaller is its slope. If the length goes to infinity, the average slope of the road goes to zero. Since there are at least two cluster points, the road has to go between them infinitely many times. It follows that the length of the road is infinite. Hence, the slope approaches zero at the place near the cluster points. The slope of the road is the directional derivative along the search direction. So, the above conclusion means that the limitation of the search direction cannot be an ascendant direction.

Suppose that (a) the algorithm finds the search direction continuously and (b) the algorithm can find a search direction for every nonstationary point. Then we can conclude that the cluster points must be stationary points.

The above idea has resulted in a lemma (section 5.1). The lemmas of this type are called slope lemmas [46] since they are motivated from observations on slope. The discussions in chapter 4-7 were essentially based on the slope lemmas. We showed how the lemmas bring a new look to the convergence theory of feasible direction methods. Since we were interested in theoretical aspects, we discussed only simple and basic algorithms in the literature. It should be noticed that these algorithms may not be the most efficient. However, they are serving as roots of some most efficient algorithms.

A measure for computational efficiency of nonlinear optimization method is the convergence rate (chapter 8). All methods described in the above are linearly convergent. In chapters 8-10, our attention was turned to methods which are quadratically or superlinearly convergent.

The first quasi-Newton method was initially proposed by Davidon [29] in 1959, but his work was published nine year later [30]. The public attention on Davidon's work was largely due to the introduction of Flectcher and Powell [61] in 1963. This method is called the DFP method today. Since then, many variations and generalizations of quasi-Newton methods [13, 14, 73, 159, 88, 171] appeared in the literature (chapter 9). There are three issues on the convergence of quasi-Newton methods: quadratic termination, global convergence, and convergence rate. All quasi-Newton methods have quadratic termination under certain condition [15, 16]. However, the global convergence is a difficult problem for quasi-Newton methods. Powell [137] in 1971 established the first convergence theorem that if the objective function is twice continuously differentiable and uniformly concave, then the DFP method with exact line search generates a sequence convergent to a stationary point. Dixon [33, 34] at the same period found that all methods in Broyden's family with exact line search actually generate an identical sequence. This discovery extended Powell's result to all members in Broyden's family. It is interesting to point out that although Powell in 1971 obtained a convergence theorem, he did not believe that DFP method in general is globally convergent. Thus, he conjectured in 1971 that there exists a twice continuously differentiable objective function such that DFP method generates a sequence convergent to a point which is not a stationary point. However, one year later he changed his mind. In fact, after eighteen months he worked on his own conjecture, but he did not find such a counterexample. Instead, he [138] proved that his conjecture in 1971 is false for objective functions with two variables and that the uniform concavity in his convergence theorem can be replaced by concavity and an upper bound on the objective function (section 10.2). Therefore, since 1972, it was considered that

DFP method with exact line search is indeed globally convergent. Today, it is so-called Powell's conjecture that DFP method with exact line search is globally convergent for continuously differentiable objective functions.

Pu and Yu [146] made an important progress on Powell's conjecture (section 10.2). They showed that if the objective function is twice continuously differentiable and the generated sequence is convergent, then the sequence converges to a stationary point. This settled Powell's initial conjecture in 1971. However, Powell's conjecture (in 1972) is still open.

Powell [140] also showed a global convergence theorem for BFGS method with inexact line search (Wolfe test) for twice continuously differentiable and concave objective functions with a upper bound. Ritter [151] tried to generalize Powell's result to whole Broyden's family. However, the restriction on search step in line search restricts success of his result. Byrd, Nocedal, and Yuan [19] successfully generalized Powell's result to all members in Broyden's family except DFP method. It is a very interesting open problem whether DFP method with Wolfe test has the same global convergence. With a modified Wolfe test and a suitably smooth objective function, Pu [145] showed that if a member in Broyden's family generates a convergent sequence, then the sequence converges to a stationary point.

For the convergence rate, Powell [137] showed that if the objective function is uniformly concave and its Hessian matrix satisfies the Lipschitz condition, then DFP method with exact line search is superlinearly convergent. This can be extended to all members in Broyden's family by Dixon theorem. Similarly, Powell [140] also showed that if the objective function is uniformly concave and its Hessian matrix satisfies Lipschitz condition, then BFGS method with Wolfe test is superlinearly convergent. Byrd, Nocedal, and Yuan [19] generalized this result to Broyden's family except DFP method under the condition that the Hession matrix is Hölder continuous. Pu [144] showed that with the modified Wolfe test, DFP method is one-step superlinearly convergent.

Combining Rosen's method with variable metric methods, Goldfarb [71] (section 10.3) and Murtagh and Sargant [118] obtained two efficient algorithms. Powell [139] showed that these two algorithms are actually equivalent. The convergence of such algorithms is still an open problem. However, several convergent variations [78, 97, 112, 149, 150] have been established. The convergence of quasi-Newton methods for nonlinearly constrained optimization can be found in [142, 143, 122, 27].

There is a long history of mathematical research on the shortest networks. Possibly, Fermat is the first person who studied this kind of problems. Given three points on the Euclidean plane, Fermat raised the problem of finding a point such that the sum of three distances from this point to three given points achieves minimum. There are two ways to generalize Fermat problem. The first generalization is the problem of finding a point such that the sum of n distances from the point to n given points achieves minimum. This generalization is still called *Fermat problem.* The second one is the problem of finding a shortest network interconnecting n given points on the Euclidean plane. Clearly, the second problem is called the Steiner minimum tree problem, which was actually proposed by Gauss in a letter to his friend [158].

Fermat's problem and the Steiner tree problem coincide only on three points. In general, their solutions are very different. In fact, Fermat problem is polynomial-time computable [99]. But, the Steiner tree problem is unlikely polynomial-time solvable [65]. Why the problem is named after Steiner instead of Gauss? A story is as follows: Although no contribution from Steiner has been found, Courant and Robbins [26] in their famous 1941 book "What Is Mathematics" referred to it as the Steiner problem. The popularity of their book has been the main reason that the misnomer "the Steiner problem" has struck. This story does not explain where Courant and Robbins got the information. In fact, they also referred to the Fermat problem as the Steiner problem. If they got both problems from Steiner's works, then Steiner is really the first person who studied the Steiner tree problem.

Although the Steiner tree problem is an old problem, it is well-studied only after 1950's. Gilbert and Pollak [68] did a lot of fundamental work. Melzak [115] gave a method to construct a Steiner tree with a given topology.

The Steiner ratio was then verified by Pollak [136] for $n = 4$, by Du, Hwang and Yao [38] for $n = 5$ and by Rubinstein and Thomas [157] for $n = 6$. Along another line, the lower bound of ρ for general n has been pushed up from 0.5 (by Moore as reported in [68]) to 0.57 by Graham and Hwang [76], to 0.74 by Chung and Hwang [24], to 0.8 by Du and Hwang [36] and to 0.824 by Chung and Graham [23]. Finally, Du and Hwang [37, 35] proved the conjecture by employing a new approach. This approach is based on their new discovery in minimax theory (section 11.1). They proposed a necessary optimality condition for a class of minimax problems.

The semidefinite programming is a fast growing area in optimization. It initially appeared for the study of graph optimization problems [106] in 1979. But, it became a really active area starting from Alizadeh's work [1] in 1991. He found the first polynomial-time algorithms and a lot of properties of semidefinite programming similar to those of linear programming. So far, more than a hundred of research papers on semidefinite programming have been published. They are basically distributed in three directions. The first is in the study of geometric and algebraic properties of spectrahedra, including duality of semidefinite programming. The second is on interior point methods and ellipsoid methods for semidefinite programming [1, 2, 3, 4, 6]. The one included in section 14.2 belongs to [93]. The third is in applications to combinatorial optimization [174, 11, 62, 69, 120, 185, 182, 63]. For a more complete list of references and survies, the reader may refer to [131, 128].

Bibliography

[1] F. Alizadeh, Combinatorial optimization with interior point methods and semidefinite matrices, *Ph. D. Thesis*, Computer Science Department, University of Minnesota, Minneapolis, Minnesota, 1991.

[2] F. Alizadeh, Interior point methods in semidefinite programming with applications to combinatorial optimization, *SIAM J. Opt.* 5 (1995) 13-51.

[3] F. Alizadeh, J.-P. A. Haeberly, and M. Overton, Primal-dual interior point methods for semidefinite programming, manuscript, 1994.

[4] F. Alizadeh, J.-P. A. Haeberly, and M. Overton, Complementarity and nondegeneracy in semidefinite programming, *Methematical Programming* 77 (1997) 111-128.

[5] E. Allen, R. Helgason and J. Kennington, A generalization of Polyak's convergence result for subgradient optimization, *Mathematical Programming* 37 (1987) 309-318.

[6] L. T. H. An, P. D. Tao, and L. D. Muu, A combined D.C. optimization-ellipsoidal branch-and-bound algorithm for solving nonconvex quadratic programming problems, *Journal of Combinatorial Optimization* 2 (1998) 9-28.

[7] L. Armijo, Minimization of functions having Lipschitz continuous first partial derivatives, *Pacific Journal of Mathematics* 16 (1966) 1-3.

[8] M. Avriel, *Nonlinear Programming: Analysis and Methods* (Prentice-Hall, Englewood Cliff, New Jersey, 1976).

[9] M. S. Barnes, A variation on Karmarkar's algorithm for solving linear programming problem, *Mathematical Programming* 36 (1986) 174-182.

[10] M. S. Bazaraa and C.M. Shetty, *Nonlinear Programming: Theory and Algorithms* (John Willy & Sons, New York, 1979).

[11] D. Bertsimas and Y. Ye, Semidefinite relaxations, multivariate normal distributions, and order statistics, *Handbook of Combinatorial Optimization (Vol 3)*, D.-Z. Du and P. M. Pardalos (eds.) (Kluwer Academic Publishers, 1998) pp. 1-19.

[12] R. G. Bland, New finite pivoting rules for the simplex method, *Mathematics of Operations Research* 2:2 (1977) 103-107.

[13] C. G. Broyden, A class of methods for solving nonlinear simultaneous equations, *Math. Comp.* 19 (1965) 577-593.

[14] C. G. Broyden, Quasi-Newton methods and their application to function minimization, *Math. Comp.* 21 (1967) 368-381.

[15] C. G. Broyden, The convergence of a class of double-rank minimization algorithms 1: general consideration, *J. Inst. Math. Appl.* 6 (1970) 76-90.

[16] C. G. Broyden, The convergence of a class of double-rank minimization algorithms 2: the new algorithms, *J. Inst. Math. Appl.* 6 (1970) 222-231.

[17] A. Buckley, An alternate implementation of Goldfarb's minimization algorithm, *Mathematical Programming* 8 (1975) 207-231.

[18] R. H. Byrd, An example of irregular convergence in some constrained optimization methods that use the projected Hessian, *Mathematical Programming* 32 (1985) 232-237.

[19] R. H. Byrd, J. Nocedal, and Y. Yuan, Global convergence of a class of quasi-Newton method on convex problems, *SIAM J. Numer. Anal.* 24 (1987) 1171-1190.

[20] R. H. Byrd and G. A. Shultz, A practical class of globally convergent active set strategies for linearly constrained optimization, unpublished manuscripts.

[21] P. H. Calamai and J. J. More, Projected gradient methods for linearly constrained problems, *Mathematical Programming* 39 (1987) 93-116.

[22] F.R.K. Chung and E.N. Gilbert, Steiner trees for the regular simplex, *Bull. Inst. Math. Acad. Sinica*, 4 (1976) 313-325.

[23] F.R.K. Chung and R.L. Graham, A new bound for Euclidean Steiner minimum trees, *Ann. N.Y. Acad. Sci.*, 440 (1985) 328-346.

[24] F.R.K. Chung and F.K. Hwang, A lower bound for the Steiner tree problem, *SIAM J.Appl.Math.*, 34 (1978) 27-36.

[25] F. H. Clarke, *Optimization and Nonsmooth Analysis* (John Wiley, New York, 1983).

[26] R. Crourant and H. Robbins, *What Is Mathematics?*, (Oxford Univ. Press, New York, 1941).

[27] T. F. Coleman and A. R. Conn, On the local convergence of a quasi-Newton method for the nonlinear programming problem, *SIAM J. Numerical Analysis* 1 (1984) 755-769.

[28] G. B. Dantzig, Linear programming, in J. K. Lenstra, A. H. G. Rinnooy Kan, and A. Schrijver (eds.), *History of Mathematical Programming* (North-Holland, 1991) pp.19-31.

[29] W. C. Davidon, Variable metric method for minimization, AEC Research and Development Report ANL-5990 (Rev.), November, 1959.

[30] W. C. Davidon, Variable metric method for minimization, *Computer J.* 10 (1968) 406-410.

[31] A. Dax, The gradient projection method for quadratic programming, Report of the Institute of Mathematics of the Hebrew University (Jerusalem, 1978).

[32] R. S. Dembo and J. G. Klincewicz, Dealing with degeneracy in reduced gradient algorithms, *Mathematical Programming* 31 (1985) 257-363.

[33] L. C. W. Dixon, Quasi-Newton algorithms generate identical points, *Math. Prog.* 2 (1972) 383-387.

[34] L. C. W. Dixon, Quasi-Newton algorithms generate identical points II, the proof of four new theorems, *Math. Prog.* 3 (1972) 345-358.

[35] D.Z. Du and F.K. Hwang, The Steiner ratio conjecture of Gilbert-Pollak is true, *Proceedings of National Academy of Sciences*, 87 (1990) 9464-9466.

[36] D.Z. Du and F.K. Hwang, A new bound for the Steiner ratio, *Trans. Amer. Math. Soc.*, 278 (1983) 137-148.

[37] D.Z. Du and F.K. Hwang, An approach for proving lower bounds: solution of Gilbert-Pollak's conjecture on Steiner ratio, *Proceedings 31th FOCS* (1990) 76-85.

[38] D.Z. Du, F.K. Hwang and E.N. Yao, The Steiner ratio conjecture is true for five points, *J. Combinatorial Theory, Ser.A*, 32 (1982) 396-400.

[39] D.Z. Du and P.M. Pardalos, A continuous version of a result of Du and Hwang, *Journal of Global Optimization* 5 (1994) 127-129.

[40] D.Z. Du and P.M. Pardalos (Editors), *Minimax and Applications* (Kluwer Academic Publishers, 1995).

[41] D.Z. Du, J.Gu and P.M. Pardalos (Editors), *Satisfiability Problem: Theory and Applications* (DIMACS Series Vol. 35, American Mathematical Society, 1997).

[42] D. -Z. Du, A modification of Rosen-Polak's algorithm, *Kexue Tongbao* 28 (1983) 301-305.

[43] D. -Z. Du, A family of gradient projection algorithms, *Acta Mathematicae Applicatae Sinica, English Series* 2 (1985) 1-13.

[44] D. -Z. Du, A gradient projection algorithm for convex programming with nonlinear constraints, *Acta Mathematicae Applicatae Sinica* 8 (1985) 7-16 (in Chinese).

[45] D. -Z. Du, Changing of point-to-set maps and point-to-set map families for continuities, *Acta Mathematicae Applicatae Sinica* 8:2 (1985) 142-150 (in Chinese).

[46] D.-Z. Du, Rosen's method and slope lemmas, in P. M. Pardalos (ed.) *Advances in Optimization and Parallel Computing* (Elsevier Science Publisher, 1992) 68-84.

[47] D. -Z. Du and J. Sun, A new gradient projection method, *Mathematicae Numerica Sinica* 4 (1983) 378-386 (in Chinese).

[48] D. -Z. Du, J. Sun and T. -T. Song, A counterexample for Rosen's gradient projection method, *Mathematical Assortment* 1 (1980) 4-6 (in Chinese).

[49] D. -Z. Du, J. Sun and T. -T. Song, Simplified finite pivoting processes in the reduced gradient algorithms, *Acta Mathematicae Applicatae Sinica* 7 (1984) 142-146 (in Chinese).

[50] D. -Z. Du and X. -S. Zhang, A convergence theorem of Rosen's gradient projection method, *Mathematical Programming* 36 (1986) 135-144.

[51] D. -Z. Du, Remarks on the convergence of Rosen's gradient projection method, *Acta Mathematicae Applicatae Sinica, English Series* 3 (1987) 270-279.

[52] D. -Z. Du, *Gradient Projection Methods in Linear and Nonlinear Programming* (Hadronic Press, Boston, 1988).

[53] D. -Z. Du and X. -S. Zhang, Global convergence of Rosen's gradient projection methods, *Mathematical Programming* 44 (1989).

[54] D. -Z. Du and X. -S. Zhang, Notes on a new gradient projection method, *System Science and Mathematical Science* 2 (1989).

[55] D. -Z. Du, F. Wu and X. -S. Zhang, On Rosen's gradient projection methods, *Annals of Operations Research* to appear.

[56] D. -Z. Du and X. -F. Du, A convergent reduced gradient algorithm without using special pivot, *Mathematicae Numerica Sinica* to appear.

[57] D.Z. Du, E.N. Yao, and F.K. Hwang, A short proof of a result of Pollak on Steiner minimal trees, *J. Combinatorial Theory, Ser. A*, 32 (1982) 396-400.

[58] J. C. Dunn, Global and asymptotic convergence rate estimates for a class of projected gradient processes, *SIAM Journal on Control and Optimization* 19 (1981) 368-400.

[59] J. C. Dunn, On the convergence of projected gradient processes to singular critical point, *Journal of Optimization Theory and Applications* 55 (1987) 203-216.

[60] R. Fletcher, *Practical Methods of Optimization, Unconstrained Optimization*, (John Wiley and Sons, New York, 1987).

[61] R. Fletcher and M. J. D. Powell, A rapidly convergent descent method for minimization, *Computer J.* 6 (1063) 163-168.

[62] A. Frieze and M. Jerrum, Improved approximation algorithms for max k-cut and max bisection, *Proc. 4th IPCO*, 1995, pp. 1-13.

[63] M. Fu, Z.-Q. Luo, and Y. Ye, Approximation algorithms for quadratic programming, *Journal of Combinatorial Optimization* 2 (1998) 29-50.

[64] M. R. Garey and D. S. Johnson, *Computers and Intractability: A Guide to the Theory of NP-completeness*, (Freeman, San Francisco, 1979).

[65] M.R. Garey, R.L. Graham and D.S. Johnson, The complexity of computing Steiner minimal trees, *SIAM J. Appl. Math.*, 32 (1977) 835-859.

[66] L. Gibbons, D. Hearn and P.M. Pardalos, A continuous based heuristic for the maximum clique problem, In *Clique, Graph Coloring, and Satisfiability: Second DIMACS Implementation Challenge* (D.S. Johnson and M.A. Trick, Editors), DIMACS Series Vol. 26, American Mathematical Society (1996) 103-124.

[67] L. Gibbons, D. Hearn, P.M. Pardalos, and M. Ramana, A continuous characterization of the maximum clique problem, *Math. of Oper. Res.* 22, 3 (1997) 754-768.

[68] E.N. Gilbert and H.O. Pollak, Steiner minimal trees, *SIAM J. Appl. Math.*, 16 (1968) 1-29.

[69] M. X. Goemans and D. P. Williamson, Improved approximation algorithms for maximum cut and satisfiability problems using semidefinte programming, *Journal of ACM* 42 (1995) 1115-1145.

[70] D. Goldfarb, A family of variable metric methods derived by variational means, *Mathematics of Computation* 24 (1970) 23-26.

[71] D. Goldfarb, Extension of Davidon's variable metric method to maximization under linear inequality and equality constraints, *SIAM Journal on Applied Mathematics* 17 (1969) 739-764.

[72] D. Goldfarb and L. Lapidus, Conjugate gradient method for nonlinear programming problems with linear constraints, *Industrial and Engineering Chemistry Fundamentals* 7 (1968) 142-151.

[73] J. Goldstein, Variations on variable-metric methods, *Math. Comp.* 24 (1970) 1-22.

[74] C. C. Gonzaga, Conical projection algorithms for linear programming, *Mathematical Programming* 43 (1989) 151-174.

[75] J. Goodman, Newton's method for constrained optimization, *Mathematical Programming* 33 (1985) 162-171.

[76] R.L. Graham and F.K. Hwang, Remarks on Steiner minimal trees, *Bull. Inst. Math. Acad. Sinica,* 4 (1976) 177-182.

[77] J. Gu, Q. P. Gu, and D.-Z. Du, Convergence property of optimization algorithms for the SAT problem, *IEEE Transactions on Computers* 45 (1996) 209-219.

[78] X. -Y. Gui and D. -Z. Du, A superlinearly convergent method to linearly constrained optimization problems under degeneracy, *Acta Mathematicae Applicatae Sinica, English Series* 1 (1984) 76-84.

[79] S. -P. Han, A successive projected method, *Mathematical Programming* 40 (1988) 1-14.

[80] A. Haraux, How to differentiate the projection on a convex set in Hilbert space: Some applications to variational inequalities, *Journal of Mathematics Society Japan* 20 (1977) 615-631.

[81] G. -Z. He, Proof of convergence of Rosen's gradient projection method, *Acta Mathematicae Applicatae Sinica* 1 (1990).

[82] J. Herstovits, A two-stage feasible direction algorithms for nonlinear constrained optimization, *Mathematical Programming* 36 (1986) 19-38.

[83] R. Horst and P.M. Pardalos, *Introduction to Global Optimization* (Kluwer Academic Publishers, 1995).

[84] R. Horst and P.M. Pardalos (Editors), *Handbook of Global Optimization* (Kluwer Academic Publishers, 1995).

[85] X. -D. Hu, Nonlinear programming: a unified approach and the upper bound of partial concentrator, Ph. D. Thesis (Institute of Applied Mathematics, Chinese Academy of Sciences, Beijing, 1989).

[86] L. K. Hua, *Optimal Selection Method,* (Science Press, 1981).

[87] F.K. Hwang, D.S. Rechards, and P. Winter, *Steiner tree problems* (North-Holland, Amsterdam, 1992).

[88] H. Y. Huang, Unified approach to quadratically convergent algorithms for function minimization, *J. Optimization Theory and Appl.* 5 (1970) 405-423.

[89] P. Huard, Optimization algorithms and point-to-set maps, *Mathematical Programming* 8 (1975) 308-331.

[90] P. Huard, Extensions of Zangwill's theorem, *Mathematical Programming Study* 10 (1979) 98-103.

[91] P. Kalfon, G. Ribiere and J. C. Sogno, A method of feasible directions using projection operators, *Information Processing* 68 (Proc. IFIP congr. 68) Vol. 1 (North-Holland, Amsterdam, 1969).

[92] N. Karmarkar, A new polynomial-time algorithm for linear programming, *Combinatorica* 4 (1984) 373-395.

[93] E. de Klerk, C. Roos, and T. Terlaky, Polynomial primal-dual affine scaling algorithms for semidefinite programming, *Journal of Combinatorial Optimization* 2 (1998) 51-70.

[94] E. de Klerk, C. Roos, and T. Terlaky, Initialization in semidefinite programming via a self-dual skew-symmetric embedding, *OR Letters* 20 (1997) 213-221.

[95] B.N. Khoury, P.M. Pardalos, and D.W. Hearn, Equivalent formulations for the Steiner problem in graphs, in D.Z. Du and P.M. Pardalos (eds.) *Network Optimization Problems* (World Scientific, 1993) pp.111-123.

[96] K. C. Kiwiel, A survey of bundle methods for nondifferentiable optimization, in M. Iri and K. Tanabe (eds.) *Mathematical Programming, Recent Developments and Applications* (KTK Scientific Publishers Tokyo, 1989) 263-282.

[97] V. Klee and G. J. Minty, How good is the simplex algorithm? in O. Shisha (ed.) *Inequalities III* (Academic Press, New York, 1972).

[98] H. -Y. Kwei (X. -Y. Gui), F. Wu and Y. -L. Lai, Extension of a variable metric algorithm to a linearly constrained optimization problem: a

variation of Goldfarb's algorithm, In K. B. Haley (ed.) *O. R. 1978* (North-Holland Publishing company, 1979).

[99] H.W. Kuhn, Steiner's problem revisited, in *Studies in Optimization*, G.B. Dantzig and B.C. Eaves (editor), (Math Assoc. of America, 1975) 98-107.

[100] H.W. Kuhn, Nonlinear programming: a historical note, in J. K. Lenstra, A. H. G. Rinnooy Kan, and A. Schrijver (eds.), *History of Mathematical Programming* (North-Holland, 1991) pp.82-96.

[101] Y. -L. Lai, An algorithm and its convergence for nonlinear constrained convex programming, *Acta Mathematicae Applicatae Sinica* 3 (1980) 322-331 (in Chinese).

[102] Y. -L. Lai, Fang Wu and X. -Y. Gui, A reduced variable metric algorithm for linearly constrained convex programming, *Scientia Sinica, Series A* 26 (1983) 785-794.

[103] M. L. Lenard, A computational study of active set strategies in nonlinear programming with linear constraints, *Mathematical Programming* 16 (1979) 81-97.

[104] E. S. Levitin and B. T. Polyak, Constrained minimization methods, *USSR Computational Mathematics and Mathematical Physics* 6:5 (1966) 1-50.

[105] D. G. Luenberger, *Introduction to Linear and Nonlinear Programming* (Addison-Wesley, Reading, Mass. , 1973).

[106] L. Lovász, On the Shannon capacity of a graph, *IEEE Transactions on Information Theory* IT-25 (1979) 1-7.

[107] L. Lovász and A. Schrjiver, Cones of matrices and set functions and 0-1 optimization, *SIAM J. Opt.* 1 (1991) 166-190.

[108] L. Lovász, Combinatorial optimization: some problems and trends, *DIMACS Tech Report*, 92-53, 1992.

[109] D. G. Luenberger, The gradient projection method along geodesies, *Management Science* 18 (1972) 620-631.

[110] N. Maculan and G. Galdino de Paula, A linear-time median-finding algorithm for projecting a vector on the simplex of R^n, *Operations Research Letters* 8 (1989) 219-222.

[111] G. P. McCormick, Anti-zigzagging by bending, *Management Sciences* 15 (1969) 315-320.

[112] G. P. McCormick, A second order method for the linearly constrained nonlinear programming problem, in J. B. Rosen, O. L. Mangasarian, and K. Ritter (eds.) *Nonlinear Programming* (Academic Press, New York, 1990).

[113] G. P. McCormick, The variable reduction method for nonlinear programming, *Management Science* 17 (1970) 146-160.

[114] G. P. McCormick and R. A. Tapia, The gradient projection method under mild differentiability conditions, *SIAM Journal on Control and Optimization* 10 (1972) 93-98.

[115] Z.A. Melzak, On the problem of Steiner, *Canada. Math. Bull.*, 4 (1961) 143-148.

[116] R. R. Meyer, The validity of a family of optimization methods, *SIAM Journal on Control and Optimization* 8 (1970) 41-54.

[117] R. R. Meyer, Sufficient conditions for the convergence of monotonic mathematical programming algorithms, *Journal of Computer and System Sciences* 12 (1976) 108-121.

[118] B. A. Murtagh and R. W. H. Sargent, A constrained minimization method with quadratic convergence, in R. Fletcher (ed.) *Optimization* (Academic Press, London, 1969).

[119] B. A. Murtagh and M. A. Saunders, Large-scale linearly constrained optimization, *Mathematical Programming* 14 (1978) 14-72.

[120] Y. E. Nesterov, Semidefinite relaxation and nonconvex quadratic optimization, *Optimization Methods and Software* 9 (1998) 141-160.

[121] V. H. Ngugen and J. J. Strodiot, A linearly constrained algorithm not requiring derivative continuity, *Engineering Structures* 6 (1984) 7-11.

[122] J. Nocedal and M. L. Overton, Projected Hessian updating algorithms for nonlinearly constrained optimization, *SIAM J. Numerical Analysis* 1 (1985) 821-850.

[123] K. Ollerenshaw, Minimum networks linking four points in a plane, *Inst. Math. Appl.*, 15 (1978) 208-211.

[124] E. R. Panier and A. L. Tits, A superlinearly convergent feasible method for the solution of inequality constrained optimization problems, *SIAM Journal on Control and Optimization* 25 (1987) 934-950.

[125] E. R. Panier, An active set method for solving linearly constrained non-smooth optimization problems, *Mathematical Programming* 37 (1987) 269-292.

[126] P. M. Pardalos, On the passage from local to global in optimization, In *Mathematical Programming: State of the Art 1994* (J.R. Birge & K.G. Murty, Editors), The University of Michigan (1994), 220-247.

[127] P. M. Pardalos, *Continuous Approaches to Discrete Optimization Problems*, In *Nonlinear Optimization and Applications* (G. Di Pillo & F. Giannessi, Editors), Plenum Publishing (1996), 313-328.

[128] P. M. Pardalos and M. Ramana, Semidefinite programming, in *Interior Point Methods of Methematical Programming* (Kluwer Academic Publishers, 1997) pp. 369-398.

[129] P. M. Pardalos and J.B. Rosen, *Constrained Global Optimization: Algorithms and Applications* (Springer-Verlag, Lecture Notes in Computer Science 268, 1987).

[130] P. M. Pardalos and J.B. Rosen, Methods for Global Concave Minimization: A Bibliographic Survey, *SIAM Review* 28, 3 (1986) 367-379.

[131] P. M. Pardalos and H. Wolkowicz, *Topics in Semidefinite and Interior-Point Methods* (American Methematical Society, 1998).

[132] R. R. Phelps, Metric projections and the gradient projection method in Banach space, *SIAM Journal on Control and Optimization* 23 (1985) 973-977.

[133] H. Phillips, *My Best Puzzles in Logic and Reasoning* (Dover Publications Inc., New York, 1961).

[134] E. Polak, On the convergence of optimization algorithms, *Revue Francaise d'Informatique et de Recherche Operationelle* 3 (1969) 17-34.

[135] E. Polak, *Computational Methods in Optimization* (Academic Press, New York, 1971).

[136] H.O. Pollak, Some remarks on the Steiner problem, *J.Combinatorial Theory, Ser.A*, 24 (1978) 278-295.

[137] M. J. D. Powell, On the convergence of the variable metric algorithm, *J. Institute Mathematics and Its Applications* 7 (1971) 21-36.

[138] M. J. D. Powell, Some properties of the variable metric algorithm, in F. A. Lootsma (ed.) *Numerical Methods for Nonlinear Optimization* (Academic Press, 1972).

[139] M. J. D. Powell, Unconstrained minimization and extensions for constraints, in P. L. Hammer and G. Zoutendijk (eds.) *Mathematical Programming Theory and Practice* (North-Holland Publishing Company, 1974) 31-79.

[140] M. J. D. Powell, Some global convergence properties of a variable metric algorithm for minimization without exact line searches, in R. W. Cottle and C. E. Lemke (eds.) *Nonlinear Programming* (American Mathematical Society, Providence, R. I. , 1976).

[141] M. J. D. Powell, An example of cycling in a feasible point algorithm, *Mathematical Programming* 20 (1981) 253-257.]

[142] M. J. D. Powell, The convergence of variable metric methods for nonlinearly constrained optimization calculations, in *Nonlinear Programming 3*, O. Mangasarian, R. Meyer, and S. Robinson (eds.) (Academic Press, New York, 1978) 27-63.

[143] M. J. D. Powell and Y. Yuan, A trust region algorithm for equality unconstrained optimization, *Math. Prog.* 49 (1991) 189-211.

[144] D. Pu, A class of DFP algorithm without exact linear search, *Asia-Pacific J. of Oper. Res.* 9 (1992) 207-220.

[145] D. Pu, The convergence of Broyden algorithms without convexity assumption, *Systems Science and Mathematical Sciences* 10 (1997) 289-298.

[146] D. Pu and W. Yu, On convergence property of DFP algorithm, *Journal of Qüfu Normal University* 14:3 (1988) 63-69.

[147] K. Ritter, A superlinearly convergent method for minimization problems with linear inequality constraints, *Mathematical Programming* 4 (1973) 44-71.

[148] W.R. Pulleyblank, Two Steiner tree packing problems, *Proc. of 27th STOC*, 1995, pp. 383-387.

[149] K. Ritter, A method of conjugate directions for linearly constrained nonlinear programming problems, *SIAM Journal on Neumerical Analysis* 12 (1975) 272-303.

[150] K. Ritter, Convergence and superlinear convergence of algorithms for linearly constrained minimization problems, In L. C. W. Dixon, E. Spedicado and G. P. Szego (eds.) *Nonlinear Optimization: Theory and Algorithms, Part II* (Birkhauser Boston, 1980) 221-251.

[151] K. Ritter, Global and superlinear convergence of a class of variable metric methods, *Math. Prog. Study* 14 (1981) 178-205.

[152] J. B. Rosen, The gradient projection method for nonlinear programming, Part I: linear constraints, *SIAM Journal on Applied Mathematics* 8 (1960) 181-217.

[153] J. B. Rosen, The gradient projection method for nonlinear programming, Part II: nonlinear constraints, *SIAM Journal on Applied Mathematics* 9 (1961) 514-553.

[154] J. B. Rosen and J. Kreuser, A gradient projection algorithm for nonlinear constraints, in F. A. Lootsma (ed.) *Numerical Methods for Nonlinear Optimization* (Academic Press, 1972).

[155] J. B. Rosen, Two-phase algorithm for nonlinear constraint problems, in O. L. Mangasarian, R. R. Meyer and S. M. Robinson (eds.) *Nonlinear Programming* (Academic Press, New York, 1978) 97-124.

[156] J.H. Rubinstein and D.A. Thomas, A variational approach to the Steiner network problem, *Proceedings of NATO Workshop on Topological Networks*, Copenhagen, Denmark, 1989.

[157] J.H. Rubinstein and D.A. Thomas, The Steiner ratio conjecture for six points, to appear in *J. of Combinatorial Theory.*

[158] P. Schreiber, On the history of the so-called Steiner Weber problem, *Wiss. Z. Erust-Monitz-Arndt-Univ. Greifswald, Math.-nat.wiss. Reihe* 35:3 (1986).

[159] D. F. Shanno, Conditioning of quasi-Newton methods for function minimization, *Math. Comp.* 24 (1970) 647-656.

[160] D. F. Shanno, Globally convergent conjugate gradient algorithms, *Mathematical Programming* 33 (1985) 61-67.

[161] D. F. Shanno, Computing Karmarkar projections quickly, *Mathematical Programming* 41 (1988) 61-72.

[162] S. E. Sutanthavibul, S. E. Shragowitz, and J. B. Rosen, An analytical approach to floorplan design and optimization, *Proc. of the 27th ACM/IEEE Design Automation Conference*, 1990.

[163] M. J. Todd, Recent developments and new directions in linear programming, in M. Iri and K. Tanabe (eds.) *Mathematical Programming, Recent Developments and Applications* (KTK Scientific Publishers, Tokyo, 1989) 109-154.

[164] C. Y. Wang, Simplifications of a new pivoting rule and Levitin-Polyak gradient projection method and their convergent property, *Acta Mathematicae Applicatae Sinica* 4 (1981) 37-52 (in Chinese).

[165] C. Y. Wang, On convergence property of an improved reduced gradient method, *Kexue Tongbao* 28 (1983) 577-582.

[166] S. Wright, Convergence of SQP-like methods for constrained optimization, *SIAM Journal on Control and Optimization* 27 (1989) 13-26.

[167] P. Wolfe, Methods of nonlinear programming, in R. L. Graves and P. Wolfe (eds.) *Recent Advances in Mathematical Programming* (1963).

[168] P. Wolfe, Convergence conditions for ascent methods, *SIAM Review* 11 (1969) 226-235.

[169] P. Wolfe, Convergence theory in nonlinear programming, in J. Abadie (ed.) *Integer and Nonlinear Programming* (North-Holland Publishing Company, Amsterdam, 1970).

[170] P. Wolfe, On the convergence of gradient methods under constraints, *IBM J. Research and Development* 16 (1972) 407-411.

[171] F. Wu and X. Gui, a class of variable metric methods with $n+1$ parameters, *Acta Mathematica* 24 (1981) 921-930.

[172] C.R. Wylie Jr, *101 Puzzles in Throught and Logic* (Dover Publication, New York, 1957).

[173] S. J. Xie, A gradient projection method for solving linearly constrained convex programming, *Journal of Guangxi University* 1 (1982) 94-101.

[174] Y. Ye, A .699-approximation algorithm for Max-Bisection, manuscript, 2000.

[175] M. Yue and J. Han, A new reduced gradient method, *Scientia Sinica* 22 (1979) 1099-1113.

[176] M. Yue and J. Han, A unified approach to the feasible direction methods for nonlinear programming with linear constraints, *Acta Mathematicae Applicatae Sinica, English Series* 1 (1984) 63-73.

[177] W. I. Zangwill, Convergence conditions for nonlinear programming algorithms, *Management Science* 16 (1969) 1-13.

[178] W. I. Zangwill, *Nonlinear Programming: A Unified Approach* (Prentice-Hall, Englewood Cliffs, New Jersey, 1969).

[179] X. -S. Zhang, An improved Rosen-Polak method, *Acta Mathematicae Applicatae Sinica* 2 (1979) 257-267 (in Chinese).

[180] X. -S. Zhang, Discussion on Polak's algorithm of nonlinear programming, *Acta Mathematicae Applicatae Sinica* 4 (1981) 1-13 (in Chinese).

[181] X. -S. Zhang, On the convergence of Rosen's gradient projection method, *Acta Mathematicae Applicatae Sinica* 8 (1985) 125-128 (in Chinese, also in *English Series* 3 (1987) 280-288).

[182] Q. Zhao, S. E. Karisch, F. Rendl, and H. Wolkowicz, Semidefinite programming relaxations for the quadratic assignment problem, *Journal of Combinatorial Optimization* 2 (1998) 71-109.

[183] H. S. Zhou, A reduced gradient method with superlinear rate of convergence, *Acta Mathematicae Applicatae Sinica* 6 (1983).

[184] G. Zoutendijk, *Methods of Feasible Direction Methods* (Elsevier, Amsterdam, 1960).

[185] U. Zwick, Outward rotations: a tool for rounding solutions of semidefinite programming relaxations, with applications to Max Cut and other problems, *Proc. 10th SODA*, 1999.

Index